Productivity Management Frontiers - I

Productivity Management Frontiers - I

Refereed Papers Presented at the First International Conference on Productivity Research, February 4–6, 1987, Miami, FL, U.S.A.

Edited by

David J. Sumanth

Director, Productivity Research Group, Department of Industrial Engineering, University of Miami, FL, U.S.A.

ELSEVIER

Amsterdam – Oxford – New York – Tokyo 1987

ELSEVIER SCIENCE PUBLISHERS B.V.
Sara Burgerhartstraat 25
P.O. Box 211, 1000 AE Amsterdam, The Netherlands

Distributors for the United States and Canada:

ELSEVIER SCIENCE PUBLISHING COMPANY INC.
52, Vanderbilt Avenue
New York, NY 10017, U.S.A.

ISBN 0-444-42741-4

Printed in The Netherlands

PREFACE

Productivity Management Frontiers - I is a result of the First International Conference on Productivity Research (with an exclusive emphasis on "Productivity Engineering" and Productivity Management). This conference was co-sponsored by the University of Miami's Productivity Research Group (in the Department of Industrial Engineering) and the Department of Conference Services. This conference, held at the University of Miami Conference Center/James L. Knight International Center in Miami, Florida, has provided a forum for an international exchange of experience about some of the latest developments in "productivity engineering and management".

From a Call for Papers, over 80 abstracts were received and sent for independent review. Out of these, 64 were accepted. Full papers for these 64 abstracts were subjected to a rigorous blind refereeing process. About two-thirds of the 64 papers were accepted for inclusion in the book. Some of the authors whose full papers were accepted in the refereeing process could not make the publishing deadline or chose not to, for their own reasons.

This book is organized under the following main headings, covering the spectrum of Productivity Measurement, Productivity Evaluation and Planning, Productivity Improvement, and Productivity Management Implementation:

A. Importance of Productivity Management
B. Productivity Measurement at Company/Division/Plant/Product Levels
C. Total and/or Partial Productivity Measurement
D. White-Collar/Knowledge-Work Productivity Measurement
E. Productivity Planning
F. Productivity Improvement Approaches
G. Productivity Improvement Through Participation and Gainsharing
H. Implementation of Productivity Management

Productivity measurement and improvement at the international, national, and industry levels have the research concern of mostly economists. Many international conferences and congresses on productivity have been addressing the concerns at these levels. This book narrows the focus at or below the Company/Enterprise level where the action for producing a product or delivering a service begins.

The conference brought together researchers and practitioners from more than 25 countries. The participation truly reflected the international, intellectual flavor of this conference.

It is earnestly hoped that this book will stimulate enough interest among the researchers and practitioners of Productivity Engineering and Management, so that the second International Conference on Productivity Research will reflect further maturity in this evolving discipline.

The reader's constructive comments are most appreciated, and will form a basis for an even more enlightening *Frontiers of Productivity Management*.

Special gratitude goes to all the conference Presentors and particularly those whose papers are included in this book. A separate Acknowledgement section follows.

David J. Sumanth
Editor

ACKNOWLEDGEMENTS

The success of this conference was due to the sponsors, organizers, committees, participants, and attendees. While all of them are recognized for their generosity of their time and talents, some of them must receive special mention.

The following persons of the University of Miami provided the financial and/or administrative support for the conference:

Dr. Norman Einspruch	Dean, College of Engineering
Dean Dennis Tarr	Dean, School of Continuing Studies
Dr. Tarek Khalil	Chairman, Department of Industrial Engineering

The Advisory Committee, Local Conference Committee, Speakers Arrangement Committee, and several other sub-committees made it possible to review abstracts and full papers, to plan the logistics, and to moderate conference sessions. The help rendered has been most unselfish and generous:

ADVISORY COMMITTEE

Dr. Everett Adam Jr.	University of Missouri-Columbia (USA)
Mr. Edmond A. Cahill	Irish Productivity Center (Ireland)
Dr. Byoung Kyu Choi	Korean Advanced Institute of Science and Technology (South Korea)
Dr. M. Zia Hassan	Illinois Institute of Technology
Dr. Warren Hauck	Bowling Green Productivity & Gainsharing Institute (USA)
Dr. Ostap Hawaleshka	University of Manitoba (Canada)
Dr. Albert T.Y. Kuo	Tunghai University (Taiwan)
Dr. R.N. Lehrer	Georgia Institute of Technology (USA)
Dr. Keith E. McKee	Manufacturing Productivity Center (USA)
Professor Rintaro Muramatsu	Waseda University (Japan)
Dr. Krish Pennathur	Indian Productivity Academy (India)
Dr. M.R. Ramsey	NEPEAN Productivity Center (Australia)
Dr. William A. Ruch	Arizona State University (USA)
Dr. Li-Yen Shu	National Sun Yat-Sen University (Taiwan)
Dr. D. Scott Sink	Virginia Productivity Center (USA)
Dr. Alan Stainer	Middlesex Polytechnic (United Kingdom)

Dr. Ing. H.J. Warnecke	Stuttgart University (West Germany)
Professor Tanaka Yoshihiko	Waseda University (Japan)

LOCAL CONFERENCE COMMITTEE

Mr. Angelo Alonso	The Productivity Center, Inc.
Dr. Shihab S. Asfour	University of Miami
Dr. Howard Gitlow	University of Miami
Mr. Jose Gregorio	Burger King Corporation
Dr. Tarek M. Khalil	University of Miami
Dr. Tomislav Mandakovic	Florida International University
Dr. Jay Spechler	American Express Company
Dr. William B. Werther	University of Miami

SPEAKER ARRANGEMENTS COMMITTEE

Dr. Oscar Adaniya	University of Miami
Dr. Keebom Kang	University of Miami
Dr. A. Rabie	University of Miami
Dr. Vinent Omachonu	University of Miami
Dr. Bulent Bayraktar	University of Miami

Others who helped with the conference include several students.

Dr. Martin Lydon (Former President, Lowell University) provided his encouragement and support at the time when most needed

Dr. Marcy Ullom (Department of Conference Services at the University of Miami), as the coordinator of the conference took on enormous responsibilities and her painstaking efforts are most sincerely appreciated. Finally, grateful acknowledgement is made to Dr. David Sumanth (Director of the Productivity Research Group at the University of Miami) for conceiving, initiating, and chairing the conference.

TABLE OF CONTENTS

KEYNOTE ADDRESS

Productivity Management Frontiers — I, edited by D.J. Sumanth
Elsevier Science Publishers B.V., Amsterdam, 1987 — Printed in The Netherlands

PITFALLS ON THE PATH TO PRODUCTIVITY

Marvin E. Mundel, Ph.D., P.E.
M.E. Mundel & Associates, 821 Loxford Terrace, Silver Spring, MD 20901, U.S.A.

ABSTRACT
Mundel, M.E., 1987. Pitfalls on the Path to Productivity. Int. Conference on Productivity Research, Miami, FL, U.S.A., Feb.4-6, 1987.

Eleven pitfalls on the path to productivity improvement are described. In many cases one or more of these pitfalls may not only impede the improvement of productivity but may actually result in lowering it. The discussion of these pitfalls will be followed by some brief suggestions as to how to achieve an increase in productivity.

1. THE ELEVEN PITFALLS

1.1 THE WRONG MIND-SET

In England, 1811-16, the Luddites, bands of workmen, were organized to destroy manufacturing machinery, in the belief that its use diminished employment.

One would have thought, in view of the experience of the 150 years following, that this viewpoint would have died away. However, I assure you that I am still frequently asked the question, "But doesn't improved productivity reduce employment?" I am not sure that my explanation is accepted; "That without increased productivity the goods or services being produced become so expensive that no one can afford them; that unemployment really results from a failure to improve productivity."

Another wrong mind-set is in the failure to perceive that if we really improve productivity there will be various, concomitant, short-term results, particularly when the change is spectacular. There may be some short term redundancy which management must plan to alleviate.

For instance, in one plant a change was introduced that enabled one worker to do work that previously required five people. Management promised all the workers on the operation that no discharges would be made. However, the workers were also told that they would be assigned to other jobs, with the same pay as previously, as soon as vacancies occurred, but they would be expected to learn new skills. With normal labor turnover the vacancies came rapidly. This was not altruism on the part of management. A satisfactory worker, adjusted to plant procedures, represents a considerable investment; retaining the workers was good business as well as good thinking. Unfortunately, not all managers think this way; they frighten workers with productivity changes which result in immediate layoffs.

In other cases it is the workers who impede productivity improvement by insisting on the maintenance of jobs even though the need has disappeared. For instance, the typesetters have long fought for the maintenance of typesetting jobs on newspapers, not just to tide them over when automation had taken their place, but for perpetuity! After years of giving in, the newspapers have more recently been fighting.

In other cases we have managers who reject productivity measurement systems with the excuse "I want something more sophisticated." They give me the definite impression that they really don't want any system.

To sum up the right mind-set:
Productivity must be measured. If you do not measure you have no datum from which to measure change.

1.2 FAILURE TO ENLIST PARTICIPATION

It may be accepted as axiomatic that people will not willingly participate in an activity that they do not think they understand. Further, if they do not understand, they cannot contribute. Even the original Henry Ford had the concept that the mechanics in his plant could help improve the product and the process although, as time went by, he seemed to forget this.

1.3 OVERLY SIMPLISTIC COUNTING

We frequently delude ourselves into thinking we have developed a means of measuring productivity by having overly simplistic methods of counting our outputs. To give you a sample of the examples of I have seen:

a. A credit card firm measuring the productivity of local offices by counting "the number of resident accounts maintained."
b. A weather bureau's productivity measured by "number of typhoons tracked."
c. Army depots measured by "tons maintained in storage per staff year."
d. OSHA's productivity (under Eula Bingham) measured by "dollars of fines levied."
e. Police forces measured by "traffic tickets per officer per day."
f. Grants offices measured by "dollars of grants per staff year."
g. A bus line measured by "bus seat miles per month."

This list could be endless. In summary, all too often a simple count is used. We should be aware the word "simple" has another meaning which is probably more appropriate here.

1.4 OVERLY DETAILED "OUTPUTS"; MISTAKING ACTIVITIES FOR OUTPUTS

Of all the personnel offices I have worked with, all but one attempted to climb into this pit. All but one proposed outputs such as:

a. Personnel file pulled

b. Personnel file posted

c. Personnel file re-filed, and so forth.

More appropriate outputs would have been "Promotion processed"; "Pay change processed", and so forth.

With the original list you can do any amount of 'a' and 'c' and accomplish nothing but kill time!

1.5 POOR WORK MEASUREMENT METHODS AND ATTITUDES

The faults committed here range from secret hidden readings on stop watches, carried in one's pocket, as advocated by some purveyors of what they call "short interval scheduling", to micromeasurement of tasks with varying content. The Veterans Administration spent 57 man years of effort using detailed MTM to set the standards for employees of the Dept. of Medicine and Surgery; employees such as Ward Secretaries, Mail delivery persons, Medical Records Typists, and so on. When the task was done it indicated 1600 employees too many when in reality the hospitals were badly understaffed.

1.6 OVERLY COMPLEX REPORTING SYSTEM

When work goes through a series of steps, with different employees, there is no need to count the production at each step, particularly when the final step becomes a recorded fact in a management information system. Further, additional work may inadvertantly be introduced into a system. I have seen several cases where documents were worked on in sequence. In order to reduce the time to move the work it was ordered that "work shall move in batches of 200." Subsequently, the workers spent considerable time, at each step of the work, counting the number of items in a batch to make sure that they were passing on 200!

1.7 FAILURE TO EMPLOY PC'S IN THE PRODUCTIVITY REPORTING SYSTEM

Some productivity systems are abandoned because of the inordinately large amount of time required to prepare the necessary reports. We just have not fully learned what the PC can do for us. (I could devote this whole paper to this topic.)

1.8 LOOKING FOR A UNIVERSAL "BUCKET OF IMPROVEMENTS"

Unfortunately, too many people expect to find a "bucket of improvements" in a book or at a seminar. Being told "where and how to look" is not enough for them. Unfortunately, they do not seem to realize that what is an improvement in one situation may be a dis-improvement in another. Even a suggestion to replace labor with equipment, even in the office, is not specific enough for these "lazy brains."

1.9 "WHO ELSE HAS DONE IT?"

This pitfall is the assumption that if no one else has done it, it is not worth trying to do it, when the reverse may be the real truth. When I was a consultant for a large meat packer my client lost a large order for sausage when a competitor underbid him. The sausage master claimed the competitor was loosing money. I claimed that they probably were using different cuts of meat. The sausage master claimed this was impossible. I suggested using linear programming to find the cheapest legal mixture. The Chairman of the Board asked if it had ever been done before. I admitted that I knew of no case.

"Get to work at it, and get to work fast," he said. "But stay mum!"

We increased our profit on the first sausage to which we applied the approach by 2800 percent! We had the industry "by the tail" for years.

1.10 NEGATIVE MOTIVATORS

Hospitals, for years, have measured their outputs, and been reimbursed essentially by "patient bed-days."

In 1971 we worked with two small Indian Health Service hospitals. We found that if you both measured productivity and budgeted the hospitals in terms of "weighted patients restored to health," that productivity went up 22 percent in the first year, and they usually had empty beds.

The use of patient-bed days for reimbursement, a few years ago, was driving Medicare costs through the ceiling. Medicare introduced DRG's, diagnostic related groupings, with a standard bed-day value for each category, as a basis for payment. A hospital bed ceased being an automatic money making machine. As a result of the use of the DRG's there are loads of empty hospital beds in areas with heavy Medicare populations.

The Veterans Administration hospitals used to be so full you had to be on a waiting list to get in. The VA applied a procedure similar to the DRG to control hospital costs. Most VA hospitals now have empty beds, lower cost and shorter ALOS's (average length of stay.) Each bed in the VA is no longer an automatic money making machine.

For another case, a client came to me with the complaint that his factory designs were resulting in overly expensive factories. Their measure of the productivity of the design department was "the difference between design cost and 10 percent of the factory construction costs."

Would the high cost of their factories surprise you?

Another great way to motivate a design department in the wrong direction is to count the number of drawings made.

It should surprise no one to learn that is it very easy to make an enormous number of views of anything!

1.11 THE FAILURE TO "STAY WITH IT"

This is the last pitfall I want to talk about. American management seems to have taken on one new direction after another, and then dropped each idea in place of a new fad.

To mention a few, we've had "statistical quality control", "job enlargement", "job enrichment", "zero defects","the management grid", and now, "just in time", "zero inventory", and "quality circles"; in government we've had "workload based budgets", "planning, programming, budgeting", "zero based budgets", "productivity improvement", and so forth.

I am not saying that some of these may not have been good ideas, or good programs, but to pursue them for too short a period to really see what they can do, and then chase after something else, only to drop that shortly afterwards, does not appear to me to be the way to get things done.

2. BEING SUCCESSFUL AT IMPROVING PRODUCTIVITY

2.1 "LUCKING OUT"

One way would be to "luck out." It is possible that an organization can, by chance, stumble on to a highly productive way to work. However, this cannot go on time after time. Some systematic approach is needed.

2.2 A SYSTEMATIC APPROACH

All of the successful attempts to improve productivity that I have seen have resulted from an approach characterized by the following 11 steps, or some variant of these steps, but with essentially the same content.

(a) General reconnaisance and missionary work

This step consists of two things. One, finding out what is really going on, and two, making certain that those who will participate know the objectives and ground rules.

I learned this step the hard way. I took on a job to improve the productivity of a leather tanning plant's shipping department; a plant that tanned leather for men's shoe uppers. I did not do a preliminary reconnaisance. I PRESUMED that leather would be heavy for work shoes and light for dress shoes; that it would be black or brown. You can imagine my consternation when I discovered that there were over 200 shades of black; that leather could be shipped as one of more than 1,000,000 specifications!

(b) Identifying the outputs

If the organization is making hard product one starts with the catalogs anddrawings. However if working with the indirect part of the manufacturing organization, or with a service organization, a formal methodology for identifying the outputs is needed. They are usually not readily visible.

We use "work-unit analysis"(1); a systematic method for proceeding from the objectives, hierarchically down to the final units of output. The approach has been used successfully with organizations as diverse as the U.S. Secret Service, The Animal Health Division of the U.S.D.A. and the Social Security Administration.

(c) Selecting work measurement methods

One has to use care to select a work measurement technique that is economical and of appropriate accuracy. A wide variety of techniques are available.

(d) Tentatively designing the productivity measurement system

This is sort of a pre-test. You check to see that if you did apply work measurement to the outputs you identified, could you use these data to support a productivity measurement system.

(d) Rough design of a PC support system

With the PC at its present stage of development and price, I cannot imagine running any productivity reporting system manually. One should plan early on an appropriate use of the use of the PC and the format of the requisite program.

(e) Additional missionary work

Now that preliminary plans are completed they should be explained to all on whom they will impinge.

(f) Applying the work measurement methods

It should be noted that in the manufacturing sector, this is the most time consuming step. In the indirect or service area this may be a small step. Step 2, identifying the outputs, may take much more time.

(g) Reducing the work measurement data to useable form (and in the service sector, adjusting the values to give 100% productivity for the base period).

In the manufacturing plant, one needs to integrate the work measurement standards into the planning and control system. The standards in this setting should have a known degree of attainability.

In the service sector the measure of productivity is a relative measure rather than an absolute one. The raw standard times may indicate a performance below 100%. One cannot allow such data to reach a legislative body; it is too easy to misinterpret. We have always adjusted the standard times so that performance in the base period is 100%.

(h) Finalizing the PC reporting system

It will be necessary to assign reporting requirements to various individuals as well as assigning responsibility for operating the system.

(i) Applying the productivity measurement system

Get it up and running.

(j) Making change the order of the day

Note, we are back to where we started; MIND-SET, but we need a "mind-set for change." Without change, productivity stagnates; organizational failure is almost inevitable. Managers must accept the encouragement of change as one of their vital responsibilities.

Some organizations have used productivity gain sharing to stimulate workers; some Quality Circles; others have professional "change agents" (industrial engineers). Some have had to make changes in the "climate" to enable I.E.'s to succeed.

For instance, in 1964 the U.S. Naval shipyards were experiencing tremendous turnover rates among their industrial engineers. The reason was not hard to find. The shipyards, for years, had no budgets for capital equipment. The industrial engineers would propose cost reduction (productivity improvement) suggestions only to have them turned down for lack of budget; in professional disgust they would quit.

The change introduced was to allow the shipyards, and other Defense Department industrially funded installations, to buy anything, without budget money, that would pay for itself in two years or less. The turnover rate plummeted; the engineers could take pride in their accomplishments.

3. SUMMARY

In summary, the message is:

3.1 MEASURE PRODUCTIVITY IN A RATIONAL FASHION

Without a datum from which to measure the effect of all factors affecting productivity there is no way of knowing what is happening.

3.2 CREATE A CLIMATE FOR CHANGE

Create a climate that is appropriate for the organization; a climate that stimulates the desire to change.

3.3 KEEP ON HELPING TO CHANGE THE WAY WORK IS DONE

Provide training, as may be required, in the technologies that help examine alternative ways of doing things.

REFERENCES

(1) "Improving Productivity and Effectiveness," M.E.Mundel, Prentice-Hall and the Asian Productivity Organization, Tokyo, 1983.

(2) "Total Productivity Measurement for Manufacturing Organizations - Algorithms and PC Programs," M.E.Mundel, The Asian Productivity Organization, Tokyo, 1986, and Kraus Thomson Organization, White Plains, N.Y.

A. IMPORTANCE OF PRODUCTIVITY MANAGEMENT

Productivity Management Frontiers — I, edited by D.J. Sumanth
Elsevier Science Publishers B.V., Amsterdam, 1987 — Printed in The Netherlands

MEETING THE NEED FOR PRODUCTIVITY EDUCATION IN COLLEGES OF BUSINESS: A DELPHI STUDY

F. B. Green
Associate Professor
of Decision Sciences

Ahmad Tashakori
Associate Professor
of Management

ABSTRACT

Productivity management is generally acknowledged as being essential to the economic well being of any business. However, many business managers are unaware of productivity improvement methodologies. A primary reason for this imbalance resides within the nation's colleges of business. Despite recent efforts to improve productivity education, there are very few colleges of business preparing managers for production and operations related fields.

This paper presents the results of a Delphi study designed to develop a production and operations management (POM) curriculum to be used as a model for institutions offering major programs in the emerging field of productivity management.

INTRODUCTION

Productivity is currently being encouraged as an integral part of business school curricula by the American Assembly of Collegiate Schools of Business. Several workshops have been conducted on the teaching of productivity. However, most colleges of business still emphasize the traditional areas of marketing, finance, accounting, and economics. New programs in management information systems and decision sciences have been developed, but only 15 of 240 accredited institutions have established undergraduate programs in fields related to production and operations management (survey based on 1984-85 catalogs of accredited schools and colleges of business). It is believed that an expanded emphasis on the efficient management of production activities is vital if business school graduates are going to play a role in improved national productivity.

The primary objective of this research effort has been to develop and design an undergraduate program of study that would be responsive to the future needs of U.S. industry. More specifically, the study attempts to determine the content of a desired production and operations management curriculum, and the degree of emphasis that should be placed on each POM topic. In order to achieve the objectives of the study, the Delphi survey technique has been employed. It is a technique that uses several rounds of questionnaires, with feedback from previous rounds, in an attempt to reach a consensus among a panel of experts regarding a particular issue.

<u>Note</u>: Both authors contributed equally on this paper.

BACKGROUND

Colleges of business are somewhat belatedly responding to a national resurgence of interest in productivity. During the late 70's and early 80's, the business community began seeking more graduates with specific skills in production, problem solving, industrial relations, and productivity enhancement. Some universities developed specific courses to meet these needs [4]. However, very few institutions developed actual majors for their undergraduate business students.

A 1981 study of management programs at 350 colleges and universities concluded that traditional approaches had failed to produce meaningful POM curricula [2]. Even when programs are developed to fill the gap in productivity education, many people sense that college of business programs fail to convey what is currently happening in industry. Too many colleges are teaching quantitative methods, operations research, and sophisticated theories as though they play a larger role in the workplace than they really do [7]. The gap that continues to exist between practitioners and academics in the POM field has been well documented [3].

To help bring industry needs and academic perceptions into closer alignment, a number of professional associations are offering assistance in curriculum development and placing considerable emphasis on industrial productivity, Japanese manufacturing techniques, computer aided decision making, management controls, and human performance. Organizations such as the American Production and Inventory Control Society, American Productivity Center, Operations Management Association, and American Institute of Industrial Engineers have education committees sponsoring research projects and developing educational materials for colleges and universities as well as their own members [8]. Dale Sauers of York College, Pennsylvania, developed a production management activities chart which ties together a number of POM activities into a practical framework which has served as the basis for curriculum development at several institutions [6].

The challenge remains, however, to assure the participation of industry practitioners along with professional educators in the process of designing a curriculum that includes the development of specific skills along with a general business education. Universities now face the challenge of implementation: What should be included in a production and operations management curriculum to meet the future needs of a nation in a highly competitive industrial world market?

A DELPHI APPROACH

The Delphi method was chosen for the purpose of developing an undergraduate curriculum. It has been shown that the Delphi technique, using a panel of experts, can lead to consensus decisions that are superior to individual decision-making, majority votes, leader decisions, or averaged individual decision-making.

While there are variations in design and content of a Delphi survey, the following are the steps that are usually needed once the survey topic has been determined: (1) selection of a panel of experts, (2) design and preparation of a survey questionnaire, and (3) a procedure for conducting successive rounds in which the questionnaire is reiterated until a consensus is reached [5]. Each successive round consists of a composite panel response from the previous round and the opportunity for each panel member to revise earlier responses in view of overall panel responses.

In order to form a panel of experts that would be able and willing to participate in all rounds, a target panel composed of both academicians and industry practitioners was contacted. A total of 70 individuals (46 academicians and 24 practitioners) expressed their willingness to participate. This formed the panel of experts for the Delphi project.

To determine the content of the desired POM curriculum through a series of questionnaires, the first step was to find out what is presently taught at U. S. universities. A search of university catalogs and POM related textbooks brought forth nineteen (19) production and productivity related topical areas, each including several subtopics reflecting specific concepts or techniques. This list basically shaped the three rounds of questionnaires discussed as follows:

ROUND I: The first round questionnaire listed the nineteen topical areas and related subtopics. Panel members were asked to 1) for each topical area listed add other subtopics that they felt should be included, and 2) list additional topical areas (and related subtopics) for consideration. A total of 58 responses to the first round resulted in an expanded list in terms of both topical areas and related subtopics.

ROUND II: The second round questionnaire was designed to allow all 70 members of the panel to 1) reevaluate the expanded list from Round I and 2) solicit their opinion as to the relative importance of the topical areas in an undergraduate POM curriculum. For this purpose the 23 topical areas were grouped into eight major categories of POM subject areas (Table 1). Panel members were asked to assign a total of 24 available semester hours of course work among the eight major categories to indicate the relative weight that each should receive in a desired POM curriculum. Panelists were also asked to rate topical areas listed under each category (1=no importance, 5=high importance). A total of 42 responses were received from Round II.

TABLE 1

Comparative Statistics From Round II and Round III

Categories and Topical Areas	Mean Round II	Mean Round III	Standard Deviation Round II	Standard Deviation Round III
PRODUCTION SYSTEMS PLANNING	2.7	2.8	1.04	0.82
Product development & design	2.9	3.0	1.12	0.88
Design of production system	3.7	4.0	1.25	0.90
Capacity requirements planning	3.9	3.7	1.23	1.17
OPERATIONS STRATEGY & BUDGETING	2.9	2.9	1.12	0.80
Job-cost control & budgeting	2.9	2.7	1.15	0.97
Logistics and distribution	3.3	3.4	1.09	0.79
*Operations strategy	3.8	4.2	1.37	0.81
PRODUCTION PLANNING & OPERATIONS	3.4	3.4	1.08	0.92
*Forecasting	3.8	3.8	1.12	0.81
Intermediate master planning	3.8	3.9	1.39	0.86
Operations scheduling	3.9	4.0	1.37	0.78
INVENTORY MANAGEMENT	4.2	4.0	1.44	0.94
Materials requirement planning	4.2	4.3	1.26	0.82
Materials inventory control	4.1	4.3	1.35	0.66
*Purchasing	3.0	3.2	1.29	0.75
QUALITY AND PRODUCTIVITY	3.8	3.8	1.60	1.22
Production activity control	3.8	4.0	1.21	0.80
Quality control	4.1	4.5	1.39	0.65
Automated manufacturing	3.5	3.8	1.37	0.85
*Just-in-time manufacturing	3.5	3.9	1.48	0.99
OPERATIONS RESEARCH IN POM	2.1	2.1	1.07	0.85
Project management	3.3	3.4	1.46	1.00
LP and optimization	2.4	2.6	1.34	1.02
Decisions under uncertainity	2.4	2.6	1.27	1.00
DECISION SUPPORT SYSTEMS	2.0	2.0	0.84	0.76
Modeling and simulations	3.0	3.2	1.16	0.75
Computerized information systems	3.5	3.9	1.42	1.01
HUMAN ASPECTS	2.8	2.9	1.82	1.70
Industrial relations	3.2	3.7	1.46	0.93
Service sector operations management	3.3	4.0	1.72	1.00

Values in category rows correspond to semester hours.
Values in topical area rows correspond to rankings of importance.
*Topical areas added as a result of Round I Questionnaire
(Complete list with subtopics available from the authors)

ROUND III: The third round questionnaire served as both feedback from Round II and as a survey instrument to solicit panel members revised responses. Therefore, a summary sheet provided descriptive statistics (mean, standard deviation, median, mode, and the interquartile range $Q_1 - Q_3$) from Round II, for categories as well as their related topical areas. An additional column was added to include individual panelist's response from Round II. Respondents were asked to reevaluate their responses, in view of the composite panel results, and enter their new responses. A total of 40 responses were received from Round III. Table 1 illustrates summary statistics from Rounds II and III.

INTERPRETATION OF RESULTS

A comparison of summary statistics of Rounds II and III clearly indicates that the Delphi process did in fact move panelists towards a consensus regarding the relative importance of the POM subject areas for an undergraduate curriculum. Measures of central tendency, measured by mean, median, and mode, changed little. Yet standard deviations showed considerable reduction in the third round indicating that the third round reduced variability considerably. This reduction is larger for topical areas than for categories.

Not only did variability of responses from the mean decline, the interquartile ranges of all 8 categories showed reductions also. These changes provide support for the contention that the third round questionnaire brought panelists much closer to a consensus on the subject.

In view of the above observations, it could be concluded that many panelists with non-comfirming responses to the second round, actually modified their third round responses toward the middle values. In other words, the third round questionnaires resulted in a near consensus among panel members regarding 1) what should be included in the curriculum, and 2) how much coverage should be received by these POM concepts and techniques.

IMPLICATIONS FOR CURRICULUM DESIGN

Colleges and universities generally apportion material to be taught in courses with credit-hours or semester-hours. A typical college course meeting three hours a week for fifteen weeks would be given a three semester-hour designation.

The results of the Delphi survey are most revealing in terms of implication for a POM major. Table 2 indicates the mean number of semester hours allocated to each category (or course) in descendimg order. It should be recalled that panel members were constrained to make these allocations within a 24 semester hour curriculum.

Two major categories of topical areas emerged as being of considerable importance in terms of semester hours allocated and rankings of their contents. These are Inventory Management and Quality and Productivity.

TABLE 2

A POM Curriculum: Courses and Topics Recommended for an Undergraduate Major of 24 Required Semester Hours

	Mean Number of Semester Hours	Percent of Program Devoted to Topics	Number of Semester Hr. (Rounded)
INVENTORY MANAGEMENT	4.0	16.8	4
Materials requirement planning (MRP)			
Materials management & inventory control			
Purchasing			
QUALITY AND PRODUCTIVITY	3.8	15.8	4
Production activity control			
Quality control			
Automated manufacturing			
Just-in-time manufacturing			
PRODUCTION PLANNING AND OPERATIONS SCHEDULE	3.4	14.1	3
Forecasting			
Intermediate master planning			
Operations scheduling			
OPERATIONS STRATEGY AND CAPITAL BUDGETING	2.9	12.0	3
Job-cost control and budgeting			
Logistics and distribution			
Operations strategy			
HUMAN ASPECTS OF PRODUCTION AND SERVICE	2.9	12.0	3
Industrial relations/human resources			
Service sector operations management			
PRODUCT DESIGN AND PRODUCTION SYSTEM PLANNING	2.8	12.0	3
Product development and design			
Design of the production system			
Capacity requirement planning			
OPERATIONS RESEARCH IN POM	2.1	8.8	2
Project management			
Linear programming and optimization			
Decision making under uncertainty			
DECISION SUPPORT SYSTEMS	2.0	8.3	2
Modeling and simulation			
Computerized information systems			
		100.0%	24

Inventory Management includes the topical areas of material requirements planning, materials management, inventory control, and purchasing. All except purchasing were given ratings of high importance. Panel members indicated that Inventory Control should constitute 16.8 percent of the major, or approximately one sixth of the overall program. Many colleges of business do not currently offer a course in inventory control or materials management.

The second category encompassing Quality and Productivity includes the topical areas of production activity control, quality control, automated manufacturing, and just-in-time manufacturing. Of the four topical areas, all were rated above average in importance with quality control leading the list. According to the survey, Quality and Productivity should constitute 15.8 percent of the overall program. Again, it should be noted that few colleges of busines offer courses in quality control or in the productivity topics mentioned.

Next in line was Production Planning and Operations Scheduling with a percentage allocation of 14.1; followed by Operation Strategy and Capital Budgeting, Human Aspects of Production and Service, Product Design and Production System Planning, each with an allocation of approximately 12 percent. Operations Research in POM and Decision Support Systems each received allocations of less than 9 percent.

This last observation is particularly interesting since many colleges of businesses do offer courses in operations research (or related courses in management science and quantitative methods); and the various topics under decision support systems are available as courses also. It appears as though business college courses available to support a POM major exist in inverse proportion to the perceived requirements of the major. This would seem to support the findings of the Bowling Green study cited earlier in this paper [2].

Since the purpose of this paper is to present an undergraduate curriculum for a POM major field of study, Table 2 includes a column in which the mean semester-hours have been rounded to the nearest whole number. The eight major categories could be offered as courses which include the topical areas listed in Table 2. However, colleges generally prefer to offer courses consisting of three semester-hours. Consequently, the topics could be rearranged so that all courses would be three semester-hours each and constitute a major with a total of 24 semester hours.

COMMENTS OF THE DELPHI PANEL

Much of the rationale for the curriculum resulting from this study may be determined from the comments of Delphi panel members.

They were usually handwritten in the remarks section of the questionnaire. To enable the reader to get the flavor of varying opinions, some of the more representative comments are presented below:

> "Quality and Productivity are the <u>most</u> important segments of today's industry in the U. S. With these under control, U. S. industry can reduce costs and increase the worldwide competitive position. The subjects listed under this category are the keys to today's industry staying alive."

> "One factor that is important to determining the amount of time required for a topical area is the span of topics and the time needed to teach it. Inventory Management is a case in point: to teach MRP, inventory control, materials management, and purchasing requires much time."

> "The impact of human relations on the manufacturing macrosystem is almost universally under-realized. A thorough study of these factors will give any practitioner a competitive advantage."

> "In Operations Strategy, we have a chance to get away from tool orientation - teach students about the big picture. That has been one of our shortcomings."

> "Decision Support System - I believe over the careers of the students in your program we'll see the integration of a whole new level of support tools, based on improvements in processing economics, artificial intelligence, data collection, etc. Students enter this arena with their eyes open."

> The 'people' component in OM is often ignored; yet any plant manager will tell you that managing people is his or her toughest job."

> "Decision making under uncertainty is done all the time by production managers. I know; I was one for 20 years, but mathematical models really weren't useful to me."

> "While I will agree that extensive rumination on linear programming is unnecessary, it does force the student to mathematically consider objectives and constraints. It poses a methodology for establishing an analytical approach to problem solving."

> "I'm not convinced that just-in-time manufacturing in and of itself is particularly novel -- it's an execution philosophy or approach. It needs to be identified and discusses for information, people should not set out to imitate a given model or approach. They should seek new ways to accomplish the objective."

Other comments reflected similar sentiments. There were also comments pertaining to topical areas that people thought should be included under different major categories, in courses apart from the major, or de-emphasized all together. Some of the items listed as topical areas cross over into other disciplines, making any kind of categorization a difficult task at best. Neither the authors nor the Delphi participants feel bound by a particular scheme of categorization.

CLOSING COMMENTS

This study was successful in bringing about a degree of consensus among participants and producing a document that may well serve to guide colleges of business in the development of an undergraduate curriculum for production and operations majors. Productivity issues were discussed, and the relative value of specific topical areas was determined with some success. Few universities offer programs resembling that which has been developed as a result of this study. Curriculum development is a continuous process and much remains to be done.

Some very important issues were not addressed in this study: What kind of background is required of faculty teaching in these areas? Are universities going to be able to find professors with the necessary qualification? What alternatives to the clasroom setting should be used to teach these topics. To what extent should industry be involved? Some of these issues are being addressed at meetings, seminars, and conferences; such as the annual Academic Practitioner Liaison Conference sponsored by the American Production and Inventory Control Society [8], and periodic meetings of the American Assembly of Collegiate Schools of Business at the American Productivity Center [1]. Whatever avenue is followed, business schools and colleges must expand the scope of their education offerings to include programs that directly affect our nation's productivity. The curriculum guidelines contained in this study may help in that endeavor.

REFERENCES

[1]American Productivity Center, "Productivity in the Business School Curriculum," Productivity Brief, Number 46, March 1985.

[2]Chan, K. H., et al, "Production and Inventory Management Educational Requirements Report for APICS Phase II Study," College of Business Administration, Bowling Green State University, Bowling Green, Ohio, May 1981.

[3]Green, F. B., "A Commentary on the Major Influences Affecting POM Course Content and Orientation," Proceedings of the 14th Annual Meeting, Southeast AIDS, Savannah, Georgia, February 1984, pp. 210-212.

[4]"Hard Times Push B-Schools into Basics," Business Week, August 30, 1982, pp. 23-24.

[5]Linstone, Harold A., and Murray Turoff (editors), The Delphi Method: Techniques and Applications, Reading, Massachusetts: Addison-Wesley Publishing Company, 1975.

[6]Sauers, Dale G., "New Styles of Managing Production: A Proposal for a New Education; Collegiate News and Views, Fall-Winter 1984, pp. 21-26.

[7]Wight, Oliver, Production and Inventory Management in the Computer Age, Boston, Massachusetts: Cahners Book International, 1974.

[8]Woolsey, Robert, E. D., "Report of the 1985 Academic Liaison Committee Workshop" Production and Inventory Management, Fourth Quarter, 1985, pp. 127-140.

Productivity Management Frontiers — I, edited by D.J. Sumanth
Elsevier Science Publishers B.V., Amsterdam, 1987 — Printed in The Netherlands

PRODUCTIVITY MANAGEMENT, AN IN-DEPTH STUDY OF SELECTED U.K. ORGANISATIONS

W. N. SHAW
Department of Business Organisation, Heriot-Watt University, 31 Grassmarket, Edinburgh EH1 2HT, Scotland, U.K.

ABSTRACT

This paper reports some of the key findings from an in-depth study of the internal productivity/efficiency departments (often labelled industrial engineering in the U.S. but generally called management services in the U.K.) in 34 U.K. commercial, industrial and public sector organisations. The results of the study are set within the context of management services in the U.K., its history, development, complexity and diversity. The findings reported here are first the characteristics of internal productivity/efficiency departments and second the characteristics of the managers of such departments.

INTRODUCTION

It has been argued by a number of commentators (Baker & Schaffer, 1969; Gruber & Niles, 1972 and Child, 1977), that as modern organisations become more complex there is a proliferation of groups of specialists within them. Hence there is an increasing need to fuse the knowledge of these specialists or experts with the operating experience of line managers. This should result in organisations being led by knowledge responsive management teams who are much more able to cope with complex, non-programmed kinds of decisions (Mumford & Pettigrew, 1975).

One such group of specialists that has, as yet, undergone little objective study is the internal productivity experts. The chief executive of a local authority, interviewed by the author of this paper suggests one reason why this may be so,

> "There are, first of all, very few internal productivity units that have a high profile. If they have then I think they tend to encounter difficulties and I think also that they would not last very long. So from that point of view they are operating within organisations and not much evidence of their activities is seen by the outside world".

The rationale or justification for an organisation having an internal productivity unit is generally to ensure that a group exists, within the organisation, which is exclusively committed to improving the efficiency, effectiveness, productivity and resource utilisation of that organisation. Within the U.K. the internal productivity experts are often located in management services departments or their equivalents, e.g. industrial engineering, productivity services, organisation and methods or internal consultancy.

Based on the writer's research (Shaw, 1984a) it has been estimated that there may well be 95,000 plus, internal productivity experts throughout industry, commerce and the public sector. Their location ranges from the prestigious Efficiency Unit of the Cabinet Office to the time and motion departments of small manufacturing companies in the clothing industry. There are very few U.K. industries in which management services is not practiced. Management Services practitioners can and do become involved in a wide range of activities in different functional areas and at various levels in the organisational hierarchy. Increasingly over the last six years within public sector organisations (Holtham and Stewart, 1981) including the Civil Service (Command 8293, 1981), application of the concepts of value for money, economy and efficiency have received much attention. A major driving force behind current U.K. interest in the topic, especially in the public sector, has been the pressures of financial stringency brought about by the policies of the present Conservative (right wing) Government which came to power in 1979 and was subsequently re-elected in 1983. One of its policies for central and local government has been to set a financial framework with strict limits on manpower and money and to create incentives towards economy and efficiency. This is very similar to President Reagan's pronouncements in 1986 on a productivity improvement programme for Federal government agencies. It is important to remember, however, that many U.K. industrial and commercial organisations like their counterparts in the U.S.A. have been working in the productivity improvement area for many years. Seeking improvements in the utilisation of an organisation's resources is not purely a U.K. phenomena, it is a topic of intense interest in many countries throughout the world. There are productivity organisations and networks at international, "continental" and national levels. For example, at the international level there is the World confederation of Productivity Science, within Europe there are two bodies, the European Association of National Productivity Centres and the European Federation of Productivity Services,whilst in Asia, there is the Asian Productivity Organisation. Many countries have their own productivity associations. In the United States for example there are, The American Productivity Management Association having a membership of 200 plus major corporations and The National Productivity Network which is a consortium of state, private and university productivity organisations whose prime objective is to help small to medium sized organisations. Over the years individual countries have mounted productivity campaigns, some of the most recent being in Denmark 1983, Norway 1982 and India 1982. A common feature of many of these campaigns has been the involvement of representatives from employers, trade unions and government. All this, is of course, in addition to the efforts continuously being made to improve resource utilisation at the level of the individual organisation. Despite the apparently large number of internal productivity experts in the U.K. there has been remarkably little empirical research carried out into the work activities of this occupational specialism. Of the few research studies that have specifically examined this area two stand out, Lansbury's (1978) views about occupational specialisation are based on a case study of management services staff in British Airways. Whereas Page's (1982) research provides a useful insight into the role of the internal consultant in a division of a multi-national company.

There have been, however, within the U.K. in recent years an increasing number of research studies of other occupational specialisms, for example personnel management (Watson, 1977; Tyson, 1979), accountancy (Hooper, 1980), data processing (Mumford, 1972) and production/operations management (Lawrence, 1983; Voss, 1983). The aim of the writer's research project was to expand our extremely limited knowledge of the activities of the internal productivity experts.

RESEARCH DESIGN

The research reported in this paper has been seeking to answer questions such as; How do specialists and in particular the internal productivity experts influence organisational decision making? What similarities and differences are there in the activities carried out by internal productivity units in different organisations? What technical and personal qualities should productivity practitioners possess? How can organisations learn to more effectively utilise their productivity units?

The research had objectives at two levels. At the more theoretical level the objective was to add to our knowledge and understanding of the concepts of roles, relationships, structures and processes within organisations. At a more practical level it was hoped that the research would lead to the establishment of guidelines for line management to more effectively utilise staff specialists whilst ensuring that staff spcialists give a more meaningful and effective service to line management. In particular, the research focused on issues such as occupational specialisms, the consultant/client relationship; the identification of problems; the implementation of solutions; power, influence and politics within organisations and their links with an organisation's objectives, structure and technology. Although only selected results from the research are described here the interested reader may find out about other aspects of the study in other papers written by the author which are cited later.

Access was gained to 34 organisations, of these 4 were used for a pilot study and 2 disbanded their productivity units. Thus most of the fieldwork took place in the remaining 28 organisations. Their industrial classification is shown in Table 1. Each of the organisations was an autonomous institution though some of them were related by ownership to larger organisations. Although their size, as measured by number of employees, ranged from 250 to 30,000, the average was 4,600 with some 52% having between 1,000 to 5,000 employees. It can be seen from the Table that a wide range of products and services are provided by the organisations.

TABLE 1

Survey organisations by industrial classification

Sic class no.	Label	Commercial	Industrial	Public Sector
16	Energy distribution			1
22	Metal manufacture		1	
25	Chemicals		2	
31	Manufacture of metal goods		1	
33	Manufacture of office machinery & DP equipment		1	
34	Electrical & Electronic engineering		2	
37	Instrument engineering		1	
41/42	Food and drink manufacturing		2	
48	Processing of rubber & plastics		1	
61	Wholesale distribution	1		
79	Postal Services & Telecommunications			1
81	Banking & Finance	3		
82	Insurance	2		
91	Public administration			6
93	Education			1
95	Medical & other health services			2
	Total	6	11	11

The units selected for study were not intended to be representative of any defined population. The main factors determining their selection were; that access could be gained; that the internal productivity units were worthwhile researching and that the units covered a reasonable spread of different industries. Information was obtained from four groups, namely the heads of the productivity units, productivity practitioners, clients (e.g. line managers) and senior management (e.g. directors and chief executives). However the key respondents were the directors or managers of the productivity units. The bulk of the data was gathered by means of structured interviews and biographical questionnaires.

In addition significant documentary data was supplied by many of the organisations, for example project reports, departmental control indices, job descriptions and organisation charts. Over one hundred, two hour structured interviews were carried out and nearly forty biographical questionnaires received. Full details of the research design and methodology are reported by Shaw (1982) in an earlier paper.

CHARACTERISTICS OF THE INTERNAL PRODUCTIVITY UNITS

Background information about the units studied was obtained, this included when the productivity units were first established, the location of such units within organisational structures, the size of units expressed as the ratio of productivity staff to total employees and the range of work activities undertaken by the units.

It was found that the majority (86%) of units visited were established between 1956 and 1972. The location of productivity units in an organisation's hierarchy varied tremendously, although two units reported directly to the Chief Executive Officer the majority (89%) reported at senior management level whilst a small number reported to middle management. A system for classifying internal (and external) productivity units has been developed and described elsewhere (Shaw 1984b). Like location the size of productivity units varied considerably, the smallest unit visited had a total complement of five whilst the largest had a staff of some 140. These figures, however, disguise the fact that 72% of the units had a staff of twenty or less. A more useful basis for comparison was found to be the ratio of productivity staff to total employees. It can be seen from Table 2 that there was a significant variation from a low of 1:14 to a high of 1:1342 for all organisations. The industrial organisations in the sample had the lowest average ratio, in other words they had quantitatively, though not necessarily qualitatively, the best staffed internal productivity units whereas the public sector organisations had quantitatively the worst staffing levels.

TABLE 2

Ratio of internal productivity staff to total number of employees

Ratio	Commercial	Organisational group Industrial	Public sector	All
Average	1:211	1:122	1:627	1:304
Range	1:82 to 1:400	1:14 to 1:233	1:49 to 1:1342	1:14 to 1:1342

One possible explanation for the differences in ratios amongst the three organisational groups could well be the location of the productivity units. In general, the industrial organisations in the sample operated on the basis of decentralised productivity units. In other words the main units were located at operational sites with sometimes, but not always, a small co-ordinating central productivity unit located at head office. This differed from the majority of commercial and public sector organisations which operated centralised productivity units based at head office that supplied a service as and when required to outlying departments or locations.

What then are these units called in the U.K.? From the small sample of this study it can be seen (Table 3) that a variety of titles were used. This confirms the findings of earlier studies (Harrison, 1979 and Minter, 1984) of much larger samples which, in addition, clearly established that management

TABLE 3

Titles of Units

Title	Commercial	Industrial	Public Sector	No.	%
Industrial engineering emphasis	0	8	1	9	32
e.g. Industrial engineering Production engineering Site services Technical services					
Management services emphasis	2	0	7	9	32
e.g. Management services Personnel & MS Finance & MS Administration					
Organisation & methods emphasis	4	0	1	5	18
e.g. O & M					
Work study emphasis	0	3	2	5	18
e.g. Work study Work study & O & M					

Source: Questionnaire

services (MS) is by far the most frequently used title in the U.K.

The next Table (4) shows, in overall rank order, the wide range of work activities undertaken by the 28 management services units. Not all the units were involved in every activity, 82% of them were undertaking work associated with new technology whilst only 14% of the units made use of operations research techniques such as simulation, modelling and forecasting. It was found that the mix of work activities varied over time. Although traditional activities such as work study and O & M figures prominently it should be noted that increasing amounts of organisation development type work such as organisational and management reviews and internal consultancy are being undertaken particularly within public sector organisations. Other interesting differences between the three groups were first, the comparative lack of involvement of industrial MS units in setting staffing and manning levels. Second, it was rather surprising to find that so few industrial MS units appear to participate in project groups. Third, none of the commercial units indicated that they were involved in bonus or incentive schemes. It should be noted however that most of them do make use of work measurement mainly as a technique for setting and controlling staffing levels. Certain activities were undertaken by so few units that they have not been recorded in the Table. However it is worthwhile noting that three units, all within industrial organisations, were involved in operating managerial controls over capital expenditure. Two units, belonging to public sector organisations had a specific remit, albeit a small one, to aid the introduction of new ideas and innovations into their organisation. Finally two units, one industrial and one public sector had recently been given an energy conservation role to play.

TABLE 4

Work Activities of MS units

Activity	Organisational Group Commercial (N=6)	Industrial (N=11)	Public Sector (N=11)	Total	%
Introducing new technology	6	7	10	23	82
Traditional work study	1	11	7	19	68
Advising management	3	5	10	18	64
Setting staffing/manning levels	6	4	7	17	61
Traditional O & M	5	3	8	16	57
Operating systems & procedures	4	5	5	14	50
Participating in project groups	5	3	6	14	50
Organisational & Management reviews	4	2	8	14	50
Providing information to management	2	4	5	11	39
Bonus/incentive schemes	0	4	7	11	39
Job evaluation	3	5	2	10	36
Training	3	2	4	9	32
Acting as internal consultant	1	1	6	8	29
Acting as project co-ordinator	1	1	4	6	21
Negotiating with TU's	1	2	3	6	21
Cost reduction programmes	0	4	1	5	18
Operations research	0	3	1	4	14

Source: Interviews

There are some interesting omissions from the list, no units were specifically involved in quality work or job design, nor were any of the industrial units doing work in the CAD/CAM field. The management services managers stated that their units had multiple missions. These were,

- being productivity innovators
- bringing about change
- advising management
- providing information to management
- being efficiency auditors

CHARACTERISTICS OF PRODUCTIVITY MANAGERS AND PRACTITIONERS

The study sought to identify some of the key characteristics of both managers and practitioners. As one would expect the most important work activity of the management services managers was found to be managing their departments. More specifically this involved organising and controlling their staff by means of allocating work and resources, planning and co-ordinating and motivating (Table 5). Given the nature of management services it was to be expected that dealing with line management would also be an important activity. It is perhaps a little surprising that staff training and develop-

TABLE 5

Main work activities of MS managers (N = 27)

Main work activities	No. of mentions
Organising and controlling staff - Supervising staff - Allocating and co-ordinating work - Planning and co-ordinating - Motivating staff - Allocating resources of unit - Managing unit - Ensuring target dates met	27
Dealing with line management - Advising management - Liaising with senior managers	20
Working on specific projects	14
Training staff	8
Keeping up to date	7
Highlighting inefficiencies in organisation	6
Attending committees	4

Source: Interviews

ment received comparatively few mentions.

The MS managers were asked to identify the attributes required of a person doing their job (Table 6). Again there was considerable agreement as to the main attributes. These were, in order of importance, good inter-personal skills, extensive organisational experience and appropriate experience/knowledge of technical aspects of the work. In the list of required attributes it is interesting to note the relative importance accorded to "experience" whether it be of an organisational or technical nature and the comparatively low importance given to handler of organisational politics, promoter of change or being a salesman.

TABLE 6

Attributes required of the MS manager

Attribute	No. of cases
Good inter-personal skills	20
Extensive organisational experience	19
Experience/knowledge of technical aspects of the work	17
Communication skills	11
Management skills	11
Ability in solving problems	10
Able to handle organisational politics	10
Able to motivate staff	7
Reasonable numeracy	7
Promoter of change	5
"Salesman"	5

Source: Interviews

A number of respondents added comments about the personal qualities that MS managers should possess. These included qualities such as commonsense, patience, self-motivated, sense of humour, thick skinned and even "kangaroo mind"!

The MS managers were then asked about the positive and negative aspects of their job. A wide range of views were expressed about the positive features namely, "the variety in the job, being at the centre of things, dealing with a wide variety of people, introducing change and being innovative". The main criticism of the job was "the frustrations of being in a service role".

Other findings from the study were that about 40% of the MS managers in the sample had been in their present job for 6 or more years. In contrast, however, over 60% of the group had worked for 2 or more employees. There was a wide disparity in formal education qualifications amongst the managers. The highest level of formal education for 27% of the sample was a high school qualification whilst at the other end of the spectrum some 38% had a first degree or postgraduate qualification.

CONCLUSIONS

This paper has provided, by necessity, a brief overview of selected results from a study of U.K. internal productivity units. The findings outlined in the paper are;

- The present U.K. government has brought resource utilisation issues to the forefront of the organisational agenda, especially in the public sector.
- The commonest description for the productivity/efficiency specialism is industrial engineering, if a worldwide perspective is taken, or management services if considered only in a U.K. context.
- Management services as a specialism is complex, exceedingly diverse and rarely has any clear cut boundaries within which it operates.
- The type of work undertaken by management services may vary significantly from industry to industry and even vary between organisations in the same industry.
- The main missions for management services are as productivity innovators and change agents.
- Management services work activities range from the traditional areas of work study and O & M to the developing area of new technology.
- The main personal attributes and qualities required of an MS manager are seen to be inter-personal and managerial skills and organisational experience allied to appropriate technical knowledge.

REFERENCES

Baker, J.K., & Schaffer, R.H., 1969. Making Staff Consulting more Effective. Harvard Business Review, Vol. 47, No. 1: 62-71.

Child, J., 1977. Organisation, London, Harper & Row.

Command Paper 8293, 1981. Efficiency in the Civil Service, London, HMSO.
Harris, N., 1979. Management Services in the U.K. Management Services, Vol. 23, No. 12: 18-24.

Holtham, C., & Stewart, J., 1981. Value for money: A Framework for Action, INLOGUV., University of Birmingham.
Hopper, T.M., 1980. Role Conflicts of Management Accountants and Their Position Within Organisation Structures. Accounting, Organisations & Society, Vol. 5, No. 4: 401-411.
Lansbury, R.D. 1978. Professionals and Management, University of Queensland Press.
Lawrence, P.A., 1983. Operations Management: Research and Priorities. Report to the Social Science Research Council.
Minter, A.L., 1984. The Technical Board Survey - Membership. Management Services, Vol. 28, No. 1: 26-31.
Mumford, E., 1972. Job Satisfaction, A Study of Computer Specialists, London, Longman.
Mumford, E., & Pettigrew, A., 1975. Implementing Strategic Decisions, Longman.
Page, A.J., 1982. The Role of the Internal Consultant in a Division of a Multi-National Company, Aston University; unpublished Ph.D.
Shaw, W.N., 1982. An Ongoing Study of the Internal Organisational Efficiency Experts. Scandinavian Journal of Materials Administration, Vol. 8, No. 3: 56-72.
Shaw, W.N., 1984a. Management Services as a Profession in BENTLEY, T.J., (ed), Management Services Handbook, Holt, Reinhart & Winston, 34-48.
Shaw, W.N., 1984b. The Place of Management Services in the Modern Organisation in BENTLEY, T.J., (ed), Management Services Handbook, Holt, Reinhart & Winston, 48-63.
Tyson, S., 1979. Specialists in Ambiguity, Personnel Management as an Occupation, University of London; unpublished Ph.D.
Voss, C.A., 1983. Future Research in production/operations management, A report to the Social Science Research Council.
Watson, T.J., 1977. The Personnel Manager, London, Routledge and Kegan Paul.

B. PRODUCTIVITY MEASUREMENT AT COMPANY/DIVISION/PLANT/PRODUCT LEVELS

Productivity Management Frontiers — I, edited by D.J. Sumanth
Elsevier Science Publishers B.V., Amsterdam, 1987 — Printed in The Netherlands

A MODEL FOR PRODUCTIVITY MEASUREMENT IN A MULTI-PRODUCT ORGANISATION USING GOAL PROGRAMMING AND MULTI-ATTRIBUTE UTILITY THEORY

G.D. SARDANA* and PREM VRAT**

* General Manager, Hindustan Brown Boveri Limited,Baroda(India)

** Professor of Industrial Engineering,Mechanical Engineering Department, Indian Institute of Technology,Delhi (India)

ABSTRACT

Sardana, G.D., and Vrat Prem.,1987. A model for productivity measurement in a multi-product organisation using goal programming and multi-attribute utility theory.

A model to measure productivity of organisational system is presented. Productivity has been viewed in a broader perspective to represent the degree of accomplishment of performance objectives, reconciled optimally through Goal Programming and evaluated and converted to a common scale through the Multi-attribute Utility Theory. The model is applicable to all types of organisations both in the manufacturing and the service sector. An illustrative case of a manufacturing organisation with multiple products is included.

1. INTRODUCTION

The importance of productivity and its measurement has been well recognised. However, as Sink (1983) points out, the concept of Productivity is still vaguely defined and least understood,although it is probably the most widely discussed topic. Different meanings, definitions, interpretations and concepts have emerged as experts working in various areas of operations have looked at it from their own emphasis and perceptions. Mali (1978) says that the several views of productivity propagated have contributed to the confusion and obscurity about its nature. The measurement is even more clouded with confusion as the same is only a process of translating the concept and objectives into a methodology. Thus there is a need to clarify the concept of productivity. This paper attempts to examine productivity in its broader perspective as an index of the accomplishment of the goals of an organisation viewed as a system. A detailed methodology is developed so as to take into account the relevant goals and the total inputs needed to achieve the same. A case study of the application of the model for a multi-product organisation is presented. The case study forms the primary focus of the paper.

2. PRODUCTIVITY: A MORE MEANINGFUL PERSPECTIVE

2.1 Basic Objectives of Measuring Productivity

A productivity index cannot be viewed as an abstract number; it should convey a meaning and a message. Measurement of productivity can therefore, be recommended to concern with the following objectives:

- to help identify potential for improvement.
- to help locate areas for re-allocation and deployment of resources.
- to help provide measures which become representative of the accomplishment of the defined and the established goals in the hands of the user.

In brief, the measurement of productivity should be viewed as an exercise to obtain a more effective performance of the organisation.

2.2 Productivity - a Broader Concept

The concepts and interpretations of the term 'Productivity' have been perceived differently. Prem Vrat and Sardana(1984) have compiled twenty such definitions. A large number of concepts consider productivity as an output-input relationship relevant mostly to a production system, implying that an organisation works as a physical system with variables and their inter-relationships amenable to precise definitions. The basic reliance is on the acceptance of a stimulus-response model of causality that an input causes an output. This conceptualisation apparently creates a bias towards production function or allied activities to the exclusion of other economic as well as non-economic performance outputs, such as achieving a share of the markets, new product introduction, completion of schedules, societal goals etc. These and several other non-economic performance consume the input resources and as such should get fully projected in a model to measure productivity. Similarly, factorial productivity measures connected with input factors such as labour, capital, etc. are misleading and inadequate. Firstly, the input factors cannot be studied in isolation to one another. Improvement in one factorial productivity is generally at the cost of the other. Besides, an input factor like labour is present everywhere. Secondly, an important input like managerial resource finds no place as an input factor in such measures.

An organisation on the other hand works as a system. Its subsystems interact with each other and share the resources. Consistent with this approach, Robins (1984) shows that outputs become the ends or the performances, whereas acquisition of inputs and processing efficiencies are the means. Productivity of a system (as also of sub-systems) represents the overall performance of the system. It signifies as to how well a system is doing against the performance objectives set in advance. Sardana and Prem Vrat (1983) point out that the setting of performance objectives as goals for achievement becomes a pivotal exercise in such an approach.

The basic approach of viewing organisation as a system (and sub-systems) together with the basic objectives of measuring productivity can be used to review and define productivity in a broader concept:"Productivity as an index of system (sub-system) performance indicates the extent of actual accomplishment of performance objective(s) in relation to the attainable level in a given external environment".

2.3 PO-P approach to measure productivity

Allard et al. (1971) recommend that comparison of actual performance with a target is undoubtedly the best form of measurement, since it concentrates attention on objectives. This philosophy of measurement together with the broader interpretation of productivity as conceptualised, has been used by Sardana and Prem Vrat (1983) to develop a more pragmatic and rational model to measure productivity of an organisation. This model, termed as Performance Objectives-Productivity (PO-P) Approach to measure productivity, takes into account performance, both economic and non-economic, and is applicable to organisations of all types including those belonging to the service sector.

3 NOMENCLATURE

Following symbols/notations are used:

$(P.I)$	Productivity Index of System
$(P.I)_u$	Productivity Index of Sub-system u.
$(P.I)_{vu}$	Productivity Index of Key Performance Area v of Sub-system u.
w_u	Weightage of Sub-system u
w_{vu}	Weightage of Key Performance Area v of Sub-system u
w_{yvu}	Weightage of Performance Objective y of Key Performance Area v of Sub-system u
O^*_{yvu}	Objectivated value of Performance Objective y of Key Performance Area v of Sub-system u.
O_{yvu}	Actual value of Performance Objective y of Key Performance Area v of Sub-system u
X_j	Quantity of Product j

d_i^-	Degree of underachievement of Goal i
d_i^+	Degree of over-achievement of Goal i
P_i	Pre-emptive priority factor assigned to Goal i

3.1 The Model

Productivity Index of a System is:

$$(P.I) = \sum_{u=1} w_u.(P.I)_u \qquad (1)$$

where $\sum_{u=1} w_u = 1$

$(P.I)_u$, the Productivity Index of Sub-system u is determined as

$$(P.I)_u = \sum_{v=1} w_{vu}.(P.I)_{vu} \qquad (2)$$

where

$$\sum_{v=1} w_{vu} = 1 \text{ for all u.}$$

$(P.I)_{vu}$, the Productivity Index of Key Performance Area v of Sub-system u is given as,

$$(P.I)_{vu} = \sum_{y=1} w_{yvu}.\ O_{yvu}/O^*_{yvu} \qquad (3)$$

where,

$$\sum_{y=1} w_{yvu} = 1, \text{ for all v and u.}$$

Substituting values of $(P.I)_{vu}$ from equation (3) in equation (2), Productivity Index of a Sub-system u, $(P.I)_u$ can be rewritten as,

$$(P.I)_u = \sum_{v=1} \sum_{y=1} w_{vu}.w_{yvu}.\ O_{yvu}/O^*_{yvu} \qquad (4)$$

Value of $(P.I)_u$ from equation (4) can be substituted in equation (1) to provide (P.I), Productivity Index of System as

$$(P.I) = \sum_{u=1} \sum_{v=1} \sum_{y=1} w_u.w_{vu}.w_{yvu}.O_{yvu}/O^*_{yvu} \qquad (5)$$

4. METHODOLOGY TO MEASURE PRODUCTIVITY

The following sequential steps are suggested under this methodology.

4.1 Identification of Sub-systems

The Sub-system boundaries are defined and demarcated so as to specify 'the user'. The sub-systems chosen are the units responsible and accountable for the performance. The number and types of sub-systems identified are contingent upon the organisation under study.

4.2 Identification of Key Performance Areas (KPA's)

The analogy is drawn from MBO which recommends selection of KPA's to achieve positive performance. KPA's in each of the Sub-systems represent areas which the organisation expects to be most result oriented. KPA's may be concerning both quantitative and qualitative performance.

4.3 Setting of Performance Objectives (PO's)

This forms the pivotal exercise. Performance objectives provide information as to the accomplishment in a KPA with reference to the input resources required to be expended. The performance objectives can be quantitative as well as qualitative and draw strength from the overall objectives of the organisation. Expression in the form of ratios is recommended, as the same provides as economic perspectives to the performance objective. Qualitative performance objectives are proposed to be quantified on a common scale of 0 to 1.0 (1 to 5 or 1 to 100 etc.). The lowest and the highest levels of the scale represent the minimum and the maximum possible performance levels of the PO respectively. The quantification of intangible and heterogeneous performance objectives is another challenge. The evaluation is recommended to be carried through the application of the well known technique of Multi-attribute Utility Theory (MAUT) as developed by Keeney and Raiffa (1976).

4.4 Ranking and Weighting of Sub-system, KPA's and PO's

All the Sub-systems, KPA's and the PO's are not equal in importance. Their relative priorities vary. These are, therefore, required to be ranked and weightage-factors allocated so as to reflect the relative importance and inter-relationship. This can be determined through the technique of Paired Comparisons by involving people concerned who operate the system as has been shown by Sardana and Prem Vrat (1986).

4.5 Objectivated Output-the Concept and Determination

The PO-P approach relies on comparing the actual performance to a target performance. 'Objectivated Output' represents the optimal level of output which is possible to be attained by a system (sub-system) under the given constraints of input resources and a set of performance objectives. Determination of 'Objectivated Output' is a significant task. A low level of objectivated output shall provide a misleading figure of high productivity and vice versa. The exercise becomes all the more difficult in presence of various incompatible and incommensurable performance objectives. The issue is proposed to be resolved through Goal Programming whereby conflicting and competing multiple objectives can be included in the formulation as proposed by Lee (1972).

5. MEDIUM ELECTRICALS LTD (MEL)* : A CASE STUDY

Medium Electricals Ltd. is a medium sized organisation,engaged in the manufacturing and marketing of electric motors of a limited variety and range. Product data such as average unit sale price, unit profitability and market demands (minimum required and maximum permitted)are detailed in Table 1(a). Actual performance data in terms of quantity produced, sale value of production, quantity sold, sale value and profitability are also detailed in Table 1(a). MEL has stipulated policy norms to meet its corporate plans. These are enumerated in Appendix 1. The subsystems, KPA's and the performance objectives have been identified and the weightage-factors as determined are given in Table 2.

Performance objectives of turnover, profitability and capacity utilization are found to be in conflict. These are required to be resolved through Goal programming, the formulation and the results are detailed in Appendix 2. Values of objectivated output as derived through Goal Programming solution are noted in Tables 1(b) and 1(c). The other actual performance data and the objectivated output values based on the organisation's policy norms have been listed in Table 1(d).

'Sale Promotion effectiveness', 'motivated environment', 'effective product image', 'earnings satisfaction' and 'image in market' are qualitative performance objectives where actual performance and objectivated output values cannot be expressed in quantitative terms. In these cases objectivated output has been assumed to be 1. The actual performance values have been graded on 0-1 scale.

The performance objective of turnover has been expressed in absolute figures and not in a ratio form. The objectivated and the actual performance values have been converted to 0-1 scale through the application of MAUT assuming a linear relationship. As MEL considers turnover less than 75% of the objectivated value a totally unacceptable output, attribute value of Rs. 110,760,000 has been equated to utility value of 0. Attribute value of objectivated turnover at Rs. 147,680,000 then becomes the utility value 1. The actual value of turnover on this scale equals 0.76.

6. PRODUCTIVITY INDICES -CALCULATIONS

The PO-P model can now be applied to arrive at PI's of KPA's, Sub-systems and the System (MEL). Calculations for PI's of KPA's

* The name of the Organisation and Confidential data have been disguised.

TABLE - 1(a) ACTUAL PERFORMANCE DATA FOR MEL

(Rs. in Thousands)

Product code	H P (Horse Power) Rating	Unit Sale Price (Rs.)	Unit Profit (Rs.)	Min. Demand (Nos.)	Max. Demand (Nos.)	Unit Material Cost (Rs.)	Actual Performance Data					
							Production (Nos.)	Sale Value of Production (Rs.)	Sale (Nos.)	Sale Value (Rs.)	Profit (Rs.)	Cost of Materials (Rs.)
X_1	0.5	1.30	0.130	150	350	0.520	340	442	340	442	44.2	176.80
X_2	0.75	1.40	0.140	1,500	3,000	0.560	2,900	4,060	2,800	3,920	392.0	1,624.00
X_3	1.50	1.70	0.170	2,000	5,000	0.700	5,000	8,500	4,900	8,330	833.0	3,500.00
X_4	2.20	2.10	0.210	2,000	5,000	0.850	4,900	10,290	4,900	10,290	1,029.0	4,165.00
X_5	5.00	2.70	0.189	2,000	5,000	1.100	4,800	12,960	4,800	12,960	907.2	5,280.00
X_6	10.00	4.00	0.280	4,000	8,000	1.700	7,700	30,800	7,500	30,000	2,100.0	13,090.00
X_7	15.00	7.00	0.490	2,000	5,000	3.000	4,900	34,300	4,800	33,600	2,352.0	14,700.00
X_8	25.00	11.00	1.10	800	2,000	4.500	2,000	22,000	1,800	19,800	1,980.0	9,000.00
X_9	40.00	17.00	2.04	250	550	7.000	310	5,270	300	75,100	612.0	2,170.00
X_{10}	60.00	27.00	3.04	150	300	8.500	290	7,830	280	7,560	851.2	2,465.00
X_{11}	75.00	36.00	5.40	30	75	14.500	29	1,044	29	1,044	156.6	420.500
X_{12}	125.00	52.00	7.80	70	150	21.000	65	3,380	60	3,120	468.0	1,365.00
X_{13}	180.00	84.00	16.80	25	50	34.600	30	2,520	25	2,100	420.0	1,038.00
X_{14}	270.00	100.00	20.00	5	15	45.000	5	500	5	500	100.0	225.00
Total:								143,896		138,766	12,245.2	59,219.30

of the Production Sub-system are illustrated as under. Values of PI's of the rest of the KPA's and the Sub-systems have been listed.

6.1 PI of Production Sub-system

i) PI of KPA - Manpower Utilization:-

0.4x0.943/0.959+0.4x0.076/0.079+0.2x4.47/4.50 = 0.977

ii) PI of KPA Asset Utilization:-

0.7x0.935/0.972+0.3x0.441/0.453 = 0.965

iii) PI of KPA Material Utilization:-

1.0x0.932/0.970 = 0.961

iv) PI of KPA Schedule Completion:-

1.0x0.795/0.900 = 0.883

v) PI of KPA Quality:-

1.0x0.98/0.99 = 0.989

PI of Production Sub-system:

0.241x0.977+0.379x0.965+0.138x0.961+0.138x0.883+0.104x0.989 = 0.958

6.2 PI of Marketing Sub-system

i) PI of KPA Sales : 0.849

ii) PI of Sales Promotion and Publicity : 0.835

PI of Marketing Sub-system: 0.846

6.3 PI of Technical Sub-system 0.949

6.4 PI of Personnel Sub-system 0.958

6.5 PI of Material Sub-system

PI of KPA Purchase Management: 0.958

PI of KPA Inventory Control: 0.913

PI of Material Sub-system: 0.926

6.6 Goals and Values Sub-system

PI of KPA financial Goals: 0.931

PI of KPA Customer Satisfaction: 0.931

PI of KPA Employee Satisfaction: 0.840

PI of Supplier's Satisfaction: 0.790

PI of Goals and Values Sub-system: 0.896

6.7 PI of MEL as a System

0.237x0.958+0.289x0.846+0.105x0.949+0.105x0.958+
0.132x0.926+0.132x0.896 = 0.912.

7. CONCLUSIONS

The case study has shown how performance, both economic and non-economic as expressed in heterogeneous units can be taken into account as the total output. For MEL a total of 28 performance objectives exist with differing priorities. It is seen that all of these consume input resources of one type or the other. Through

PO-P approach it has been possible to rank the same in order of importance and develop indices for the KPA's and the Sub-systems. The primary objective of productivity measurement is to establish the potential for improvement and make people accountable for the state of productivity. PO-P model serves both objectives. It determines the gap between the actual and what is possible to accomplish (unlike subjective targetting). This sets the potential for improvement. The productivity indices calculations are carried out for KPA's and Sub-systems upwards, thus fixing the responsibility. An important contribution of the process of productivity measurement through this model is that it sets out realisable targets and goals, as objectivated output. The same are established after the performance objectives at conflict with one another have been resolved through Goal Programming. PO-P model can be applied to any type of organisation.

8. REFERENCES

Allard,R., Keal,J.E., Mount,H.W.,Samuel, P.J., Smith, G.R. and Swannack,A.R.,1971. Productivity Measurement-A Symposium for the Seventies. Institute of Personnel Management,London.

Keeney,R.L. and Raiffa,H., 1976. Decision with Multiple Objectives. John Wiley & Sons, New York.

Lee, S.M.,1972. Goal Programming for Decision Analysis.Auerback, Philadelphia.

Mali,Paul.,1978. Improving Total Productivity, John Wiley and Sons, New York.

Prem Vrat and Sardana, G.D.,1984. Concept of Productivity:A Plea for Reappraisal. UDYOG PRAGATI,7(1):16-24.

Robbins,Stephen P.,1984. Management: Concepts and Practices. Prentice-Hall,Englewood Cliffs, NJ.

Sardana, G.D. and Prem Vrat,1983. Performance Objectives-Productivity (PO-P):A Conceptual Framework and a Mathematical Model for Productivity Management. Productivity,24(3):299-307.

Sardana,G.D. and Prem Vrat,1986. Productivity Measurement of a Small Undertaking. Proceedings,Vol.II,XVIII National Convention, Indian Institution of Industrial Engineering,Bombay.

Sink,D.S., 1983. Productivity Management: The Basics.IE (Management) News, 16(1): 1-3.

APPENDIX - 1 : POLICY NORMS AT MEL

Production Sub-system:

i) Atleast 99% of production to be defect free.

ii) Production and Order booking not be exceed 300,000 Horse Power. (Permitted Production Capacity).

iii) Shop rejections not to exceed 3% of the cost of materials.

iv) In a year an attendance of 2100 Hours per workman is expected.

v) Facilities F_1, F_2, F_3, F_4 constitute critical facilities and require maximisation of their utilization.

Marketing Sub-system:

i) Average finished stock not to exceed average sale in

one month.

ii) Turnover growth rate to be atleast 10% per year.

iii) Average debt not to exceed average sales of a month.

iv) Atleast 80% of the new orders to be booked from existing customers.

v) Turnover less than 75% of the objectivated value is not acceptable.

Materials Sub-System:

i) Atleast 90% of the purchases to be made in time.

ii) A minimum of 95% of purchases to be defect free.

iii) Average inventory not to exceed average consumption of three months.

iv) Stock outs not to exceed 1% of the total indents.

v) Purchase credit not to be less than 30 days.

APPENDIX - 2 : GOAL PROGRAMMING FORMULATION

MEL has decided to accord the following priorities to goals:

P_1 Achieve Capacity Utilizationoof facilities F_1, F_2, F_3 and F_4.
P_2 Minimise under-achievement of profit.
P_3 Minimise under-achievement of turnover.

Achievement of capacity utilization of facility F_2 is two times more important than F_3. Achievement of profit is two times more important than achievement of turnover.

A Goal Programming model can now be formulated as under:-

$$\text{Minimise } Z = [P_1\{(d_1^-+d_1^+)+2(d_2^-+d_2^+)+(d_3^-+d_3^+)+(d_4^-+d_4^+)\}, 2P_2d_5^-, P_3d_6^-]$$

subject to goal constraints:

I.Capacity Constraint:

For F_1:- $0.10X_1+0.10X_2+0.10X_3+0.12X_4+0.12X_5+0.14X_6+0.19X_7+0.2X_8+0.24X_9+1.15X_{10}+1.25X_{11}+1.8X_{12}+3.0X_{13}+3.0X_{14}+d_1^--d_1^+ = 6000.$

For F_2:- $0.19X_1+0.19X_2+0.20X_3+0.26X_4+0.27X_5+0.36X_6+0.49X_7+0.88X_8+1.45X_9+2.28X_{10}+2.65X_{11}+3.33X_{12}+4.25X_{13}+4.60X_{14}+d_2^--d_2^+= 13000$

For F_3:- $0.05X_1+0.05X_2+0.05X_3+0.06X_4+0.06X_5+0.06X_6+0.07X_7+0.14X_8$ $0.17X_9+0.19X_{10}+0.21X_{11}+0.21X_{12}+0.21X_{13}+0.21X_{14}+d_3^--d_3^+ = 2500$

For F_4:- $0.23X_1+0.14X_2+0.18X_3+0.18X_4+0.18X_5+0.18X_6+0.60X_7+0.68X_8+0.85X_9+1.14X_{10}+0.74X_{11}+0.74X_{12}+0.74X_{13}+0.74X_{14}+d_4^--d_4^+ = 10000.$

II. Profit Constraint:

$0.130X_1+0.140X_2+0.170X_3+0.210X_4+0.189X_5+0.28X_6+0.490X_7+1.100X_8+2.040X_9+3.04X_{10}+5.400X_{11}+7.800X_{12}+16.800X_{13}+20.00X_{14}+d_5^- - d_5^+ = 14000.$

III. Turnover Constraint:

$1.300X_1+1.400X_2+1.700X_3+2.100X_4+2.700X_5+4.000X_6+7.000X_7+11.000X_8+17.000X_9+27.000X_{10}+36.000X_{11}+52.000X_{12}+84.000X_{13}+100.000X_{14}+d_6^--d_6^+ = 14000.$

IV. HP and the System Constraints:

$0.5X_1+0.75X_2+1.5X_3+2.2X_4+5.0X_5+10.0X_6+15X_7+25X_8+40X_9+60X_{10}+75X_{11}+125X_{12}+180X_{13}+270X_{14} \leq 300{,}000.$

V. Demand Constraints:

$150 \leq X_1 \leq 350$ $1500 \leq X_2 \leq 3000$ $2000 \leq X_3 \leq 5000$ $2000 \leq X_4 \leq 5000$
$2000 \leq X_5 \leq 5000$ $4000 \leq X_6 \leq 8000$ $2000 \leq X_7 \leq 5000$ $800 \leq X_8 \leq 2000$
$250 \leq X_9 \leq 550$ $150 \leq X_{10} \leq 300$ $30 \leq X_{11} \leq 75$ $70 \leq X_{12} \leq 150$
$25 \leq X_{13} \leq 50$ $5 \leq X_{14} \leq 15$

X_j d_i^-, d_i^+ for all i,j are the usual non-negativity constraints. The coefficients of X_j in capacity constraint equations are the standard hours required to process unit product quantity on the relevant facility. The right hand side values are the total capacity on the facility.

GOAL PROGRAMMING SOLUTION:

The problem has been solved on ICL 2960 system by using the programme as proposed by Lee (1972). The results are to be interpretted from three listings,

i) Analysis of the objective
ii) Slack Analysis
iii) Variable Analysis.

Capacity F_2 has been fully utilized. MEL can reach turnover of Rs. 147,680,000. There is under-achievement of capacities of facilities F_1,F_3 and F_4. 'Analysis of the Objective' provides details of under-achhievement of the 'priorities'. Priority-P_3 is achieved, whereas there is an underachievement of P_1 and P_2. Details of departure, positive or negative, as obtained from 'Slack analysis' are used to correct the goal figures so as to obtain the reconciled figures. 'Variable analysis' details achievable quantitities of decision variable, X_j. These are tabulated in Tables 1(b) and 1(c).

TABLE- 1(b)
MEL objectivated output values (Rs. in Thousands)

Product code	PRODUCT RELATED DATA		
	Product Quantity (Nos.)	Standard Hours Required (Hours)	Material cost)(Rs.)
X_1	350	525	182
X_2	3000	6000	1680
X_3	5000	22500	3500
X_4	5000	30500	4250
X_5	5000	75000	5500
X_6	8000	280000	13600
X_7	5000	200000	15000
X_8	2000	150000	9000
X_9	320	35200	2240
X_{10}	300	52500	2550
X_{11}	30	6900	435
X_{12}	70	25900	1470
X_{13}	32	17440	1107.200
X_{14}	5	4000	225
Total:		905965	60739.200

TABLE - 1(C)
Goals related data from GP (Rs. in thousands)

Item	Assumed Goal Value	Objectivated Output Value
Turnover (Rs.)	140,000	147,680
Profit (Rs.)	14,000	13,126
Capacity F_1 (hours)	6,000	5,204
Capacity F_2 (hours)	13,000	13,000
Capacity F_3 (hours)	2,500	2,268
Capacity F_4 (hours)	10,000	9,717
Capacity - $(F_1 - F_4)$	31,500	30,626

TABLE - 1(d)
Policy norms related data

(Rs. in Thousands)

Item	Unit	Objectivated Values	Actual Values
Direct Workmen	Nos.	450	445
Indirect Workmen	Nos.	100	100
Other Staff	Nos.	50	50
Employees at the beginning of the year	Nos.	600	600
Employees at the close of the year	Nos.	600	595
Attendance (Indirect Labour)	Hrs.	210,000	209,000
Attendance (Direct Labour)	Hrs.	945,000	934,500
Hours worked (Direct Labour)	Hrs.	850,000	795,000
Personnel Expenses	Rs.	11,500	11,600
Standard Hours Recovery (all facilities)	Hrs.	905,965*	881,630*
Defect free production(Sale Value)	Rs.	146,203	141,000
Capacity (Critical M/c)	Hrs.	31,500	31,500
Standard Hrs. Recovery (Critical M/c)	Hrs.	30,626**	29,455.9*
(Cost of) Materials converted	Rs.	58,917.02	55,200.00
Orders scheduled	Nos.	2000	2200
Orders completed in time	Nos.	1800	1750
Purchases (Direct Materials)	Rs.	60,739.20	61,000.00
Purchases in time (Direct Materials)	Rs.	54,665.28	51,850.00
Value of Accepted Purchases	Rs.	57,702.24	55,340.00
Av.Monthly Stock(Direct Materials)	Rs.	15,184.80	16,280.00
Purchase Payments	Rs.	60,739.20	61,000.00
Payments in Time	Rs.	60,739.20	44,530.00
Total Issue Indents	Nos.	5,000	5,700
Stock Outs	Nos.	50	62
Av.finished Product Stock(Monthly)	Rs.	12,306	14,800
Av.Debts (Monthly)	Rs.	12,306	15,000
Repeat Orders Booked	Rs.	126,400	115,475
Total Orders Booked	Rs.	158,000	155,000
Total Investment	Rs.	450,000	450,000
Capitalised Costs	Rs.	2,000	2,000

* Product Qty. X unit standard hours for processing.

** From Goal Programming Solution.

TABLE - 2 : MEL SUB-SYSTEMS, KPA's PERFORMANCE OBJECTIVES AND THEIR WEIGHHTAGES

Sub-system	Weightage Factor	Key Performance Area	Weightage factor		Performance Objectives	Weightage Factor	Objectivated Value	Actual Value
1	2	3	4		5	6	7	8
Production	0.237	Manpower Utilization	0.241	1.	Labour Utilization:	0.4		
					Standard Hours / Attendance Hrs. (Direct Labour)		905,965 / 945,000 = 0.959	881,630 / 934,500 = 0.943
				2.	Cost Effectiveness:	0.4		
					Standard Hours / Personnel Expenses		905,965 / 11,500,000 = 0.079	881,630 / 11,600,000 = 0.076
				3.	Indirect Support Ratio:	0.2		
					Attendance Hrs (Direct Labour) / Attendance Hrs (Indirect Labour)		945,000 / 210,000 = 4.5	934,500 / 209,000 = 4.47
		Asset Utilization	0.379	1.	Capacity Utilization:	0.7		
					*Standard Hrs. Production / *Capacity Machines		30,626 / 31,500 = 0.972	29,455.9 / 31,500 = 0.935
				2.	Capital Cost Control:	0.3		
					Standard Hours / Capitalised Costs		905,965 / 2,000,000 = 0.453	881,630 / 2,000,000 = 0.441
		Materials Utilization	0.138	1.	Yield (Direct Materials)	1.0		
					Useful Conversion of Materials / Cost of Direct Materials		58,917,024 / 60,739,200 = 0.970	55,200,000 / 59,219,300 = 0.932
		Schedule Completion	0.138	1.	Order Processing Efficiency:	1.0		
					Orders Completed in Time / Orders Scheduled		1,800 / 2,000 = 0.900	1,750 / 2,200 = 0.795
		Quality	0.104	1.	Index of Defect Free Production:	1.0		
					Value of Defect Free Production / Value of Total Production		146,203,000 / 147,680,000 = 0.99	141,000,000 / 143,896,000 = 0.98
Marketing	0.289	Sales	0.80	1.	Turnover:	0.30	147,680,000	138,766,000
				2.	Profitability:	0.35		
					Profit / Sales		13,126,000 / 147,680,000 = 0.089	12,245,200 / 138,766,000 = 0.089
				3.	Collection Index:	0.25		
					Average Sales (Monthly) / Average Debt (Monthly)		147,680,000 / 12x12,306,000 = 1	138,766,000 / 12x15,000,000 = 0.77
				4.	Stock Turnover:	0.10		
					Av. Sales (Monthly / Av. Product Stock (Monthly)		147,680,000 / 12x12,306,000 = 1	138,766,000 / 12x14,800,000 = 0.78
		Sales Promotion and Publicity	0.20	1.	Sales Promotion Effectiveness:	0.5	1	0.82
				2.	Effective Product Image	0.5	1	0.85

TABLE - 2 (Contd.)

1	2	3	4	5	6	7	8
Technical	0.105	Product Performance		1. Design Maturity	0.4	1	0.93
				2. Economical Cost (Design)	0.6	1	0.96
Personnel	0.105	Industrial Relations		1. Index of Useful Hours:	0.7		
				Labour Hours Worked (Direct Labour) / Attendance Hrs. (Direct Labour)		$\frac{850,500}{945,000} = 0.90$	$\frac{795,000}{934,500} = 0.85$
				2. Employee Turnover	0.3		
				No. of Employees at End of Year / No. of Employees at Beginning of Year		$\frac{600}{600} = 1$	$\frac{595}{600} = 0.99$
Materials	0.132	1. Purchase Management	0.30	1. Timeliness of Purchases:	0.6		
				Purchase in Time (Direct Materials) / Purchases (Direct Materials)		$\frac{54,665,280}{60,739,200} = 0.9$	$\frac{51,850,000}{61,000,000} = 0.85$
				2. Quality (In Purchases):	0.4		
				Value of Accepted Purchases / Total Purchases		$\frac{57,702,240}{69,739,200} = 0.95$	$\frac{55,340,000}{59,219,300} = 0.93$
		2. Inventory Control	0.70	1. Turnover Ratio:	0.6		
				Av. Material Consumption (Monthly) / Av. Stock (Monthly)		$\frac{60,739,200}{12x15,184,800} = 0.33$	$\frac{59,219,300}{12x16,280,000} = 0.30$
				2. Stockouts:	0.4		
				Issue Indents (Received) / No. of Stockouts		$\frac{5,000}{50} = 100$	$\frac{5,700}{62} = 91.9$
Goals and Values	0.132	1. Financial Goals	0.40	1. Return on Investment:	1.0		
				Profit / Investments		$\frac{13,126,000}{45,000,000} = 0.29$	$\frac{12,245,200}{45,000,000} = 0.27$
		2. Customer Satisfaction	0.30	1. Repeat Order Index:	1.0		
				No. of Repeat Orders / No. of Total Orders		$\frac{126,400,000}{158,000,000} = 0.800$	$\frac{115,475,000}{155,000,000} = 0.745$
		3. Employee Satisfaction	0.15	1. Motivated Environment:	0.6	1	0.80
				2. Earnings Satisfaction:	0.4	1	0.90
		4. Supplier Satisfaction	0.15	1. Timely Payments:	0.5		
				Payments in Time / Total Payments		$\frac{60,739,200}{60,739,200} = 1$	$\frac{44,530,000}{61,000,000} = 0.73$
				2. Image in Market:	0.5		
				Willingness to de repeat business		1	0.85

* Critical facilities only

Productivity Management Frontiers — I, edited by D.J. Sumanth
Elsevier Science Publishers B.V., Amsterdam, 1987 — Printed in The Netherlands

PRODUCTIVITY EVALUATION FOR MODERN MANUFACTURING TECHNOLOGY.

E. A. CAHILL,
Divisional Director, Business Advisory Service, Irish Productivity Centre,
IPC House, 35/39 Shelbourne Road, Dublin 4, Ireland.

ABSTRACT:
Cahill, E. A. 1987. Productivity Evaluation for Modern Manufacturing Technology.

The main advantage advanced manufacturing technology will confer will be flexibility. However, this very flexibility will render many of the traditional bases for forward loading machine systems with work inappropriate. Choosing among a variety of loading criteria calls for an overall evaluation method. In this paper one such method based on productivity measurement is discussed.

The measurement of productivity in manufacturing industry has for many years relied on measures of labour productivity and direct labour productivity in particular This approach was satisfactory as long as the following conditions remained valid:

(a) That the percentage of total cost represented by direct labour cost was significant (> 40%).

(b) That the causal relationship between output and direct labour input was consistent (output was direct labour-paced).

For many years work at a theoretical level had been carried on to extend the popular measurement systems to involve the influence of capital as a factor. While many successful models of total factor productivity measurement were devised, they did not find universal or wide spread application in practice mainly because of practical difficulties involved in satisfactorily measuring capital input.

In manufacturing particularly the partial direct labour measure was used widely and frequently formed the basis for incentive reward systems. Furthermore, in the decades when financial management and marketing assumed the mantle of key functional areas and glamour activities, manufacturing managers tended to ignore new concepts of productivity measurement.

But the environment in organisations has changed. Manufacturing has once more become the focus of attention and the requirements for successful manufacturing have altered significantly. Fast and continuously changing markets are demanding high quality, specialised products and at prices below what was readily paid for high volume standard products in the past.(1)

Manufacturing managements have responded by developing production environments based on flexibility using computerised machine and information control. New concepts of production organisation which dispense with many of the buffers to irregularity and uncertainty have had to be devised. The combination of computer-based control systems, computerised numerically controlled (CNC) machines and automated handling equipment as found in the typical flexible manufacturing system are the basic building blocks of the factory of the future.

Such production arrangements which attempt to confer the economic unit cost advantages of single product large lot size mass production systems on multi product small lot size flexible manufacture demand a series of carefully chosen criteria for determining the configurations under which they are operated.

Within the individual systems themselves criteria for scheduling and loading are required to meet a variety of objectives while at the level of the individual plant or firm, the congruity of goals for all sub-systems must promote the overall objectives of the company. Systems can be optimised to give lowest investment in work in process, minimum travelling and queuing time in the system, maximum throughput of products etc. With a variety of products and parts being made the basis for balancing the overall system becomes complex. At that stage, it becomes important to return to basics to set the rules. A productivity measurement system is required which will supply the information on which decisions can be made and which allow alternative arrangements and plans to be evaluated. For the measurement system to be effective it must first of all be applicable at the level of the individual department or section and must be aggregateable to give the productivity performance of the plant, division or firm.

The research into models for productivity measurement has given rise to a number of useful and sophisticated approaches. The multi factor productivity measurement model (MFPMM), which was devised largely by Dr. Scott Sink at Oklahoma State University and later at the Virginia Productivity Center, by refining and simplifying Hiram Davis' work in the 1950's and later work by the American Productivity Center presents an interesting and powerful tool for bringing the scope for productivity management into the open. As an intrinsic element of management it has always been dealt with implicitly. However, the MFPMM model has excluded consideration of capital on the grounds of difficulties associated with calculating an appropriate cost of capital and because techniques exist to ensure some level of satisficing in relation to capital productivity.(2)

While sympathising with the rationale and accepting the approach by the model to cultivating a degree of "user friendliness" to encourage adoption, it cannot be accepted that a realistic attempt at using productivity measurement to answer the problems set out above could ignore the issue of capital involved. This is particularly so when many of the performance criteria involve questions of fixed capital allocation and choices with regard to specific levels of working capital (inventories).

The other approach from the research literature which commends itself would be that used by Dr. David Sumanth in developing his total productivity model. This model has a product orientation in that the model calculates productivity indices for each product or product group and can then weight these to aggregate the combined results to give firm level evaluation, if necessary.(3) The analysis carried out is very similar to that in the MFPMM model but on a wider range of input parameters.

The extension which this paper seeks to explore would be the adaptation of the approaches to sub-systems of the total firm operations system.

A number of problems must be addressed in relation to whichever model is utilised where the departments and sections of the manufacturing operations are treated separately. These are:

(a) An output price must be derived to apply to each unit of production as it leaves each department. When a product is being dealt with in its totality its sales price is known

(b) An allocation method for apportioning overhead expenses among departments is needed. Traditional cost accounting methods need revision in this regard precisely because of the reduction in the general levels of direct labour, which has been the traditional basis for allocation.

(c) An approach must be adopted to provide a measure of the capital inputs to the various segments of the overall system.

Let us enlarge on each of these points.

Output Prices:

To be in a position to examine productivity change in a disaggregated structure of departments or sections it is necessary to adopt an output measure for unfinished products, components and subassemblies. Where the subassemblies or components are marketable items in their own right then the market price should be used less any allowance for packaging, etc. which may be carried out separately. For all other work in process a percentage of the finished product price should be applied at various stages of manufacture.

This provides each department with an output figure which represents the gross value of output from that department. Since this figure includes output associated with departments which have worked on products previously the figure for incremental output or nett output is the more appropriate figure for measuring productivity in the department itself.

Overhead Allocation:

The most commonly utilised method of apportioning overheads (particularly factory overheads) has been based on direct labour hours. When direct labour accounts for a low percentage of overall input then any apportionment of a larger element of input using this base will give rise to large variations in total input as a result small changes in direct labour. A better choice of base might be that of machine hours, particularly as that will tend to relate more closely with variations in production activity. There may be a tendency to treat overheads as strictly fixed and independent of changes in activity. In the example shown later the effect of fixed and variable overhead expenses can be seen.

Capital Inputs:

The allocation of fixed capital between departments and sections is generally a straight forward exercise. To make a measurement most relevant to department managers only activity which is directly involved in manufacturing should be considered. Any investment of a purely financial nature should be considered only at the corporate level. Similarly the marketing activity of the company should also be separated out. When the amount of capital involved is known for each section and department then the other element needed to compute capital input is the cost of capital. It is not intended in this paper to discuss at length the finer points of calculating the firms cost of capital. There are examples of such a calculation to be found in many tests. The only point that the author would make is that consistency of approach is most important if this is being done over time.

The issue of Working Capital is possibly of more importance as changes in loading and scheduling can significantly alter the levels of inventory and work in process in a department and therefore the level of working capital. In the example shown however, this element is treated by way of demonstrating the effect of such a change.

Evaluating Alternative Load Plans:

In order to examine the validity of using this approach to evaluate alternative load plans a simple example is given below. The detailed statements in relation to each of the plans is set out in the Appendix to this paper and the effects of the changes are tabulated using the basic parameters of Dr. Scott Sink's Multifactor Productivity Measurement Model with a full range of inputs including Capital. These results are shown in Table 1.

TABLE 1: PRODUCTIVITY EVALUATION OF ALTERNATIVE LOAD PLANS

	Load Plan 1 (a) Dept.A	Load Plan 1 (b) Dept.B	Load Plan 2 (c) Dept.A	Load Plan 2 (d) Dept.B	Weighted Change Ratios (e) Dept.A Q	(f) Dept.A P	(g) Dept.A V	(h) Dept.B Q	(j) Dept.B P	(k) Dept.B V	DEPARTMENT A Weighted Productivity Performance (l) Index	DEPARTMENT A Monetary Effect (m)	DEPARTMENT B Weighted Performance Index (n) Productivity	(o) Price Recovery	(p) Profitability	Monetary Effects (r) Productivity $	(s) Price Recovery $	(t) Profitability $
Output	7,200	6,360	4,700	8,860	.653	1.0	.653	1.393	1.0	1.393								
Inputs:																		
Labour	900	1,013	518	1,120	.576	1.0	.576	1.106	1.0	1.106	1.034	69.7	1.259	1.0	1.259	291	-	291
Materials	2,400	200	1,400	1,200	.583	1.0	.583	3.5	1.714	6.00	1.12	168	.398	.583	.232	-421.4	-500.0	-921.4
Energy	240	272	107	390	.446	1.0	.446	1.434	1.0	1.434	1.464	49.7	.971	1.0	.971	-11.15	-	-11.15
Other Factory Expenses	1,200	1,088	532	1,296	.443	1.0	.443	1.191	1.0	1.191	1.474	252	1.170	1.0	1.170	219.8	-	219.8
Admin. Expenses	800	900	800	900	1.0	1.0	1.0	1.0	1.0	1.0	.653	-277.6	1.393	1.0	1.393	353.7	-	353.7
Capital Costs	1,200	1,600	1,200	1,800	1.0	1.0	1.0	1.125	1.0	1.125	.653	-416.4	1.238	1.0	1.238	428.8	-	428.8
Total Inputs:	6,740	5,073	4,557	6,705	.676	1.0	.676	1.272	1.038	1.321	.966	-155	1.095	.962	1.054	862	-500.0	362.0

NOTES:

(e) = $\frac{\sum Q_{A2} P_{A1}}{\sum Q_{A1} P_{A1}}$

(f) = $\frac{\sum Q_{A2} P_{A2}}{\sum Q_{A2} P_{A1}}$

(g) = $\frac{\sum Q_{A2} P_{A2}}{\sum Q_{A1} P_{A1}} = \frac{(c)}{(a)}$ = (e)x(f)

(h) As above for B

(j) " " " "

(k) " " " "

(l) = $\frac{\text{Col.(e) for Outputs}}{\text{Col.(e) for Indiv. Inputs}}$

(m) = Col.(a) for each corresponding Input x (Col.(e) for Output - Col.(e) for each Input).

(n) = $\frac{\text{Col.(h) for Outputs}}{\text{Col.(h) for Indiv. Inputs}}$

(o) = $\frac{\text{Col.(j) for Outputs}}{\text{Col.(j) for Indiv. Inputs}}$

(p) = $\frac{\text{Col.(k) for Outputs}}{\text{Col.(k) for Indiv. Inputs}}$

(r) = Col.(b) for each corresponding Input x (Col.(h) for Outputs - Col.(h) for each Input).

(t) = Col.(b) for each corresponding Input x (Col.(k) for Outputs - Col.(k) for each Input).

(s) = (t) - (r)

Note: Because Col.(e) and (g) are identical the Weighted Performance Indices for Productivity and Profitability are the same for Dept. A.

Hence Monetary Effects are also the same.

Price Recovery Effects are Zero.

Example:

To illustrate the operation of the approach let us examine a simplified example of loading a three department manufacturing set up for the following requirements:

Product 4 - 100 units   Product 4 consists of an assembly of
1-Product 3 4-Product 2 2-Product 1

∴ Product 1 - 200 units
Product 2 - 400 units
Product 3 - 300 units

Capacity Available:

Dept.A	Mach.Group 1 - 100 hrs.	Mach.Group 2 - 200 hrs.
Dept.B	Mach.Group 3 - 300 hrs.	Mach.Group 4 - 100 hrs.
Dept.C	Assembly - 150 hrs.	

Machining Requirements per Product Unit:

Product						
1	.5 hr. M/C Gr.1	or	.75 hr. M/C Gr.2	1 hr. M/C Gr.3	or	.5 hr. M/C Gr.4
2			.5 hr. M/C Gr.2	or .4 hr. M/C Gr.3		.125 hr. M/C Gr.4
3				1 hr. M/C Gr.3	or	.6 hr. M/C Gr.4
4	1 hr. Assembly.					

Labour Requirement - per M/C hr.:

Machine Group	1	2	3	4
Labour Class 1	.4	.1	.1	.4
Labour Class 2	.6	.2	.18	.66
Assembly:	.5 hr. Labour Class 1 per unit		1.5 hrs. Labour Class 1 per unit	

Labour Class 3: Fixed 20 hrs.Dept.A 30 hrs.Dept.B 10 hrs.Dept.C

Material Requirements/Unit

	Raw Materials	Components
Product 1	2 Units RM1	1 Unit Comp. A
Product 2	2 Units RM1	1 Unit Comp. A
Product 3	2 Units RM2	1 Unit Comp. B

	Dept. A		Dept. B		Dept. C
M/c Grp.:	1	2	3	4	Assembly
Energy per M/C hr.units	16	4	7	12	
per hr. Assembly					40
Other Factory O/H per M/c hr units	4	1	1.4	2.4	

Alternative Load Plans: (Production Amounts)

	Dept. A		Dept. B		Dept. C
M/c Grp.:	1	2	3	4	Assembly
Plan 1: (Feasible)					
Product 1	200		200		
2		400		400	
3			17	83	100
4					

	Dept. A		Dept. B		Dept. C
M/c Grp.:	1	2	3	4	Assembly
Plan 2: (Minimum Cost)					
Product 1	33	167	100	100	
2		150	250	400	
3			100		
4					100

Analysis of Results:

Before analysing the results it should be pointed out that Table 1 confines itself to considering the effect of changes in Departments A & B, since no changes took place in Department C. Also, since the load plans are being considered as alternatives, there are no price change effects except in relation to materials which reflect shifts of materials between Departments A & B. By and large therefore, the monetary effects on profitability or cost effectiveness are all as a result of changes in productivity.

The overall evaluation, using total inputs, gives the explanation of the changes which took place between the two plans. As can be seen the overall output levels of the two departments taken together is unchanged, while the total inputs are lower in Plan 2 than Plan 1.

In terms of the change both Depts. showed an improvement in labour productivity. The combined total effect of $360 represents the improvement achieved over the level of labour input which would have been merited by output weighted changes in labour.
e.g. Labour in Depts. A & B was $1,913 in Plan 1, if it had changed in line with changes in Output Value, it would have been $1,998 in Plan 2. It was actually $1,638, an improvement of $360. This reflects the better output to labour ratio of Dept. B to Dept. A.

The changes in the usage of material in each department however, shows up a problem with the detailed analysis using the model. While it reflects accurately the effect of switching production between departments it gives a slightly exagerated view of the monetary effects.

Careful interpretation is needed of the departmental changes particularly as the nett effect overall is zero.

In Department A the productivity effect on materials is a saving of $168 which represents the shift of the lower price weighted product out of Department A. In Department B the weighted productivity performance index for material reflects the significant shift of production into Department B with its attendant increase in materials. Also, what was being added was work with a higher material to output ratio and this in turn depressed the productivity measure. The Price Recovery Effect arises from the increase in average price per unit of material (0.666 to 1.142) and the number of units used (1,050).

In order to calculate the indices for the overall operation from the departmental data. The Weighted Change Ratios can be aggregated using base period share of total input weights for input elements and base period share of total output for output. In this case the base period would refer to Load Plan 1.

e.g. $\frac{(900)}{(1,913)}$ .576 + $\frac{(1,013)}{(1,913)}$ 1.106 = 856 = $\frac{1\,638}{1,913}$

= Overall Weighted Change Ratio for Labour

The changes in energy utilisation reflect the shifts in activity between Departments A and B. The positive monetary effect in Department A more than offsets the slight negative effect in Department B.

The double gain in Department A and B on Factory Overheads shows how essentially variable overheads are reflected in the model while Administration Overheads are fixed and as expected show negative effects where output drops and positive effects where it rises.

The use of the model does show clearly that the productivity changes that have taken place and can therefore be used to evaluate different load plans. It also shows the advantage of lowering the productivity of one section in the interest of raising that of the system as a whole. Needless to add there would be considerable scope in practise to develop simulation models to evaluate productivity changes arising from changes in factory loading.

The technique described here affords the Manager a means of determining the most economically efficient way of loading between departments for any level of demand. The model has the capability of handling simulations of feasible solutions. This is particularly useful if differential price changes have occurred between output and various inputs. Normally changes in input prices would not be taken into account by reconsidering loading plans because fixed routing and loading has been a long tradition of manufacturing. Now a technique exists which can explore the effect of these changes and which can provide projection of permissible levels of change to contain effects within targets.

The model is particularly powerful in tracing the effects of changes over time. There is also considerable scope for development of the approach to cater for more complex aspects of manufacturing performance in the future.

Bibliography

(1) Edghill. J.S & Davies. A, Flexible Manufacturing Systems - The Myth and the Reality; International Journal of Advanced Manufacturing Technology Sept.1985.

(2) Sink D.S , Productivity Management: Planning Measurement and Evaluation, Control and Improvement, Wiley, New York 1985

(3) Sumanth. D.J., Productivity Measurement and Evaluation Modes for Manufacturing Companies, Unpublished Dissertation 1979.

Load Plan 1

	Dept. A			Dept. B			Dept. C		
Outputs	Q	P	V	Q	P	V	Q	P	V
Prod. 1	200	16	3,200	190	24	4,560			
Prod. 2	400	10	4,000	400	21	8,400	20	21	420
Prod. 3				100	6	600	5	6	30
Prod. 4							95	200	19,000
Total Output			7,200			13,560			19,450
Incr. Output			7,200			6,360			5,890
Inputs									
Labour 1	60	4	240	62	4	248	50	4	200
Labour 2	100	5	500	105	5	525	150	5	750
Labour 3	20	8	160	30	8	240	10	8	80
Total Labour			900			1,013			1,030
Materials									
Raw Materials	1,200	1	1,200	200	.5	100			
Components	600	2	1,200	100	1	100			
Prev. Depts.						6,740			11,813
Total Materials			2,400			6,940			11,813
Energy	2,400	.1	240	2,720	.1	272	4,000	.1	400
Other Fact O/H	600	2	1,200	544	2	1,088	700	2	1,400
Admin. O/H	400	2	800	450	2	900	500	2	1,000
Capital	12,000	.1	1,200	16,000	.1	1,600	6,000	.1	600
Total Inputs			6,740			11,813			16,243
Incr. Output			7,200			6,360			5,890
Dept. Inputs			6,740			5,073			4,430
Dept. Prod. Net			1.068			1.254			1.330
Gross			1.068			1.148			1.197

Load Plan 2

	Dept. A			Dept. B			Dept. C		
Outputs	Q	P	V	Q	P	V	Q	P	V
Prod. 1	200	16	3,200	190	24	4,560			
Prod. 2	150	10	1,500	400	21	8,400	20	21	420
Prod. 3				100	6	600	5	6	30
Prod. 4							95	200	19,000
Total Output			4,700			13,560			19,450
Incr. Output			4,700			8,860			5,890
Inputs									
Labour 1	27	4	108	70	4	280	50	4	200
Labour 2	50	5	250	120	5	600	150	5	750
Labour 3	20	8	160	30	8	240	10	8	80
Total Labour			518			1,120			1,030
Materials									
Raw Materials	700	1	700	500 200	1 .5	600			
Components	350	2	700	250 100	2 1	600			
Prev. Depts.						4,557			11,262
			1,400			5,757			11,262
Energy	1,070	.1	107	3,900	.1	390	4,000	.1	400
Other Fact O/H	266	2	532	648	2	1,295	700	2	1,400
Admin. O/H	400	2	800	450	2	900	500	2	1,000
Capital	12,000	.1	1,200	18,000	.1	1,800	6,000	.1	600
Total Inputs			4,557			11,262			15,692
Incr. Output			4,700			8,860			5,890
Dept. Inputs			4,557			6,705			4,430
Dept. Prod. Net			1.031			1.321			1.330
Gross			1.031			1.204			1.239

APPENDIX I

Productivity Management Frontiers — I, edited by D.J. Sumanth
Elsevier Science Publishers B.V., Amsterdam, 1987 — Printed in The Netherlands

A STANDARD COST BASED MODEL OF PRODUCTIVITY ANALYSIS

ADIL T. TALAYSUM, PhD and M. ZIA HASSAN, PhD
School of Business Administration, Illinois Institute of Technology, Chicago, Illinois 60616

ABSTRACT
Talaysum, A.T. and Hassan, M.Z.

There continues to be a void in accounting based analysis of productivity relationships at the plant level despite renewed interest in productivity from economics and management. To help fill this void, a model of standard cost based productivity analysis is presented in this paper which culminates in the establishment of an index aimed at measuring the relative profitability foregone due to productivity variances. The interdependencies between the determinants of productivity and profitability are highlighted within the context of the model and potential biases in the index are illustrated. The model serves to bridge the gap between physical input-output relationships and the profitability measure without relying on intertemporal analysis of productivity.

In the last couple of decades, a substantial body of literature has developed around the concept of productivity both at the aggregate economic level and at the firm level. The profession of accounting has only recently begun to pay attention to productivity studies, however, as evidenced by a few articles. A great void in the accounting based analysis of productivity still exists as Kaplan [1983, p. 693] has so eloquently expressed:

> Apparently, productivity measurement, perhaps the most basic measure of manufacturing performance over time, has not been considered part of the information that will assist managers in their decision making and control activities. Also, proposed productivity measures have not yet been subject to the scrutiny we have applied to accounting variances, one of our traditional financial-based performance measures. That is, the effects of activity volume, and of substitution among labor, capital, materials, energy and other key inputs, on proposed productivity measures, have not been the subject of accounting research... Thus, developing improved productivity measurement would seem to be a fertile field for accounting researchers.

The traditional productivity measures borrowed from economics and suggested for use by accountants, such as output-per-man-hour or Kendrick's total productivity index, have a number of theoretical limitations in guiding managerial decision making. One important limitation that has not been addressed is that these measures are aimed at tracking intertemporal movements in productivity in reference to a historical starting date (base period). Thus, we can learn from these measures the gains or losses in productivity as compared to time zero or another base period, but there is nothing to guarantee that the level of productivity at time zero was optimal, or close to optimal, in the first place. It is one thing to know that some measure of productivity has advanced by X percent over time, however, this does not tell us whether it could have by X+Y percent or whether X percent was the maximum attainable. There is no absolute yard stick against which productivity performance in any one given time period can be compared and deviations from optimality, in the sense of a precise target, can be measured. Firms set targets for return on assets (ROA) or on equity (ROE) and measure deviations from such targets; they are not satisfied with knowing that ROA or ROE has advanced by so many percentage points from last year. Why should it be different with productivity analysis?

The answer to this question may lie in the hesitancy of accountants to define a reference level of productivity that may be considered optimal. It appears to be an ambitious task reserved for the theoretical sphere of economics. The closest conception to such theoretical optimality may be Salter's [1960] best practice production function from which input-output ratios could presumably be derived. The estimation of such a production function is statistically very difficult and even when it is estimated, changes in technology are likely to render the estimated function obsolete. Furthermore, the optimal technique of production in the industry which reveals such a production function may necessitate formidably large investment expenditures for a given firm in order for it to achieve the implementation of the said technique and the consequent presumed optimality. Productivity targets, for the sake of practicality, must be based on standards more readily available to the firm.

Despite its analytical simplicity, standard cost accounting may well be the only avenue toward defining such productivity targets and tracking deviations from those targets. This paper explores this avenue. The framework presented, with a bias for the conceptual, elicity considerably further research from accounting and information systems to facilitate its day-to-day usage. It is hoped, however, that the model will generate sufficient interest from the accounting profession so that productivity measures will eventually be used alongside financial ratios to analyze the performance of the firm, and that it will form a basis for suggesting means of improving profitability through the productivity gains route.

PRODUCTIVITY MODEL BASED ON DIRECT INPUTS

Implicit in the set-up of standard cost accounting is the premise that the standard quantity of any direct input i per unit of output Q to be used in the production process is <u>perceived</u> to be the optimal utilization of that input under the operational and organizational constraints specific to the plant. While such utilization may not be optimal in the theoretical sense discussed above, it is nevertheless the target level of utilization which has been settled upon by management to reflect the best performance attainable under the constraints specific to the plant. Any positive difference between the actual quantity of input i used up and the allocated standard quantity of input i represents substandard utilization.

Let sq_i = the standard quantity of direct input i allowed for a given physical output of Q units,

aq_i = the actual quantity of direct input i used up in producing output Q,

then, the traditional partial productivity of input i can be defined as:

Q/sq_i = optimum partial productivity of direct input i,

Q/aq_i = actual partial productivity of direct input i.

Based on these definitions, we can further define a partial productivity variance (PPV_i) for input i as:

$$PPV_i = Q/(aq_i - sq_i). \tag{1}$$

Now, let ac_i = actual (per unit) price or cost of input i, then the cost overrun per unit of output Q due to substandard productivity of input i can be expressed as the cost variance (CV_i) of input i:

$$CV_i = ac_i\ (aq_i - sq_i)/Q = ac_i\ \ (PPV_i)^{-1} \tag{2}$$

Thus, cost variance is inversely proportional to productivity variance.

For a plant with N different direct inputs Q stemming from substandard productivity of all direct inputs is denoted as total cost variance (TVC):

$$TCV = \sum_{i=1}^{N} CV_i \tag{3}$$

This last result clearly sets aside the cost overruns (per unit of output Q) due to substandard utilization of direct inputs as productivity-based cost overruns even though the cost of inputs are externally determined and serve to magnify the TCV for a given level of PPV_i.

Now, let the plant produce M different products, each of quantity Q_j, then the entire cost overrun from all the products, (X) will ge given as:

$$Z = \sum_{j=1}^{M} TCV_j\ (Q_j) \tag{4}$$

Given that profit (before allocated overhead and general administrative and selling expenses) equals revenue minus cost, the pre-tax profit from the plant would have been greater by an amount X if X were zero. Essentially, this means that if productivity variances for all direct inputs used in all M products were zero, then the total cost of production would have been lower by X dollars and, consequently, the pre-tax profit would have been higher by X dollars (assuming no inventory effects). Notice that negative productivity variances in some inputs (meaning that actual utilization was less than standard utilization, or an attained level of partial productivity higher than expected) would tend to offset positive partial productivity variances in other inputs. Thus, "super" productivity of some inputs helps alleviate the cost overruns stemming from substandard productivity performance in other inputs. Clearly, productivity based cost behavior is dependent upon the entire sphere of inputs.

At this point, let,

t = the firm's marginal corporate income tax rate,

A = the total net value of the plant assets,

π = the actual after-tax profit of the plant

π/A = the actual ROA of the plant

Since the after-tax equivalent of X dollars of pre-tax profit foregone due to productivity variances will be $X(1-t)$ dollars, if productivity variances did not exist or cancelled out to zero, the after-tax profit of the plant would have been $\pi + X(1-t)$ dollars. Consequently, the ROA would have been $[\pi + X\,(1-t)]/A$.

This last result enables the establishment of a productivity-based profitability index, to be called the Z-index, which is the ratio of the actual (realized) ROA to the attainable ROA:

$$Z = [\pi/A].[(\pi + X(1-t)/A]^{-1} = \pi/[\pi + X(1-t)] \qquad (5)$$

The Z-index, thus, turns out to be simply the ratio of actual after-tax profit to the attainable after-tax profit since the asset denominators cancel out.

Clearly, the larger the productivity variances, the lower will be the Z-index, with an index value of one standing for optimum productivity performance in the plant as a whole. The index stands for a clear cut measure of the relative profitability lost due to substandard productivity of the inputs. The Z-index, in short, is the culmination of the relationship between input productivities and profitability. In a plant with a constant set of capital inputs, i.e., fixed machinery, the steps leading to this index can be used to analyze the effects of various adjustments on the quantity and quality of direct inputs on profitability. This will be undertaken in the next section. A later section will expound on the model to accomodate changes in fixed investment capital inputs of the plant.

INTERDEPENDENCIES IN THE MODEL

Determinants of profitability adjustments are numerous and are not necessarily the same ones as the determinants of productivity adjustments. While in the long-run increases in productivity will, ceteris paribus, lead to increases in profitability; in the short-run increases in some input partial productivities may be obtained at the expense of higher unit input prices and/or deteriorations in other input partial productivities which may lead to unexpected and multi-directional impacts on profitability. The many determinants of intertemporal productivity changes are usually too interwined to enable the isolation of the impact of each determinant of productivity. The standard cost productivity model alleviates some of this analytical difficulty by recognizing the impact of certain determinants before the standard quantities are established and by adjusting the standard quantities accordingly. There are, nevertheless, a number of influences on productivity adjustments that cannot be so easily incorporated into the standard quantities. These influences will carry over to the productivity and cost variances of the model resulting in biases in the determination of the Z-index. The following discussion looks at each of the major determinants of productivity adjustments at the plant level in order to facilitate the attainment of the maximum benefit from the use of the model.

1. Adjustments in Quality of Direct Inputs:

Improvements in the quality of direct inputs (for instance, in the purity of some industrial raw materials inputs) generally reduce the required quantities of these inputs, and deteriorations in quality have the opposite effect. If the quality level of direct inputs is decided upon before the standard quantities are set, any reductions or increases in the quantities of these inputs due to quality changes will presumably be reflected in the determination of the standard quantities. Thus, the model prevents these quality generated adjustments from directly appearing in productivity variances because such adjustments will be built into standard quantities.

However, quality changes may indirectly influence the productivity and cost variances in two ways. First, higher quality inputs generally facilitate better estimation of the size of the output which can be attained by using these inputs. From an engineering viewpoint, input-output relationships can be ascertained with a greater degree of accuracy if higher quality inputs are used. This means that the standard quantitites set will be more realistic and the potential difference between the standard and actual quantities will be lower. Thus, with a higher quality of these inputs, the productivity variances are likely to be smaller, and consequently are Z-index likely to be higher and vice-versa. Secondly, the higher the quality of inputs, the higher will be the cost (or price) per unit of these inputs meaning that for a given productivity variance, the corresponding cost variance would be larger, and the Z-index lower. Since these two effects operate in opposite directions, they may have a tendency to cancel each other out, leaving the Z-index relatively free of such bias. But this need not always be the case, especially in the instance of very large price increments for a given rise in the quality of the inputs; clearly in that case the second effect would dominate the first one, causing a downward bias in the Z-index.

There is a further advantage of the standard cost productivity analysis over the intertemporal productivity analysis approach in that the impact of one input's partial productivity on the partial productivities of other inputs is incorporated into the standard quantities before the standard quantities are set. Gold [1979, pp. 45-61] presents an insightful model of the interdependencies between partial productivities of different inputs. He shows that the partial productivity of any one given input will be a function of, among other things, the partial productivity of every other input and the ratio of each input's quantity utilized to the quantity of every other input. Essentially, the productivity of labor (output-per-man-hour) will be influenced by the productivity of materials (output over materials volume), the productivity of fixed assets (capacity over fixed assets) and the ratio of the size of each of the mentioned inputs to every other input. This means, for instance, that a rise in the productivity of materials would increase the productivity of labor unless the quantity of materials was reduced to compensate for its higher productivity. Similarly, more efficient machinery may lead to a higher productivity of fixed assets causing, in turn, the productivity of labor to go up if the additional capacity thus provided is utilized. In intertemporal analyses of productivity changes, it is very difficult to disentangle these interdependencies after the rise of productivity has occurred. In the model presented here, however, a reasonable estimation of the impact of these interdependencies would be undertaken in establishing the input-output ratios necessary for the estimation of the standard quantities. While such estimates will fall short of perfection, they are nevertheless more feasible than trying to delineate the magnitude of the interdependencies by statistical methods after the fact. Errors in the setting of the standard quantities will, however, translate to productivity variances, consequently underestimations of standard quantities will lower the Z-index, and overestimations will induce an upward bias in the Z-index. This is the age-old problem of the accuracy of standard cost accounting which carries over to the standard cost productivity model.

2. Changes in Capacity Utilization:

Previous research has identified a tendency for productivity to rise with an increase in output at the aggregate economic level. At the plant level, the same phenomenon is likely to prevail in the following manner: Some inputs decline less than proportionately to a reduction in the output because the firm faces rigidities in reducing the quantity of these inputs and has to employ them in larger quantities than optimal. Labor is the best example for such an input; when demand slackens, the size of the lay-offs of workers tend to be smaller than that required by the production needs. Union contracts, the cost of losing the already trained workers and having to retrain workers in the next economic upswing advocate reducing workers in less than proportion to the decline in output. Consequently, when capacity utilization declines the partial productivity of labor can also be expected to decline, other things held constant; and conversely. This is less true for the material inputs because their utilization can more readily be adjusted to production levels due to the short-term nature of most procurement agreements.

In the model presented above, this phenomenon will partially be accounted for by adjusting the standard quantities. If the standard quantity set per unit of output is not a constant but is allowed to rise to some degree with rises in output and allowed to decline with reductions in output, some of the said bias will be integrated into the standard quantities and prevented from spilling over to the productivity variances. In other words, if fifty units of the input are set as the standard quantity for one-hundred units of the output, more than twenty-five units of the input will be set as the standard quantity for fifty units of the output to account for rigidities in input lay-offs. The curvilinear character of the input-output relationship suggested is over and beyond the curvilinearity dictated by the production technique alone. The establishment of the standard quantities within this context is not easy; errors in estimation of such quantities are naturally expected to occur biasing the productivity variances and, in turn, biasing the Z-index. This is all the more reason for the firm to undertake as careful a job of standard quantity determination as possible. Every effort needs to be made to base the standard quantities on objective projections; the potential tendency of managers to overestimate the standard quantities to make themselves "look good" must be guarded against to the fullest possible extent.

3. Other Determinants:

There are a number of other causes of productivity changes that are not necessarily tied to a change in technology or to a change in the size of the plant's fixed assets. Gold [1979, pp. 85-88] classifies these as changes in the scope of operations (i.e. the degree of forward or backward integration), changes in output composition (without changing the variety or the identity of the products) and changes in inventory management, materials handling or other aspects of production management aimed at reducing input requirements by reducing the wastage and idleness of resources. The specific effect of each of these factors is too broad in scope to be described here, but it should suffice to say that many of these effects can be anticipated and therefore build into the standard quantities. Again, misspecifications in standard quantities are likely to occur, leading to biases in the Z-index. In this context, however, the biases may have to be interpreted in a different light. Suppose that management expected the standard quantity of a given input to decline by five percent due to a materials waste reduction program but the program, in fact, was capable of only a three percent reduction. This would mean that the standard quantity was underestimated by two percent. The resultant downward bias in the Z-index may not be all that undesirable because it is a signal to management that the waste reduction program needs to be further refined to attain the anticipated benefit from the waste reduction program. As such, part of the partial productivity variance is accounted for by expectations that were not fulfilled by the physical input-output relationships and is consequently congruent with the spirit of productivity variance interpretation in general. Hopefully, the causes of each observed productivity variance will carefully be searched after the fact, to uncover the magnitude of underestimation in standard quantities. For the next observation period, the firm should try to remove the causes of the mismatch between expectations and the objectively assessed capabilities of the production system. This can be done by further refining the production relationship so as to attain the previously anticipated input partial productivities, or when this is not feasible, by realistically accepting the currently available production limitations and establishing the standard quantities to reflect the input-output relationships obtainable from the existing production, procurement and control systems.

INFLUENCE OF FIXED INVESTMENT

While net fixed assests were not considered a direct input into the production process in the above presented model, changes in the size of the fixed investment of the plant and its capacity influence the productivity of direct inputs and the Z-index in a number of ways. This section will review such influences, but it is first necessary to define the productivity of fixed investment simply as the ratio of capacity over net fixed assets. While the definition is analogous to the partial productivity definitions used earlier, the choice of capacity as opposed to output in the numerator underlines the concept that increases in fixed investment provide additional capacity even though such capacity may not be fully used. Thus, the link between the productivity of fixed investment and output is capacity utilization. An increased productivity of fixed investment may not be translated into larger output if the higher capacity is not utilized. In most cases, it is not possible to reduce the size of the fixed investment if the additional capacity provided by an increase in the productivity of fixed investment is currently unnecessary. Thus, in total contrast to direct input partial productivities, neither the numerator nor the denominator of the productivity of fixed investment can readily be adjusted to accomodate the fluctuations in the demand for the plant's output. At times, large investment expenditures are undertaken to bring in new technologies aimed at improving productivity (as measured by output-per-man-hour), only to find that the larger fixed costs associated with the additional investment have not been offset by the expected gains from productivity because the additional capacity has not been fully utilized. It is entirely possible for profitability to deteriorate when productivity of fixed investment improves but capacity utilization does not keep pace. Even when the additional capacity is not fully utilized, however, the productivity to direct inputs and the Z-index are significatnly affected by changes in technology and the size of the fixed investment. These influences are briefly analyzed below in the context of the proposed standard cost productivity model.

1. Changes in Technology:

Suppose a new production technology is installed in the plant which does not increase capacity but reduces one or more of the direct input requirements for a given level output. These so-called process innovations typically increase the size of the net fixed assets and reduce manpower and/or materials volume. Since the reductions in the direct inputs will be anticipated by engineering and management estimates, they will be built into the standard quantities of these inputs. If these estimates have been realistic, in and by itself, there is no reason to expect the X(1-t) term of the Z-index to be altered because productivity variances will not be affected and because unit prices of these inputs can, in the short-run, be expected to remain constant. But the increased fixed investment may result in a greater increase in fixed expenses than the reductions in direct costs during the earlier years of the project leading to a decline in the actual profit. Since the Z-index is the ratio of actual profit to attainable profit, an increase in fixed expenses will lead to a reduction in the Z-index even though the X(1-t) term is not changed. Thus, any increase in fixed expenses, which is not offset by the reduction in direct input costs, induces a downward bias in the Z-index. If the reductions in the input requirements, which were already built into the standard quantities and prevented from appearing in the productivity variances, result in direct cost reductions that exceed the increase in fixed expenses, then the Z-index will be higher because the actual profit will be higher. Thus, if the new process technology improves actual profitability, then the Z-index will have an upward bias and vice-versa. To escape this bias, one may use the X(1-t) term as the sole and final measure of the profit impact of productivity variances, but that measure lacks a reference point; an index is typically more communicative than an absolute dollar figure. Notice that the said bias exists for any change in fixed expenses which impacts actual profit (depreciation charges would be a good example), consequently the degree of operating and financial leverage are inversely related to the Z-index at lower levels of output.

It is entirely possible, of course, that in the process technology illustrated above the unit prices of the reduced direct inputs may not remain constant. A decline demand for the input may lead to a reduction in the unit prices of the said inputs. In that case, the cost variance corresponding to any given productivity variance would be larger; essentially such an outcome would alleviate the profitability impact of productivity variances.

Let us now trun to an increase in fixed investment for a new technology which allows the plant to produce the same number of units of a more enhanced thus economically more valuable version, of one of the existing product lines. The impact on actual profitability in this case will be dependent on the relative growth of costs (fixed and variable) to revenues. If actual profitability improves, the Z-index will also improve (even though the productivity variances may remain unaffected) for the reasons explained above. The productivity variances, in turn, may well embody an upward or downward bias depending upon whether the standard quantities are established liberally or conservatively in response to such a change. It would be safe, however, to assume that the Z-index would be less reliable immediately following such a change because whatever biases may exist would appear in larger proportions given the uncertainties accompanying the change.

2. Changes in Capacity:

An increase in capacity may be attained in the "pure" case by an expansion of fixed assets while leaving the production process and the product characteristics constant. In the absence of positive or negative economies of scale, direct input requirements (per unit of output) should not change with a simple capacity expansion. If the profit margin is positive and the additional capacity is utilized, however, the actual profitability will improve causing the aforementioned upward bias in the Z-index. On the other hand, if the fixed and variable expenses added by the capacity expansion are not offset by higher revenues, the reverse will be true. Thus, the utilization level of the added capacity is the critical determinant of the direction of the bias in the Z-index; at lower levels the index will be biased downward, and conversely.

When there are psoitive economies of scale that can be attained by capacity expansion, the situation becomes more complicated. Direct input requirements (per unit of output) are likely to decline in direct relationship to the level of utilization of the added capacity, thus setting the standard quantities clearly becomes a challenge. At higher levels of utilization of the added capacity, standard quantities per unit of output are expected to be reduced, but the exact magnitude of the reductions are beyond precise determination. In this context, productivity variances will be heavily influenced by divergencies between (naive) expectations and the physical input-output relationships that are in fact embodied in the new capacity. The best that can be said in such a situation is that expost analyses of the productivity variances will strengthen the understanding of the system and will hopefully lead to the elimination of such biases in the following periods. This is especially true when we recognize that capacity expansion or technological changes do not usually arrive in the clean, easily classifiable forms illustrated for the sake of analytical simplicity in this discussion; rather a capacity expansion may embody reduced input requirements, enhanced products, scale economies, etc. further complicating the biases involved.

CONCLUDING COMMENTS

The model presented in this paper does not require any additional data collection beyond what is necessary for traditional cost accounting. It, however, puts the data in a different light by bridging some of the gap between physical production relationships and profitability measures. The Z-index, proposed for routine use, stands for a measure of the relative profitability lost due to shortfalls in input productivities. The disciplines of accounting and finance have achieved a relatively high degree of sophistication in analyzing the impact of financial changes on bottom-line measures of the firm's performance. In an age when the importance of production *per se* is being rediscovered, the approach taken in this paper may alleviate the neglect in the analysis of the impact of production changes on these bottom line measures.

The calculation of the Z-index is obviously the means of a larger end; the causes of productivity variances must be analyzed in order to enhance the understanding of the existing production relationships and to identify the weak points in the production system, whether such weaknesses are caused by human factors or by technical factors. Furthermore, the model puts a dollar benefit value X(1-t) to eliminating the causes of productivity variance. Clearly, if higher benefits can be attained by, for instance an expansion of the sales force to enhance revenues than by eliminating every weak point in the production system, the firm may well stop short of further penetrating into diminishing returns in trying to capture optimum production efficiencies. The model also has a clear advantage over intertemporal analyses of productivity: in intertemporal analyses, it is difficult to distinguish, for instance, whether the anticipated improvement in profitability was not realized because the expected level of production efficiency was not attained, or because the planned improvement in productivity could not possibly have led to increases in profitability due to the entailed higher fixed expenses, higher input prices etc. This model clearly separates the profit impact of the shortfall between expectations and results, i.e. the first possibility as the X(1-t) term, forcing management to seriously examine the second possibility before undertaking the productivity increasing change.

Among other biases mentioned above, the Z-index is heavily influenced by the unit prices of direct inputs in that they magnify productivity variances into cost variances. Consequently, a possible objection to the model may be based on its neglect of price variances, i.e. the difference between standard unit input prices and actual input prices. It is possible to arrive at a variance of the Z-index which isolates away the price variances but such a construct, in practice, becomes quite cumbersome. Furthermore, the objective of the model is to show the profitability impact of productivity variances; that means the impact of productivity variances within the actual economic environment facing the firm and within the input prices dictated by that economic environment. If the firm is able to reduce input prices and thereby raise the Z-index, that simply means that productivity variances are less economically important to the firm; at higher prices, on the other hand, productivity variances translate to larger lost profits. This message, among other things, is precisely what the model is intended to convey.

REFERENCES

Baird, Robert, "Production Functions, Productivity and Technological Change", in Gold (1977).

Erdilek, Asim, "The Concept and Measurement of Productivity and Technological Change in Input-Output Analysis", in Gold (1977).

Gold, Bela, Foundations of Productivity Analysis (Pittsburgh: University of Pittsburgh Press, 1955).

_____ (ed.), Research, Technological Change and Economic Analysis (Lexington, Mass.: Lexington Books, D.C. Heath, 1977).

_____, Productivity, Technology and Capital (Lexington, Mass.: Lexington Books, D.C. Heath, 1977).

_____, "Practical Productivity Analysis for Accountants", Management Accounting (May, 1980).

Kaplan, Robert S., "Measuring Manufacturing Performance: A New Challenge for Managerial Accounting Research", The Accounting Review (October, 1983), pp. 686-705.

Kendrick, John W. and Daniel Creamer, Measuring Company Productivity: Hand-Book with Case Studies (New York: The Conference Board, 1965).

Mammone, James L., "Productivity Measurement: A Conceptual Overview", Management Accounting (June 1980a), pp. 36-42.

_____, "A Practical Approach to Productivity Measurement", Management Accounting (July 1980b), pp. 40-44.

Salter, W.E.G., Productivity and Technical Change (Cambridge, Eng., Cambridge University Press, 1960).

Productivity Management Frontiers — I, edited by D.J. Sumanth
Elsevier Science Publishers B.V., Amsterdam, 1987 — Printed in The Netherlands

DEPARTMENT LEVEL PRODUCTIVITY MEASUREMENT IN MANUFACTURING ORGANIZATIONS

ALAN R. PHILLIPS
Pfizer Hospital Products Group Inc., 235 East 42nd St., New York, NY 10017

ABSTRACT
Phillips, Alan R., 1987. Department level productivity measurement in manufacturing organizations.

This research effort was directed toward devising an improved approach to measuring manufacturing productivity in order to enhance its utility as a management tool. A methodology has been developed to measure the productivity of the smallest organizational units or departments engaged in the production process within a manufacturing plant. All significant manufacturing inputs are considered, including labor, material, and capital resources. The measurement methodology is based upon a unique, partial-factor output concept which enables dissimilar output units to be meaningfully aggregated at each discrete level of the manufacturing process. This allows multi-factor productivity to be calculated at the department level, the product level, and the plant level.

1 INTRODUCTION

Productivity measurement in industrial enterprises has generally been applied at the firm level. Although valuable information is generated through such measurement activities, a significant problem arises in interpreting the data, and trying to determine the reasons for the changes in productivity which occur over time. This problem is largely due to the complexity of the typical multi-process, multi-product manufacturing plant. Changes in productivity indicated at the firm or plant level are difficult to trace to their sources. The more recently applied product level productivity measurement approaches do not go far enough in addressing this basic problem.

This research effort was directed toward devising an improved approach to measuring manufacturing productivity in order to enhance its utility as a management tool. A methodology has been developed to measure the productivity of the smallest organizational units or departments engaged in the production process within a manufacturing plant. All significant manufacturing inputs are considered, including labor, material, and capital resources. The measurement methodology is based upon a unique, partial-factor output concept which enables dissimilar output units to be meaningfully aggregated at each discrete level of the manufacturing process. This allows multi-factor productivity to be calculated at the department level, the product level, and the plant level.

It is desirable to consider as many of the different input resources as possible when measuring productivity (National Research Council 1979). This is due to the fact that one input resource can exert a significant influence over how effectively another input resource is utilized. A classic example of such an effect would be the impact an investment in new capital equipment might have upon labor productivity, or upon the yield of an expensive raw material into saleable product in a given manufacturing plant. When such effects are measured at or below the individual department level, valuable insight is obtained. With this added detail and insight, the causes of significant productivity changes can be traced to their sources. Problem areas that could go undetected or unexplained at the firm level can be identified and corrected within operating departments.

Department level productivity measurement adds a new dimension to traditional performance indicators by comparing current levels of performance to actual past levels of performance, and by consolidating the individual labor, material, and equipment factors into an integrated total performance measure. Department supervisors and managers will be able to understand and quantify the effect that changes in the productivity of individual input resources will have upon the total productivity of the department or organization. Upper level managers will be better able to understand the full implications of their decision making.

The application of department level measurement will significantly enhance the utility of productivity measurement as an advanced management tool in manufacturing firms. Management control and understanding of the entire manufacturing process should be improved, leading to higher productivity, better decision making, and improved profitability.

2 MEASUREMENT METHODOLOGY

In order to measure multi-factor productivity at and below the individual department level, several problems must be overcome. A significant problem is posed by the dissimilarity of output units being produced within individual departments, which must be aggregated together at very detailed levels. To overcome this problem, a partial-factor output concept has been developed. The total output value of a given process step or production operation is broken down into the separate output shares of each corresponding input resource. This effectively transforms the dissimilar units of output into common value terms at the lowest discrete levels of the manufacturing process. These outputs can then be meaningfully aggregated and summarized, allowing partial-factor and multi-factor productivity to be calculated at and below the level of individual organizational units, departments, or work centers.

In its most simple form, productivity can be expressed as the ratio of the output produced over the input resources consumed by a given organizational unit. In and of themselves, productivity ratios have little meaning or significance. What is important are the changes in these ratios that occur over time. Measuring the changes that occur in productivity over time can provide valuable insight into how effectively an organizational unit is utilizing its resources.

In order to calculate such changes, a period of time must be selected which represents baseline productivity, against which future productivity ratios can be compared. This baseline period of time is referred to as the base period (Mundel (1978)). A base period should be sufficiently long to capture a fully representative sample of production history (Hines (1978)). In most manufacturing operations, one calendar year is a convenient period of time to utilize as a base period since the entire manufacturing cycle will have been repeated many times across the complete product line during the course of one year. A secondary consideration in selecting a calendar year for the base period is that past historical production data will be summarized and readily available in an annualized format. In cases where a one year period of time is used, the base period is referred to as the base year. Any past year which has had abnormal production history, or unusually severe problems should not be selected as a base year since distorted productivity data will result (Craig and Harris (1973)).

The productivity change that has occurred over a current time period, such as a month, quarter, or year is calculated by comparing the current period productivity ratio to the base year productivity ratio. A useful way to express the change in productivity over time is as a percentage change. The percentage change in productivity over a current time period can be calculated by using the following general equation:

$$\text{\% change in productivity} = \left[\frac{\dfrac{\text{(current period output)}}{\text{(current period input)}}}{\dfrac{\text{(base year output)}}{\text{(base year input)}}} \times 100 \right] - 100 \qquad (1)$$

The concept of a percentage change is well understood by a wide cross-section of people and is quite useful as a performance indicator within manufacturing plants.

With these basic productivity measurement concepts covered, the department level productivity measurement methodology can now be developed. This methodology is first developed to allow productivity changes to be measured at the lowest discrete levels of the manufacturing process. The lowest discrete level of the manufacturing process can be defined as any individual process step or production operation that produces an identifiable number of intermediate output units. An example of a discrete level manufacturing operation would be the injection molding of a plastic component used in the manufacture of a sophisticated medical device.

In order to measure the productivity of a given manufacturing process step, one must know the quantity of all input resources to the manufacturing process, as well as the number of intermediate or finished units of output. The classical inputs to a typical discrete manufacturing operation would consist of labor, materials, and capital. Miscellaneous other input resources such as electrical energy, or expense items such as cutting oil are generally not traceable at the level of an individual processing step and can not practically be considered until the measurement methodology is applied at higher organizational levels.

The labor is represented by the total labor hours applied to the specific discrete operation. Labor hours worked can be weighted on the basis of hourly compensation paid by skill level classification (Kendrick and Creamer (1965)). As an example, an hour worked by a skilled person being paid \$10.00 per hour would be equivalent to two hours worked by an unskilled worker being paid \$5.00. In those cases where there have been no significant changes in the skill classification of workers between the base period and the current period, weighting of labor hours worked by skill classification would not be necessary. This would also be the case where a common skill base of workers exists throughout a given manufacturing plant.

The material input is represented by the dollar value of the units of material used, expressed in base year prices. Citing as an example an injection molding operation, this would be the base year dollar value of the plastic resin used to produce the yielded number of molded parts.

The capital input consists of the consumption value of the resources employed in the process, and can be expressed in terms of machine hours (Fabricant (1947)). The resources employed in the process would consist of such things as land, buildings, machinery, and equipment. At the level of the discrete manufacturing step, the capital input is represented by the number of equipment hours applied to the given production step. The proportional value of land, buildings and other non-dedicated fixed capital assets can be apportioned on the basis of the square footage being used. Working capital is not considered as a capital input below the plant or firm level since it would be difficult to apportion a share of total working capital to any given production step or manufacturing department.

The total input to a given discrete manufacturing operation can be represented by the following equation:

$$I_T = I_L + I_M + I_C \tag{2}$$

Where: I_T = total input
I_L = partial input of labor
I_M = partial input of materials
I_C = partial input of capital

It should be noted that this equation is presented here to illustrate the concept of total input. The summation of the partial input factors would require that they be expressed in common value terms such as equivalent base year dollars.

The output of a given discrete manufacturing operation consists of the number of units of intermediate or finished product produced. There are typically many different discrete steps or operations performed in a given manufacturing plant, involving many different products and many different departments. In general, the output at the discrete level will be represented by intermediate units. In measuring firm level productivity, output is generally expressed as the dollar selling value of goods and services produced at base year prices (Kendrick and Creamer (1965), Craig and Harris (1973), Taylor and Davis (1978)). This approach to output is not practical below the firm level since the concept of selling value is not workable when dealing with intermediate units of production, like an injection molded component.

Units of intermediate or finished product produced is a valid output measure at the individual discrete process level. A problem arises, however, when the output units from different discrete production steps must be aggregated, or combined since their true values will not be the same. To illustrate this, 1,000 intermediate units of tablets would not be equivalent in value to 1,000 intermediate units of capsules in a pharmaceutical plant. To overcome this problem, a partial factor output concept has been developed which enables the dissimilar intermediate output units to be transformed into equivalent valued units of the corresponding input resource. Once expressed in common value terms, these transformed outputs can be thought of as the partial factor output shares of the partial factor input resources. These outputs can be meaningfully aggregated and summarized at levels above the individual discrete process steps.

The total output of a given discrete manufacturing operation can be represented by the following equation:

$$O_T = O_L + O_M + O_C \tag{3}$$

Where: O_T = total output
O_L = partial output share of labor
O_M = partial output share of material
O_C = partial output share of capital

Again, it must be noted that this equation is presented to illustrate the concept of total output. The summation of the partial output factors would require that they be expressed in common value terms, such as equivalent base year dollars.

The partial output shares are first calculated through a partial-factor productivity approach. It will be recalled that partial-factor productivity is the ratio of output to a single input resource. At the level of an individual discrete manufacturing step, partial-factor productivity can be represented by the following equation:

$$P_{PF} = \frac{\text{Intermediate output units}}{I_{PF}} \tag{4}$$

Where: P_{PF} = partial-factor productivity
I_{PF} = partial input resource units

In this generalized expression, I_{PF} could represent either labor hours, equipment hours, or material dollars at base year prices. If we wished to calculate the partial factor productivity of labor, for example, the specific equation would be represented as follows:

$$P_{PFL} = \frac{\text{Intermediate output units}}{I_L} \tag{5}$$

Where: P_{PFL} = partial-factor productivity of labor

I_L = partial input of labor in units

The percent change in the partial factor productivity for a current period is calculated as follows:

$$\%\text{ change in } P_{PF} = \left[\frac{P_{PF}\text{ of current period}}{P_{PF}\text{ of base year}} \times 100\right] - 100 \tag{6}$$

The individual partial output shares are calculated as follows:

$$O_L = \frac{P_{PFL} \text{ of current period}}{P_{PFL} \text{ of base year}} \times I_L \text{ of current period} \quad (7)$$

$$O_M = \frac{P_{PFM} \text{ of current period}}{P_{PFM} \text{ of base year}} \times I_M \text{ of current period} \quad (8)$$

$$O_C = \frac{P_{PFC} \text{ of current period}}{P_{PFC} \text{ of base year}} \times I_C \text{ of current period} \quad (9)$$

In this context, O_L can be described as the equivalent number of labor hours that would yield the current period units of intermediate production at the base year output rate. As such, O_L represents the partial output share of labor for that individual discrete manufacturing operation. Since the values of all O_L are expressed in the common unit of labor hours, they can be meaningfully aggregated across the different discrete manufacturing operations, provided the different operations are performed by workers of a common skill base. This is often the case within individual departments or work centers.

Similarly, O_M represents the partial output share of material for the individual discrete manufacturing step, and is expressed in the common units of base year dollars. As such, all O_M can be meaningfully aggregated across the different discrete manufacturing operations.

O_C represents the partial output share of capital for the individual discrete operation and is expressed in terms of equipment hours. All O_C can be meaningfully aggregated across all discrete operations performed within the same equipment center. This is also usually the case within individual departments.

It is now possible to aggregate the individual O_L, O_M, and O_C partial output shares for each discrete manufacturing operation within a given department. By the same logic, the individual I_L, I_M, and I_C partial input resources can also be aggregated at the department level. The percent change in department level partial-factor productivity can be calculated for each input resource as follows:

$$\text{\% Change in } P_{PFL} \text{ departmental} = \left[\frac{\left(\sum_{i=1}^{n} O_{Li} \div \sum_{i=1}^{n} I_{Li} \right) \text{ current period}}{\left(\sum_{i=1}^{n} O_{Li} \div \sum_{i=1}^{n} I_{Li} \right) \text{ base year}} \times 100 \right] - 100 \quad (10)$$

Where: i = each individual discrete manufacturing operation performed in the department during the period represented.

n = the total number of discrete operations i performed in the period represented.

The bracketed denominator of the above equation, or the equivalent material or capital equations, will always be unity. This is true because O_L and I_L are always the same value in the base year. Similarly, O_M and I_M are the same and O_C and I_C are the same in the base year. The concept of the base year mandates that productivity is assumed to be constant across the entire year in order for the base year to provide a basis for comparison and measurement of future results. This allows the above equation for the partial-factor labor productivity to be simplified as follows:

$$\% \text{ change in } P_{PFL} \text{ departmental} = \left[\left(\sum_{i=1}^{n} O_{Li} \right) \text{current period} \div \left(\sum_{i=1}^{n} I_{Li} \right) \text{current period} \times 100 \right] - 100 \quad (11)$$

The departmental level material and capital partial-factor productivity are calculated by the following simplified equations:

$$\% \text{ change in } P_{PFM} \text{ departmental} = \left[\left(\sum_{i=1}^{n} O_{Mi} \right) \text{current period} \div \left(\sum_{i=1}^{n} I_{Mi} \right) \text{current period} \times 100 \right] - 100 \quad (12)$$

$$\% \text{ change in } P_{PFC} \text{ departmental} = \left[\left(\sum_{i=1}^{n} O_{Ci} \right) \text{current period} \div \left(\sum_{i=1}^{n} I_{Ci} \right) \text{current period} \times 100 \right] - 100 \quad (13)$$

It is important to note that the summarization of partial-factor productivity at the department level is a weighted aggregation of all the discrete intermediate production operations performed during the current period within the given department. Those intermediate operations which consume proportionally more input resources will exert proportionally greater influence on the summarized department level productivity result.

The measurement methodology must now be extended to provide a means of calculating the multi-factor productivity results at the department level. Multi-factor productivity is the ratio of the total output over the sum of all the inputs for a given economic unit. In this case, the given economic unit is an individual production department or work center within a manufacturing plant. Department level multi-factor productivity can be calculated as follows:

$$P_{MF}\ (\text{departmental}) = \frac{O_T\ \text{departmental}}{I_T\ \text{departmental}} \qquad (14)$$

$$= \frac{\sum_{i=1}^{n} O_{Li} + \sum_{i=1}^{n} O_{Mi} + \sum_{i=1}^{n} O_{Ci}}{\sum_{i=1}^{n} I_{Li} + \sum_{i=1}^{n} I_{Mi} + \sum_{i=1}^{n} I_{Ci}} \qquad (15)$$

In order to meaningfully sum the partial-factor output terms into total output, and the partial-factor input terms into total input, each of these terms must be converted into common value units. The common unit of value that can be used is the equivalent number of base year dollars.

In the case of labor, O_L and I_L are expressed in terms of labor hours. These factors can be converted into the equivalent number of base year dollars by multiplying each term by the appropriate base year hourly labor rate. This hourly rate should include all of the cash components of compensation, including the cost of fringe benefits.

In the case of material, O_M and I_M are already expressed in terms of base year dollars, and as such, require no further adjustments at this time.

The capital output and input terms are somewhat more difficult to handle. O_C and I_C must be converted from equipment hours into equivalent base year dollars. It will be recalled that equipment hours were selected as the measure of capital input because they best represent the true consumption of capital resources in the manufacturing process. A method of converting such a capital input into dollar terms was developed by Craig and Harris (1973). Craig and Harris refer to the dollar value of capital input as "the service value of capital". While they apply this concept at the firm level, it can readily be extended to the department level. The service value of capital is represented by the equivalent dollar cost of the actual consumption of capital assets in the manufacturing process, expressed in base year dollar terms. The concept is built on the facts that specific capital assets have a finite useful lifetime, that the dollars invested in such assets must generate a minimum rate of return to their owners, and that in addition to recovering a minimum rate of return, they must also recover their acquisition cost over their finite useful lifetimes. In engineering economics, the equivalent annual dollar cost of such assets can be calculated using the capital recovery factor. This annual dollar cost becomes the capital input to the productivity equation in the Craig and Harris model. The calculation of this capital input can best be illustrated with a simple example.

Assume that a manufacturing department consists of one machine that was purchased in the base year at a total installed cost of one million dollars. It is anticipated that the machine will have a finite useful life of ten years. The company which purchased the machine has an average annual return on assets of 15%. The machine is available for operation 2,000 hours per year. The annual capital recovery cost of this machine is calculated by multiplying the ten year 15% capital recovery factor (A/P, i %, n) times the total installed base year cost (Grant, Ireson, and Leavenworth (1965)).

$$\begin{aligned}\text{Annual Capital Recovery Cost} &= (A/P,\ 15\%,\ 10) \times (\text{Cost}) \\ &= (.19925) \times (\$1{,}000{,}000) = \$199{,}250\end{aligned} \tag{16}$$

$$\text{Hourly Capital Recovery Cost} = \frac{\$199{,}250}{2{,}000} = \$99.63$$

The capital input I_C and output O_C can now be converted into equivalent base year dollar terms in this example by multiplying the number of equipment hours representing each factor by the hourly consumption cost of 99.63 base year dollars per departmental equipment hour.

All partial input factors and all partial output factors can now be expressed in the common terms of equivalent base year dollars. The percent change in multi-factor productivity at the department level can be calculated as follows:

$$\%\ \text{change in}\ P_{MF}\ \text{departmental} = \left[\frac{\left(\dfrac{\sum_{i=1}^{n} O_{Li} + \sum_{i=1}^{n} O_{Mi} + \sum_{i=1}^{n} O_{Ci}}{\sum_{i=1}^{n} I_{Li} + \sum_{i=1}^{n} I_{Mi} + \sum_{i=1}^{n} I_{Ci}}\right)_{\text{current period}}}{\left(\dfrac{\sum_{i=1}^{n} O_{Li} + \sum_{i=1}^{n} O_{Mi} + \sum_{i=1}^{n} O_{Ci}}{\sum_{i=1}^{n} I_{Li} + \sum_{i=1}^{n} I_{Mi} + \sum_{i=1}^{n} I_{Ci}}\right)_{\text{base year}}} \times 100\right] - 100 \tag{17}$$

Once again, the bracketed lower base year denominator will always be unity. Therefore, the percent change in multi-factor productivity at the department level can be calculated by the following simplified equation:

$$\begin{array}{l}\% \text{ change in } P_{MF} \\ \quad\text{departmental}\end{array} = \left[\frac{\sum_{i=1}^{n} O_{Li} + \sum_{i=1}^{n} O_{Mi} + \sum_{i=1}^{n} O_{Ci}}{\sum_{i=1}^{n} I_{Li} + \sum_{i=1}^{n} I_{Mi} + \sum_{i=1}^{n} I_{Ci}}_{\substack{\text{current}\\ \text{year}}} \text{ X } 100 \right] - 100 \quad (18)$$

The department level productivity measurement methodology has now been fully developed.

3 SUMMARY AND CONCLUSIONS

The objective of this research effort was to develop a department level productivity measurement methodology that would enhance the utility of productivity measurement as a management tool in manufacturing organizations. A methodology has been developed that allows productivity to be measured at the lowest discrete level of the manufacturing process. A unique, partial-factor output concept was conceived in order to overcome the problem presented by the need to aggregate the dissimilar intermediate units of output produced by the many different discrete level manufacturing operations in the typical manufacturing plant.

In conjunction with firm level productivity measurement, department level productivity measurement can open new avenues of analysis and control to manufacturing managers. Such tools become increasingly important if manufacturing firms are to successfully compete in the high-technology, competitive markets of today. The application of productivity measurement in manufacturing companies can lead to better understanding, management, and control of the firm's basic resource consumption processes.

The department level productivity measurement methodology should significantly enhance the utility of productivity measurement as a manufacturing management tool. The application of this methodology should enable managers at all levels of the organization to exercise a greater level of control over the total manufacturing process. A heightened awareness of productivity, and the complex interactions of all the input resources, would be an expected outcome. Ultimately, improved productivity, competitive position, and profitability should result.

4 REFERENCES

American Productivity Center, Inc. 1979. Productivity Measurement: An Executive Overview. Houston, Texas.

Cocks, Douglas L. 1974. The Measurement of Total Factor Productivity for a Large U.S. Manufacturing Corporation. Business Economics, (September), 7-21.

Craig, Charles E. and R. Clark Harris. 1973. Total Productivity Measurement at the Firm Level. Sloan Management Review, (Spring), 13-29.

Fabricant, Solomon. 1947. An Evaluation of the Productivity Conference. In Summary of Proceedings of Conference on Productivity, October 28-29, 1946. U.S. Department of Labor, Bureau of Labor Statistics. Bulletin 913. Washington, D.C., 1-8.

Fabricant, Solomon. 1969. A Primer on Productivity. Random House, New York.

Grant, Eugene L., W. Grant Ireson, and Richard S. Leavenworth. 1965. Principles of Engineering Economy. Sixth Edition. John Wiley & Sons. New York. 36-37, 173, 594.

Greenberg, Leon. 1973. A Practical Guide to Productivity Measurement. Bureau of National Affairs, Inc. Washington, D.C.

Hines, William W. 1978. Guidelines for Implementing Productivity Measurement. In Productivity: A Series from Industrial Engineering, by M. E. Mundel. American Institute of Industrial Engineers, Norcross, Georgia. 30-33.

Kendrick, John W. and Daniel Creamer. 1965. Measuring Company Productivity: Handbook with Case Studies. Second Edition. The Conference Board. New York, N.Y.

Mundel, Marvin E. 1978. Measures of Productivity. In Productivity: A Series from Industrial Engineering, by M. E. Mundel. American Institute of Industrial Engineers, Norcross, Georgia. 27-29.

National Research Council. 1979. Measurement and Interpretation of Productivity. Panel to Review Productivity Statistics. National Academy of Sciences. Washington, D.C.

Siegel, Irving H. 1976. Measurement of Company Productivity. In Improving Productivity Through Industry and Company Measurement. National Center for Productivity and Quality of Working Life. Washington, D.C. (October), 15-24.

Siegel, Irving H. 1980. Productivity Measurement: An Evolving Art. Work in America Institute Studies in Productivity. Highlights of the Literature. Work in America Institute, Inc. Scarsdale, New York.

Sumanth, David J. and M. Zia Hassan. 1980. Productivity Measurement In Manufacturing Companies by using a Product-Oriented Total Productivity Model. Proceedings - 1980 Spring Annual Conference. American Institute of Industrial Engineers, Norcross, Georgia. 255-266.

Taylor, Bernard W. and K. Roscoe Davis. 1978. Corporate Productivity - Getting It All Together. In Productivity: A Series from Industrial Engineering, by M. E. Mundel. American Institute of Industrial Engineers, Norcross, Georgia. 34-38.

C. TOTAL AND/OR PARTIAL PRODUCTIVITY MEASUREMENT

Productivity Management Frontiers — I, edited by D.J. Sumanth
Elsevier Science Publishers B.V., Amsterdam, 1987 — Printed in The Netherlands

A TECHNOLOGY-ORIENTED TOTAL PRODUCTIVITY MEASUREMENT MODEL

JOHNSON AIMIE EDOSOMWAN
Dept. 146. Bldg. 012, International Business Machines Corporation, P.O. Box 950, Poughkeepsie, New York 12602, USA

ABSTRACT
Edosomwan, J. A., 1987. "A Technology-oriented Total Productivity Measurement Model

In recent years, technology-assisted means of production is the most common form of productivity improvement technique utilized in the work environment. The new manufacturing frontier faces massive development, introduction, and use of computers and robotics in the work environment. The impact of the various forms of technology on total productivity deserves significant attention. The main thrust of this paper offers a technology-oriented total productivity measurement model based on the development, production, and human-related variables that are likely to be affected when technologies such as computers and robotics are developed, introduced and used in the work environment. The input variables of interest are: Development expenses, labor, computer operating expenses, robotics operating expenses, energy, materials, capital, retraining expenses, and administrative expenses. The output variables are: Value of technologically conceptual ideas, partial units produced, finished units produced, and other revenue associated with units produced. Methods of allocating overhead expenses to the various inputs are suggested. The various input and output elements are identified and computed in specified technological life-cycle stages. A step-by-step implementation methodology is presented. The case studies are based on the validity of the model in a situation that involves the development, introduction and use of a computer-aided device and a robotic device. The problems encountered during implementation are discussed, as well as ways to correct them.

INTRODUCTION

In this era of technological explosion, companies are faced with the challenges of how to maintain technological leadership and the competitive edge, increase revenue, diversify products, keep abreast of the market place, and provide the best possible technology for the customer. However, the life cycle of the new technologies has continued to be shortened from about a ten-year span in 1960, to about a two-year span in 1986. How to respond to the short life cycle of technology, and manage it for productive growth is a significant issue. One way to do this is through a careful understanding and management of the input-output relationship in the technology life-cycle stages. Following this, developing and implementing a technology-oriented productivity measurement system can have the following benefits:

1. A technology-oriented productivity measurement system improves resource planning for all phases of the technology life cycle. It provides common measures and concepts to compare similar or different types of technology.
2. A technology-oriented productivity measurement systems creates the awareness required for effective supervision of all actions to be taken, and improves decision making through better understanding of the actions already taken to address a given problem in each of the phases of the technology life cycle.
3. A technology-oriented productivity measurement system provides the basis for the firm to make decisions on when to phase out a given technology. The productivity values and indices provide a way of detecting deviations from establishing standards on a timely basis so that something is done about such deviations. In situations where the performance of the work groups, technology, task, customer, departments, projects and the firm as whole are to be measured, the technology-oriented measurement system can be implemented with minor modifications on the definition of inputs, outputs, and the allocation criteria for overhead expenses.

PRODUCTIVITY MEASUREMENT HIERARCHY (PMH)

In order for a productivity measurement system to provide useful results, it is important for the level of measurement to be specified and understood within the productivity measurement hierarchy shown in Figure 1. The productivity measurement hierarchy suggests that productivity measurement should start at the basic levels (individuals, task, and technology). If input and output are clearly identified at this level, measures obtained are meaningful. It becomes less cumbersome to quantify and measure productivities in Level 2 (work group, project, product), Level 3 (departments, project group, product group) and through aggregation technique to Level 4 (plant or site unit level). The hierarchy also provides a basis to identify where the productivity issues are when viewed from a global perspective. Most importantly it shows that the productivity of any nation starts from the basic units. This paper focuses on presenting a model that is useful for the technology unit, and is also applicable to other units through aggregation techniques and slight modification in notations.

THREE TYPES OF PRODUCTIVITY MEASURES COMMONLY USED FOR THE VARIOUS LEVELS WITHIN THE PRODUCTIVITY MEASUREMENT HIERARCHY

1. Total Productivity: Is the ratio of total output to all input factors. Edosomwan (1985a), Sumanth (1979), Kendrick and Creamer (1965), Craig and Harris (1973), and American Productivity Center (1976) have proposed such measures.

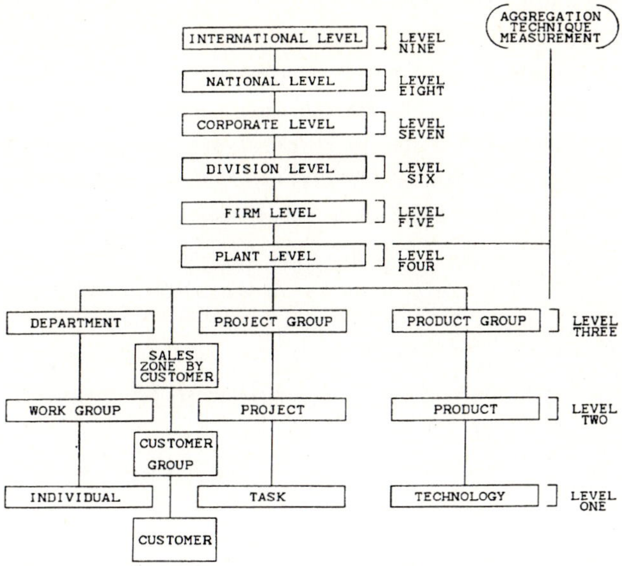

Figure 1. The Productivity Measurement Hierarchy (PMH)

2. Total Factor Productivity: Is the ratio of net output to the sum of associated labor and capital (factor) inputs. Mali (1978), Taylor and Davis (1977), and Kendrick and Creamer (1965) have recommend this measure.
3. Partial Productivity: Is the ratio of total output to one class of input. Output per man-hour (labor productivity) is commonly used. Most of the indices published by the U.S. Department of Labor (1980) are partial measures. Melman (1956) and Turner (1980) have also used such measures.

The total productivity measure is a better measure since it takes into account total output and input. The total factor measure has one major disadvantage of omitting the cost of materials, which is one of the vital input components. The partial measure only considers one input factor and does not present a total picture of the true total cost of doing business. At the basic level, only a few productivity measurement models are available for a task or project. None of the available models are concerned with technology phases measurement. A problem still exists in defining input and output components, as well as identifying the contribution to the output of a given input. Some of the input and output elements specified in the task-oriented total productivity measurement model developed by Edosomwan (1985a) provided the basis for quantifying some of the measurable input and output components of the Technology-Oriented Total Productivity Measurement Model. The works of

Kendrick and Creamer (1965), and Craig and Harris (1973) also provided the basis for quantifying some of the input and output components. The major significant difference between previous productivity models and the Technology-Oriented Total Productivity Measurement Model (TOTPMM) is that it recognizes a new technological version where development expenses, retraining expenses for new technologies, computer operating expenses, and robotic operating expenses are important additions to the input components. Technologically conceptual ideas and other associated values have also been considered as addition to output components. Also, productivity is measured in all phases of the technology for the first time.

THE TECHNOLOGY-ORIENTED TOTAL PRODUCTIVITY MEASUREMENT MODEL (TOTPMM)

The measures derived from this model are in the form of an index obtained by considering all phases of the technology life cycle shown in Table 1. The input and output components of (TOTPMM) are shown in Figures 2 and 3 respectively. The key definitions associated with TOTPMM are as follows:

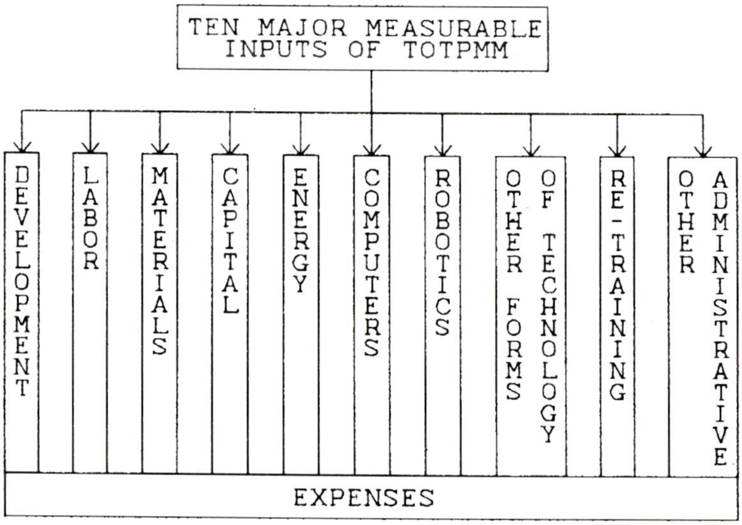

Figure 2. Input Components Considered in the Technology-Oriented Total Productivity Measurement Model (TOTPMM)
Source: Edosomwan, J. A., 1987. Reprinted with permission of Marcel Dekker, Inc. New York..

TABLE 1: Technology Life Cycle Phases Components
Source: Edosomwan, J. A., 1987. Reprinted with permission of Marcel Dekker, Inc., New York.

PHASE	1	2	3	4
DESCRIPTION	Project Proposal Study and Definition of Ideas	Design	Development	Manufacturing Verification, Production and Sales
APPROXIMATE NUMBER OF YEARS FOR ALL PHASES	← INITIAL LIFE OF TECHNOLOGY	—— VARIES FROM FIVE TO TWO YEARS ——		TO END OF LIFE OF TECHNOLOGY →
END OBJECTIVE AT EACH PHASE	Determine feasibility, requirements, initital specifications cost, quantities and estimated revenue and profits.	Refine specifications, build models, define market support and verify costs revenue and profits, commit idea and model to marketing.	Finalize specifications costs, and test manufacturability. Release for announcement and commit price and revenue.	Verify manufacturing capability, produce deliver and install continue to forcast future demand. Produce until demand indicates end of life.
SIGNIFICANT INPUT COMPONENTS	- Labor expenses - Other administrative expenses	- Development expenses - Labor, materials - Capital - Energy - Other administrative expenses	- Development expenses - Labor, capital, material - Energy - Other administrative expenses	- Labor, material - Capital, energy - Computer operating expense - Robotic operating expense - Re-training expenses - Other administrative expenses
SIGNIFICANT OUTPUT COMPONENTS	- Technologically conceptual ideas in monetary terms ($ Value)	- Finished model - Partial model in monetary terms ($ Value)	- Finished units prototype - Partial units prototype in monetary terms ($ Value)	- Partial units produced - Finished units produced - Other output associated with units produced in monetary terms ($ Value)

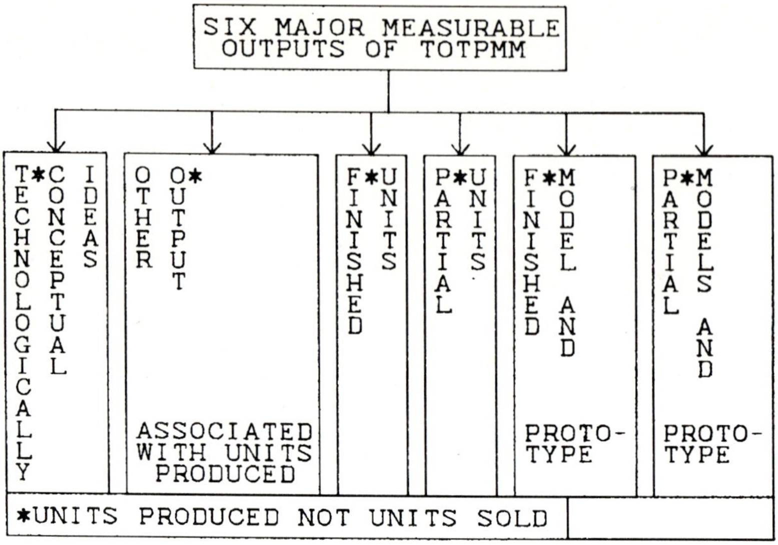

Figure 3. Output Components Considered in the Technology-Oriented Total Productivity Measurement Model (TOTPMM).
Source: Edosomwan, J. A., 1987. Reprinted with permission of Marcel Dekker, Inc., New York

NOTATIONS FOR DERIVATION OF PRODUCTIVITY VALUES AND INDEXES

Let:

i = Technology type (i=1,2,3,......m)
j = Technology life cycle phase (j=1,2,3,4)
k = Technology development/manufacturing/service site (k=1,2,3,......n)
t = Study period (t=1,2,3,......p)

TCIijkt = Total value of technologically conceptual ideas of type i, produced in phase j, in site k, in period t.

FMPijkt = Total value of technology finished model and prototype of type i, produced in phase j, in site k, in period t.

PMPijkt = Total value of partially completed technology model and prototype of type i, produced in phase j, in site k, in period t.

FUPijkt = Total quantity of finished units of technology of type i, produced in phase j, in site k, in period t.

PUPijkt = Total quantity of partial units of technology of type i, produced in phase j, in site k, in period t.

PCTijkt = Percent completion of partial units of technology of type i, produced in phase j, in site k, in period t.

SPTijkt = Base period selling price per unit for a unit of technology of type i, produced in phase j, in site k, in period t.

TOTijkt = Total quantity of output of technology of type i, produced in phase j, in site k, in period t.

DETijkt = Development expense input in monetary terms utilized to produced Technology of type i, in phase j, in site k, in period t.

LRTijkt = Labor hours input utilized to produce technology of type i, in phase j, in site k, in period t.

LRRijkt = Labor rate per hour utilized to produce technology of type i, in phase j, in site k, in period t.

METijkt = Materials and purchase part expense utilized to produce technology of type i, in phase j, in site k, in period t.

CETijkt = Capital related expense (includes fixed and working capital such as cash, accounts receivable, tools, plant, buildings, etc.). Utilized to produce technology of type i, in phase j, in site k, in period t. (Capital is computed using lease value concept.)

EETijkt = Energy related expense (includes electricity, solar energy, water, coal, gas, etc.) input utilized to produce technology of type i, in phase j, in site k, in period t.

CREijkt = Variable computer related expense input utilized to produced technology of type i, in phase j, in site k, in period t.

RREijkt = Variable robotics related expense input utilized to produce technology of type i, in phase j, in site k, in period t.

OFTijkt = Variable expense from other forms of technology utilized to produced technology of type i, in phase j, in site k, in period t.

RETijkt = Retraining expense utilized for technology of type i, in phase j, in site k, in period t.

OETijkt = Other administrative expense utilized to produced Technology of type i, in phase j, in site k, in period t.

TBT = Base period time which is the reference period to which output and input in monetary terms is reduced. Thus, the total productivity of a task expressed as dollar output/dollar input in constant dollars.

TPTijkt = Total productivity of technology of type i, in phase j, in site k, in period t.

TFTijkt = Total factor productivity of technology of type i, in phase j, in site k, in period t.

PPTijkt = Partial productivity of technology of type i, in phase j, in site k, in period t.

OOTijkt = Other output associated with units produced of technology of type i, in phase j, in site k, in period t.

TITijkt = Total input utilized to produced technology of type i, in phase j, in site k, in period t.

The total and partial productivities are expressed as follows:

$$TPTijkt = \frac{\text{Total measurable output produced of technology of type i, in phase j, in site k, in period t.}}{\text{Total measurable input utilized to produce technology of type i, in phase j, in site k, in period t.}} \quad (X1)$$

$$TPTijkt = \frac{TOTijkt}{TITijkt} \quad (X2)$$

$$TPTijkt = \frac{FMPijkt + TCIijkt + PMPijkt + (FUPijkt)(SPTijkt) + (PUPijkt)(PCTijkt)(SPTijkt) + OOTijkt}{DETijkt + (LRTijkt)(LRRijkt) + METijkt + CETijkt + EETijkt + CREijkt + RREijkt + OFTijkt + RETijkt + OETijkt} \quad (X3)$$

Partial productivity considers only one input factor. Total factor productivity considers only labor and capital input factor. For example, partial productivity with respect to labor input is expressed as follows:

$$\underset{[LABOR]}{PPTijkt} = \frac{TCIijkt + FMPijkt + PMPijkt + (FUPijkt)(SPTijkt) + (PUPijkt)(PCTijkt)(SPTijkt) + OOTijkt}{(LRTijkt)(LRRijkt)} \quad (X4)$$

ALLOCATION CRITERIA FOR OVERHEAD EXPENSES

Edosomwan (1985b, 1987) and Kendrick (1984) pointed out that allocation criteria used for output and input elements could vary depending on the type of environment, cost elements, accounting information and managerial preferences. The following allocation criterion are suggested:

1. Complexity Factor Criteria: The complexity of the technology/ product/task is used to determine a proportional contribution to the total quantity produced. Complexity in this case means items such as, number of assemblies within a given technology, insertion rate, kitting time, etc..
2. Quantity Produced Proportionality Criteria: The proportional contribution to total quantity produced is used to allocate overhead expenses. This is applicable in situations where there is more than one type of technology, product or task.
3. Direct Hours Allocation Criteria: This approach requires daily direct labor recording and the derivation of burden rate from net expense and direct hours. Therefore, cost of product includes labor and burden. Allocation of overhead is done using direct hours and one burden rate.

IMPLEMENTATION METHODOLOGY

The implementation steps for the Technology-Oriented Total Productivity Measurement Model (TOTPMM) are briefly described in Figure 4. A team work approach among departments, managers and employees is recommended to facilitate better productivity measurement. The implementation steps presented in Figure 4 are based on the two case studies presented in this paper. Additional work may be required depending on the type of technology, organizational environment, and the type of data collection system already available.

TWO CASE STUDIES

The TOTPMM was used to measure and compare the productivities of types of technology in an organization that produces technology to customer orders. The first case study involved a situation of a computer technology monitored throughout its life cycle. The second case study involved a situation of a robotic device technology monitored throughout its life cycle. In both case studies, two phases of each technology type were analyzed and studied. Historical data on input and output was used, especially in the development and project proposal phases. The implementation methodology specified in Figure 4 was followed.

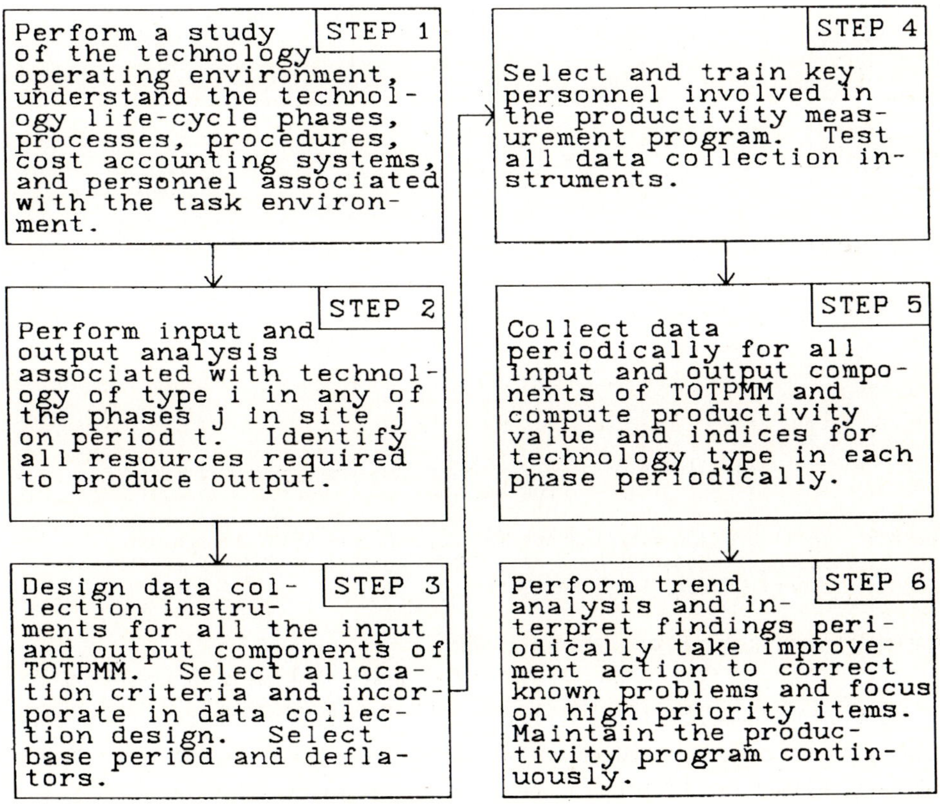

Figure 4. Implementation Steps for TOTPMM

RESULTS AND DISCUSSION

As shown in Figures 5 and 6 respectively, TOTPMM enabled us to compare the total productivity indexes of two types of computers and robotic devices in different stages of their life cycles. By examining the total productivity trend of model C212JX computer over its life cycle, improvement actions were put in place to improve the performance trend in model C212JXY. Similar actions were taken for the robotic device models NM-70X and NM-70X, respectively. In both technologies, the series of engineering changes (EC) and specification revisions had significant negative impact on productivity. For example, as shown in Figure 5, Phase 4, the performance of Computer C212JXY deteriorated because of a major EC change. Phase I of C212JX and C212JXY was significantly impact by labor productivity. In both technologies, energy expense, computer operating expenses, robotic operating expense and other administrative expenses impacted the total productivity in all phases.

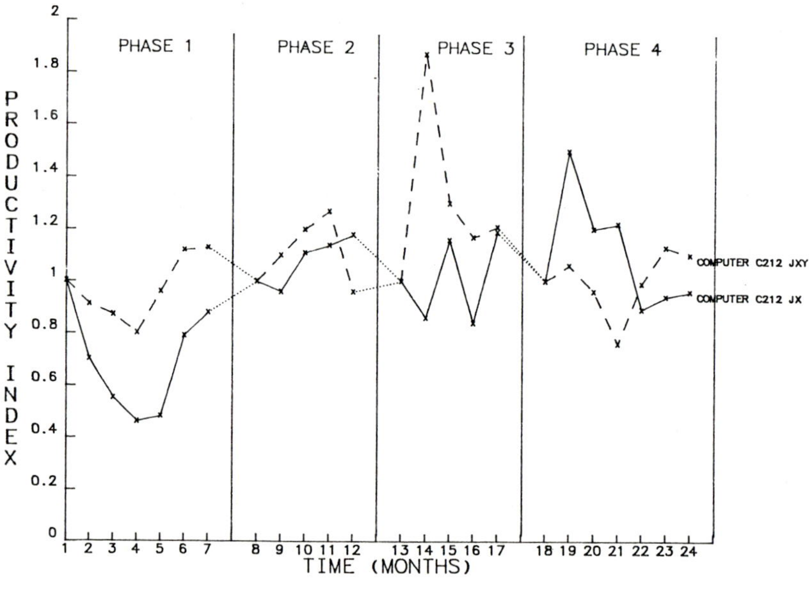

Figure 5. Total Productivity Indices for C212 JX and C212 JXY Computers
Source: Edosomwan, J. A., 1987. Reprinted with permission of Marcel Dekker, Inc., New York.

PROBLEMS ENCOUNTERED DURING IMPLEMENTATION

Data collection and tracking was the greatest problem encountered during implementation. Because of the various phases involved in the technology life cycle, a large data base had to be set up to collect all the input and output components needed for TOTPMM. A computer program was written to compute the values needed for the computation of productivity. In the proposal design and development phases, it was a bit difficult to identify the contribution of a given input to the output obtained. This problem was resolved through an appropriate tracking mechanism implemented to record input and output elements. Resistance to the productivity measurement program problem was overcome through a teamwork approach. Sessions were held to explain the purpose and benefits of the study to everyone. It must be pointed out that the problems encountered during implementation could vary from one organization to the other depending on technology types, organizational climate and other factors.

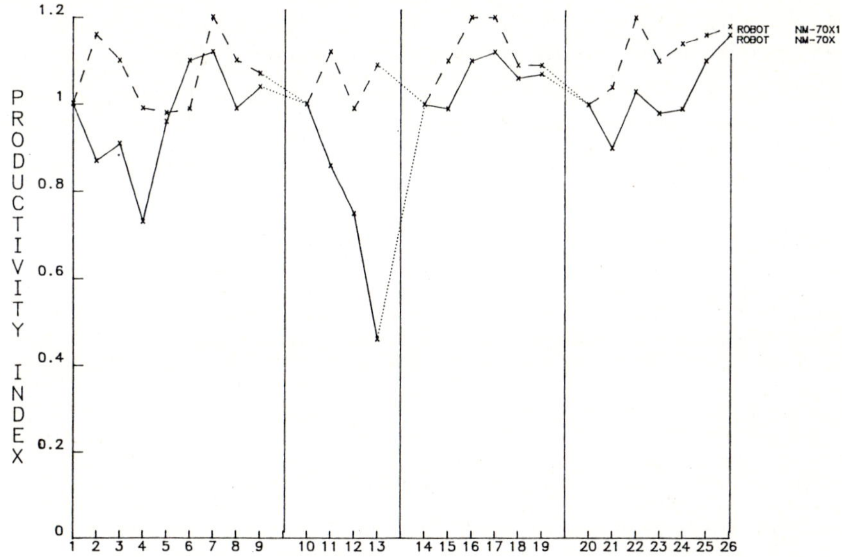

Figure 6. Total Productivity Indices for Robot NM-70X and NM-70Y
Source: Edosomwan, J. A., 1987. Reprinted with permission of Marcel Dekker, Inc. New York.

CONCLUSION

Perhaps this is a first-time attempt to develop and implement a technology-oriented total productivity measurement model that recognizes all phases of technology life-cycle stages, and total measurable input and output components. The model has also been proven to be sensitive in the two cases presented. For existing companies trying to implement this type of measurement system, it is suggested for them to start with detailed analyses of technology input and output and historical data for at least two years. New companies have the advantage of proceeding immediately to appropriate tracking of the technology input and output elements. The TOTPMM is a powerful model with strong potential for application to various product groups, task and project, in both manufacturing and service organizations. For each potential implementation case, minor modifications may have to be made in data collection instruments. TOTPMM has also been used for product and technology equipment selection as shown in the works of Edosomwan (1985b, 1986a, 1986b, 1985c). Application calls for an on-going data collection, computation, comparison monitoring, and analysis of productivity trend patterns. Only in such situations do firms realize optimum benefits for its use in a productivity measurement program.

REFERENCES

American Productivity Center, "Productivity and the Industrial Engineer," Region VIII AIIE Conference, Chicago October 27, 1978 (Seminar Notes).

Craig, C.E. and Harris, C.R. (1973). "Total Productivity Measurement at the Firm Level," Sloan Management Reviews 14(3):13-29.

Edosomwan, J.A. (1985a). "A Task-Oriented Total Productivity Measurement Model for Electronic Printed Circuit Board Assembly," First International Electronics Assembly Conference Proceedings Santa Clara, California. October 7-9, 1985.

Edosomwan, J.A. (1985b). "A Methodology for Assessing the Impact of Computer Technology on Productivity, Production Quality, Job Satisfaction, and Psychological Stress in an Assembly Task," "Doctoral Dissertation, Department of Engineering Administration, the George Washington University, Washington, D.C. 20052. Research Sponsored by IBM and Social Science Research Council (SSRC) (US Department of Labor).

Edosomwan, J.A. (1986b). "The Impact of Computer-Aided Manufacturing on Total Productivity," Eighth Annual Conference on Computers and Industrial Engineering Journal Proceedings. March 19-21, Orlando, Florida.

Edosomwan, J.A. (1986a). " A methodology for assessing the impact of robotics on total productivity," Proceeding for Annual International Industrial Conference, May 11-15, 1986, Dallas, Texas.

Edosomwan, J.A. (1985c). "A Methodology for Assessing the Impact of Computer-aided Manufacturing on Productivity, Job Satisfaction, and Psychological Stress in an Assembly Task," Annual International Industrial Engineering Conference Proceedings, Chicago, Illinois, December 7-11, 1985.

Edosomwan, J.A. Productivity and Quality Management, Marcel Dekker Publishing Company, New York, 1987 in Press.

Kendrick, J.W. (1984) Improving Company Productivity Handbook with case studies in collaboration with the American Productivity Center. The Johns Hopkins University Press, Baltimore, Maryland.

Kendrick, J.W. and Creamer, D. (1965) "Measuring Company Productivity: Handbook with Case Studies," Studies in Business Economics, No. 89, National Industrial Conference Board, New York, N.Y.

Mali, P. (1978) Improving Total Productivity: MBO Strategies for Business, Government, and Non-Profit Organizations, John Wiley and Sons.

Melman, Seymour (1956) Dynamic Factors in Industrial Productivity, John Wiley and Sons, New York.

Sumanth, D.J. August (1979) Productivity Measurement and Evaluation Models for Manufacturing Companies. Ph.D Dissertation, Department of Industrial Engineering IIT Chicago.

Taylor, B.W. (III) and Davis, R.K. (1977) "Corporate Productivity - Getting It All Together," Industrial Engineering 9(3):32-36.

Turner, Jon A. (1980) Computers in Bank·Clerical Functions: Implication for Productivity and the Quality of Working Life. Ph.D Dissertation, Columbia University.

US Department of Labor, Bureau of Labor Statistics. Monthly Labor Review, January, 1980, pp.40-43.

Productivity Management Frontiers — I, edited by D.J. Sumanth
Elsevier Science Publishers B.V., Amsterdam, 1987 — Printed in The Netherlands

A TOTAL PRODUCTIVITY MODEL FOR LONGWALL UNDERGROUND COAL-MINING FOR STRATEGIC ANALYSIS AND PLANNING

Alan Ivor STAINER, PhD
Middlesex Business School, Middlesex Polytechnic, London, England

ABSTRACT

Stainer, A.I., 1987. A total productivity model for longwall underground coal-mining for strategic analysis and planning

The paper provides a critique of the present state of the art of productivity measurement in underground coal-mining. Weaknesses of the partial measure of labour productivity as well as critical observations for improvement are revealed.

The subject is put into a technological and economic perspective with special emphasis on mechanisation. An output model is built up, founded on the number of, as well as physical features of longwall coal faces. The problems pertaining to the statistical information are discussed and validated via the physical equations of the output model. The effects of mechanisation are also highlighted relating to machine available time and output. An input model is also shown, with a discussion on the aggregation of various inputs, based on a bench-mark year to allow for inflation and international currency exchange rate movements, with particular emphasis on the problem of capital input measurement.

Utilising the input and output models, a total productivity model is developed, based on inter-linking ratios, to reflect the major factors influencing total productivity.

A case study of a 20-year span in the UK underground coal-mining industry is used to illustrate how major trends can be highlighted and thence utilised as a tool for strategic management and forward planning.

1. PRODUCTIVITY

The word productivity is too commonly used in a narrow sense as synonymous with labour productivity. More properly, productivity should be regarded as the relationship between output and all its inputs. The majority of these input components are under the control of the industry's management, either directly or indirectly.

As mining becomes more mechanised and automated, the partial measure of labour productivity becomes far less relevant. Management must ask itself whether the yardstick of productivity is to be a high tonnage per manshift or a low cost per ton. The true answer is a combination of the many factors both physical and economic affecting productivity. The weakness of measuring productivity, in purely labour terms, is that it ignores the other important inputs, especially that of capital.

For instance, one can consider output per manshift underground (O.M.S.) as the ratio

$$\frac{\text{tons of coal}}{\text{manshifts underground}}$$

This has a clear meaning and is obviously connected with productivity. It is a ratio which can be computed for a single mine, as well as for the whole mining industry. It is in a class of ratios, which should be followed with interest by the management of a mine and by those responsible for planning mining policy, for instance on a national scale. However, the question is whether one can say anything more about this ratio, which may depend on factors such as geological conditions, equipment used, skill of the workers and their efforts. If all elements, except one, are constant, the indicator will, of course, measure the changes due to the residual factor. It can therefore, depending on the circumstances, be taken as a measure of :

(i) the geological difficulty in a particular mine;

(ii) the efficiency of the equipment used;

(iii) the effect of training programmes for the workers.

As such it may be useful, but "it all depends".

No single ratio can be considered in isolation. Thus a model is necessary built upon interlinking ratios to reflect those parameters considered the most important by management to monitor and control. These ratios culminate in a figure measuring total productivity, that is :

$$\text{total productivity} = \frac{\text{total output}}{\text{total input}}$$

Total productivity can be briefly described as the overall measure of economic effectiveness on the basis of real output per unit of resource, or resources, utilised. This effectively means that all economic measures must be kept at a base to allow comparison and to isolate inflation.

2. THE MEASUREMENT OF OUTPUT

The only sensible way to measure output would be in tonnage, and the measure chosen is market or saleable tonnage. That is after being cleaned and prepared for the market. It is true this does not allow for the quality of the coal, but to measure output in calorific values or some form of hard coal equivalent would, at the present state of the art, only cloud the issue.

For longwall production, equations can be built up for management purposes based on expected pithead tonnages (EPT):

EPT = height extracted x length of face x advance x density of extraction.

No measurement is meaningful without a time scale. Thus, if the advance per day is taken, the formula would then give the expected daily output per face in pithead tonnages. To obtain market tonnage or its equivalent, the pithead tonnages must be converted via a vend factor, where

$$\text{vend} = \frac{\text{market tonnage}}{\text{EPT}}$$

By adapting this, a basic model can be built up, starting at the face, and expanding to reflect various aggregates, such as the mine or the industry, as follows:

$$\overline{DOF} = \bar{c} \cdot \bar{k} \cdot \bar{s} \cdot \bar{l} \cdot \bar{a} \qquad \text{Equation I}$$

where: $\overline{DOF}$ = average daily output per longwall face in market tonnage

$\bar{c}$ = average vend (market tonnage/pithead tonnage)

$\bar{k}$ = weighted average density of extraction

$\bar{s}$ = weighted average of extracted seam thickness, including dirt bands, and allowing for operational losses

$\bar{l}$ = average length of longwall face, including stable holes

$\bar{a}$ = average advance of longwall face per working day

Equation I is an expansion and sophistication of the EPT calculation.

It introduces an average vend (c), converting the pithead tonnages into market tonnages, these being the prepared coal as taken by the market. Also, average seam thickness is appropriately weighted for overall statistical accuracy.

The annual output from longwall faces in market tonnage can be shown as:-

$$O = \bar{d} \cdot \bar{n} \cdot \overline{DOF} \qquad \text{Equation II}$$

where: O = output per annum from longwall faces in market tonnage

$\bar{d}$ = average number of days worked per longwall face during the year

$\bar{n}$ = average number of longwall faces per working day during the year

Thus, combining Equations I and II:

$$O = \bar{d} \cdot \bar{n} \cdot \bar{c} \cdot \bar{k} \cdot \bar{s} \cdot \bar{l} \cdot \bar{a} \qquad \text{Equation III}$$

Although labour productivity is not considered by any means to be the be-all or end-all, it is considered that, as traditionally so much emphasis has been placed on it, a further equation (Equation IV) would be developed incorporating it, not so much for its own sake but to show its inter-relationships:-

$$\frac{O}{MY_u} = \overline{DOF} \cdot \frac{MS_u}{MY_u} \cdot \frac{1}{MS_u/dn} \qquad \text{Equation IV}$$

where: $\frac{O}{MY_u}$ = average output per man-year for underground longwall workers during the year

$\frac{MS_u}{MY_u}$ = average manshifts per man-year for underground longwall workers

MS_u/dn = average underground manshifts per longwall face-day worked

The increase in mechanisation throughout Europe has played a major role in the advancement of productivity. Equation III can be applied to mechanised faces exclusively, that is faces where a machine either loads prepared coal or cuts and loads coal simultaneously; coal felled in the stable holes of mechanised faces is included in mechanised output.

A further sophistication is added relating mechanised longwall output to an equation, based on the machine rather than on the man:-

$$O_m = \bar{d}_m \cdot \bar{n}_m \cdot \frac{MAT}{McS} \cdot \frac{O_m}{MAT} \cdot \frac{McS}{dn_m} \qquad \text{Equation V}$$

where: O_m = output per annum from mechanised longwall faces in market tonnage

$\bar{d}_m$ = average number of days worked per mechanised longwall face during the year

$\bar{n}_m$ = average number of mechanised longwall faces per working day during the year

$\frac{MAT}{McS}$ = average machine available time in minutes per machine shift at mechanised longwall faces during the year

$\frac{O_m}{MAT}$ = average mechanised longwall output per minute of machine available time during the year

$\frac{McS}{dn_m}$ = average number of machine shifts per mechanised longwall face-day worked during the year

Utilising the equations, longwall output can be measured and analysed. In this way, the various physical factors affecting output can be studied.

3. THE MEASUREMENT OF INPUT

All productive inputs to the longwall system must be included. That is all inputs in the production of longwall hard coal up until it is in a prepared state in market tonnage ready to be distributed. Thus all costs are included, except interest payments and marketing costs. The latter are isolated as they are a separate process from production.

The only practical way of aggregating inputs must be in economic terms, expressed in currency. The economic ills of exchange rates and inflation must be isolated, by expressing prices and exchange rates at base-year value. 1960 has been chosen as a bench mark year for this purpose.

There are a multitude of inputs, but these can be broken down into three major areas : manpower, capital and supplied/overheads. Each of these must be expressed in 1960 prices. Once this has been achieved, they can be aggregated to form the total input, which can be converted for international comparative purposes in US dollars at the 1960 exchange rate.

4. TOTAL PRODUCTIVITY

The ultimate total productivity measure should be :

$$\frac{\text{longwall hard coal output}}{\text{total productive inputs}} = \frac{O}{TI}$$

An integrated model is therefore built on interlinking ratios and illustrating the important elements of the total productivity of hard coal longwall mining. The ratios comprise three sectors, incorporating underground manpower, longwall mechanisation and underground energy utilisation. Obviously, many more features can be included but, if the model is to be of any practical use, it should be as uncluttered as possible. Hereafter, the model would be referred to as the Total Productivity Model, where all inputs are expressed in US dollars, at 1960 prices and exchange rates, and all physical measures are metric:

$$\frac{O}{TI} = \frac{O}{MY_u} \cdot \frac{CS_m}{MI_m} \cdot \frac{MI_m}{MI} \cdot \frac{MI}{TI} \cdot \frac{O_m}{CAP_m} \cdot \frac{CAP_m}{CS_m} \cdot \frac{O}{O_m} \cdot \frac{MY_u}{kWh_u} \cdot \frac{kWh_u}{O}$$

4.1 Manpower Ratios

$\frac{O}{MY_u}$ Output per man-year for underground longwall workers

This ratio best reflects underground labour productivity related to man-years, as it is the most reliable figure achieved in European labour statistics.

$\frac{CS_m}{MI_m}$ Capital stock in longwall face mechanisation to mechanised longwall face manpower input

This ratio reflects the input of manpower working direcctly at the mechanised longwall faces, with the capital machinery and equipemnt at and held for those faces, at 1960 prices, exchange rates and technology, and expressed in US dollars. In this way, the trends and effects of the proportion of manpower input to capital stock can be analysed.

$\frac{MI_m}{MI}$ Mechanised longwall face manpower input to total manpower input

This monitors the trends in manpower input at mechanised longwall faces, compared with the total production manpower input of the whole longwall mining industry. Both inputs are at 1960 prices and exchange rates, expressed in US dollars.

$\frac{MI}{TI}$ Manpower input as a proportion of total input

The important proportion of labour to total input is mirrored by this ratio, holding prices and exchange rates constant at 1960, expressed in US dollars.

4.2 Longwall Mechanisation Ratios

$\frac{O_m}{CAP_m}$ Mechanised longwall face capacity utilisation

This monitors the longwall mechanised output in relation to the capacity that could reasonably be expected to be achieved from the faces.

$\frac{CAP_m}{CS_m}$ Mechanised longwall face capital stock efficiency

This ratio matches up the capacity of the mechanised longwall faces with a capital stock of the equipment, directly used or being held for those faces. Thus, it is a measure of the efficiency of that capital stock with prices, exchange rates and technology held constant at 1960, expressed in US dollars.

$\frac{O}{O_m}$ Mechanised longwall output factor

This is a simple measure of the total longwall output in relation to the mechanised longwall output.

4.3 Energy Ratios. The concluding two ratios balance the model. The use of electrical energy underground, though not forming a major economic element, reflects the trends in the man/machine relationship.

$\frac{MY_u}{kWh_u}$ Man-years underground to underground electrical energy, for longwall output

This relates manpower to electrical energy, both working underground in longwall mining. It shows the physical relationship between man and machine.

$\frac{kWh_u}{O}$ Electrical energy underground per ton

This mirrors the input of underground electrical energy in physical terms per ton of all longwall output.

4.4 Total Productivity. The combination of the nine component ratios results in the total productivity measure.

$\frac{O}{TI}$ Total productivity measure

This is the relationship between the output of longwall hard coal and the total productive inputs, expressed in US dollars and held at 1960 prices and exchange rates.

The majority of the nine elements of the Total Productivity Model, which are the bases of the components, are self-explanaroty. However, three areas need further clarification because of their complexity or controversial nature.

4.5 Capital stock in longwall face mechanisation (CS_m)

This must measure in 1960 prices and exchange rates, the capital stock of all equipment being held for or being used at mechanised longwall face areas. Not only prices, but technological specification should be held at the 1960 base; this is because it has been found over a period of time considerable expenditure has been made over and above inflation on specification enhancement. Thus, the excess paid for inflation and technological improvement can be isolated.

4.6 Mechanised longwall face manpower input (MI_m)

This must measure in 1960 terms, the amount of money paid to mechanised longwall face manpower during the year. Only manpower employed on productive work within the face area itself should be included, and those working in roadways and headings excluded. This is so that this input can be matched against the capital stock in longwall face mechanisation (CS_m).

4.7 Capacity of mechanised longwall face system in market tonnage (CAP_m)

Of all the parameters measured or estimated in mining, capacity is possibly the most difficult and subjective. However, a meaningful measure of capacity can be based upon the mechanised longwall faces themselves as it is matched up with the mechanised capital stock and manpower inputs to those faces. Thus, assumption is made that the skill of men, shaft capacity, washery intake, haulages, etc. are sufficient. Therefore, if one concentrates on the capacity of mechanised longwall faces, then it would have four basic ingredients:

(a) the area of the seam exposed;

(b) the dirt and density of the seam;

(c) the time available for mining the seam; and

(d) the effectiveness of technology and organisation, monitored in mechanised longwall face advance per unit time.

These are reflected in Equation III, and it can be seen that the relationships between the four are multiplicative. Thus,

Capacity $\propto$ area exposed x dirt and density x time worked x effectiveness

In any reasonable time span for planning, the area, dirt and density factors are fixed, making capacity proportional to the time worked and the effectiveness of technology and organisation. Mechanised longwall face capacity was therefore based on Equation V, using obtainable values.

5. THE ANALYSIS

Using the Total Productivity Model and supporting equations, the United Kingdom's longwall hard coal mining industry can be analysed for the 20-year period 1960 to 1980. All economic terms are based on 1960 and expressed in US dollars at the 1960 exchange rate; the latter is carried out for international comparative purposes. All physical measures are in metric tons and metres.

5.1 Total longwall output in million tons and daily output per face in tons are calculated based on Equations I and III. They are self-checking devices to alert statistical inaccuracies in the make up of total longwall output, showing several important trends. Whilst daily output per face more than trebled in the period to 620 tons, the actual number of faces fell from 3603 to 742 with a drop of total output from 174M to 107.7M tons. Far fewer faces with much larger face area and more rapid daily advances were being utilised. This is a trend throughout Europe.

5.2 Output per man-year underground in tons is based on Equation IV and shows the relationship between underground labour productivity and daily output per face. The latter has steadily increased whilst the former peaked in 1972 at 698 tons, falling to 602 tons in 1980. The interesting feature is the steady increase in mashifts underground per day-face; in other words, the faces are more intensely worked in manshift terms

5.3 Mechanised longwall output in million tons and daily output per mechanised face in tons are based on Equations I and III, and apart from being a self-checking devices, show that even mechanised faces have declined in number, more than halving since 1964 to 662 in 1980. Here again, the increased daily output per face has been mainly a function of the use of larger face areas, and to a lesser extent the daily advance per face.

5.4 Mechanised longwall output in million tons and mechanised advance in centimeters related to machine available time are founded on Equations III and IV. They illustrate the fact that machine available time per machine shift has been steadily falling as faces get further away from the mine shaft. However, output per machine available time has steadily increased mainly due to larger face areas whilst advance per machine available time has only increased moderately.

5.5 Total input per ton in US dollars at 1960 prices and exchange rate show the total input per ton in real terms. In money terms it is always argued that manpower is the main input, in real terms manpower input has been steadily falling due to a more efficient and smaller labour force, from 60.4% in 1960 to 35.6% in 1980.

5.6 Total Productivity Model (see Table).

The manpower ratios show that underground labour productivity peaked in 1972 and thence declined, whereas total productivity was at its height between 1967 and 1969. Also, due to increased mechanisation and labour cut-backs, mechanised capital stock has steadily increased in relation to mechanised longwall face manpower.

The longwall mechanisation ratios show that the industry was working at 62.7% capacity in 1980 compared with 60.3% in 1960. Also, during the period 1960 to 1980, the mechanised longwall face capacity has been outstripping the mechanised capital stock available and the percentage mechanised output was doubled to 1/1.01, i.e. 99%.

The energy ratios highlight the use of more powerful equipment and the reduction in the labour force. The kilowatt-hours underground per ton increased from 6.56 to 18.5 in the period whilst the man-years underground to kilowatt-hours underground decreased from 354 to 99.

6. STRATEGIC ANALYSIS AND PLANNING

Strategic analysis and planning since 1980 in the United Kingdom coal industry has not been based on labour productivity but upon total productivity figures or their inverse of real cost per ton. Utilising the Total Productivity Model and the supporting equations, forecasts have been developed founded on historical trends, highlighting the inter-dependencies of both economic and physical factors.

In this way, the industry is now in the total productivity era and only utilises labour productivity in a supporting role. This approach can be used at all levels of longwall mining and can relate more particularly at the level of the mine to achieve an integrated management control system.

Thus, in strategic management, scenarios are built up, particularly for long range planning purposes, and are used as a basis for policy making at the national level, as well as a basis for analysing the economic viability of individual mines or groups of mines. It has become a fact of life that forward planning must be based on total productivity allowing for the considerable capital expenditures required in modern longwall mining.

TOTAL PRODUCTIVITY MODEL
UNITED KINGDOM

where $\frac{O}{TI}$ is expressed in kilogrammes per US dollar at 1960 prices and exchange rates

$$\frac{O}{TI} = \frac{O}{MY_u} \cdot \frac{CS_m}{MI_m} \cdot \frac{MI_m}{MI} \cdot \frac{MI}{TI} \cdot \frac{O_m}{CAP_m} \cdot \frac{CAP_m}{CS_m} \cdot \frac{O}{O_m} \cdot \frac{MY_u}{kWh_u} \cdot \frac{kWh_u}{O}$$

	$\frac{O}{TI}$	$\frac{O}{MY_u}$	$\frac{CS_m}{MI_m}$	$\frac{MI_m}{MI}$	$\frac{MI}{TI}$	$\frac{O_m}{CAP_m}$	$\frac{CAP_m}{CS_m}$	$\frac{O}{O_m}$	$\frac{MY_u}{kWh_u}$	$\frac{kWh_u}{O}$
1960	85.18	431	2.45	6.37	.604	.603	.674	2.22	354	6.56
1961	85.61	445	2.22	8.54	.587	.605	.685	1.85	340	6.60
1962	89.68	491	2.20	11.03	.568	.605	.710	1.51	301	6.76
1963	92.76	514	2.17	12.75	.553	.607	.744	1.34	287	6.78
1964	95.24	533	2.28	14.40	.549	.609	.702	1.23	272	6.89
1965	95.42	528	2.46	15.57	.537	.611	.663	1.14	244	7.77
1966	92.93	548	2.77	17.06	.510	.611	.583	1.08	223	8.18
1967	96.80	580	3.31	18.66	.504	.613	.475	1.07	204	8.46
1968	99.80	627	3.51	18.40	.475	.615	.502	1.05	186	8.58
1969	97.56	680	3.80	17.83	.459	.616	.488	1.04	158	9.31
1970	94.43	691	3.96	17.85	.439	.617	.478	1.03	146	9.90
1971	77.52	598	4.69	17.67	.421	.525	.411	1.03	136	12.25
1972	90.42	698	4.89	17.47	.410	.618	.406	1.03	136	10.56
1973	70.67	565	5.38	16.23	.394	.512	.393	1.02	120	14.71
1974	78.67	691	5.38	16.37	.363	.618	.392	1.02	116	12.47
1975	76.28	662	5.65	15.81	.361	.621	.374	1.02	116	13.05
1976	72.41	637	5.62	15.92	.363	.621	.354	1.01	113	13.89
1977	67.93	636	5.87	14.11	.362	.625	.351	1.01	111	14.48
1978	64.60	568	6.33	13.30	.360	.625	.348	1.01	109	15.67
1979	61.92	578	6.79	11.51	.358	.627	.340	1.01	105	16.95
1980	60.61	602	7.13	10.21	.356	.627	.335	1.01	99	18.50

7. REFERENCES

Adam, E.E., Hershauer, J.C. and Ruch, W.A., 1981, Productivity and Quality : Measurement as a Basis for Improvement, Prentice-Hall

Bailey, D. and Hubert, T., 1980, Productivity Measurement, Gower

Cahen, L., 1961, Measuring the Total Productivity of a Sector of Industry - The French Coal-Mining Industry, Productivity Management Review, No.25

Craig, C.E. and Clark Harris, R., 1973, Total Productivity at the Firm Level, Sloan Management Review

Domar, E.D., 1962, On Total Productivity and All That Journal of Political Economy, Vol. Y70, Part VI

Eilon, S. and Teague, J., 1973, On Measures of Productivity, Omega, Vol. 1, No. 5

Eilon, S., Gold, B. and Soesan, J., 1976, Applied Productivity Analysis for Industry, Pergamon

Fenske, R.W., 1965, An Analysis of the Meaning of Productivity, Productivity Measurement Review, No. 42

Greenberg, L., 1973, A Practical Guide to Productivity Measurement, Bureau of National Affairs

Ksansik, J.E., 1974, Measuring Productivity, Managerial Planning

Lawlor, A., 1985, Productivity Improvement Manual, Gower

Malhotra, R., 1975, Factors Responsible for Variation in Productivity of Illinois Coal Mines, Illinois State Geological Survey

Mali, P., 1978, Improving Total Productivity, John Wiley

Norman, R.G. and Bahiri, S., 1972, Productivity Measurement and Incentives, Butterworths

Risser, H.E., 1966, Coal Mine Productivity - Some Things the Averages don't Tell, Illinois State Geological Survey

Smith, I.G., 1973, The Measurement of Productivity, Gower

Stainer, A.I., 1979, Productivity in Longwall Mining of Hard Coal, Loughborough University of Technology, PhD Thesis

Thorelli, H.B., 1960, Productivity - Notes on a Tantalizing Concept, Productivity Measurement Review, No.22-23

Wearly, W.L., 1975, The Yardstick of Productivity, Coal Age, Vol. 80, No. 8

Productivity Management Frontiers — I, edited by D.J. Sumanth
Elsevier Science Publishers B.V., Amsterdam, 1987 — Printed in The Netherlands

TOTAL AND PARTIAL PRODUCTIVITY MEASUREMENT AT THE PLANT LEVEL: EMPIRICAL EVIDENCE FOR LINERBOARD MANUFACTURING

KONSTANTINOS P. TRIANTIS
Department of Industrial Engineering and Operations Research, Virginia Polytechnic Institute and State University, Northern Virginia Graduate Center, 2990 Telestar Court, Falls Church, VA 22042 (U.S.A.)

ABSTRACT

Triantis, Konstantinos P., 1987. Total and Partial Productivity at the Plant Level: Empirical Evidence for Linerboard Manufacturing

The objective of this paper is to empirically measure and evaluate total and partial productivity ratios for three linerboard manufacturing facilities. Due to the existence of data at the process level for one of the mills, it was possible to specify total and partial productivity ratios for both the mill and its main production processes. The performance of these measures is compared to the derived technical efficiency measures which are obtained using statistical production function and linear programming specifications. In this way, the informational content of the total and partial productivity ratios is investigated. Data are obtained from the mills' operating statements which report mill and process input/output production and cost accounting information.

1 INTRODUCTION

Management uses a number of performance measures to evaluate its past operations and to make strategic decisions concerning the future. Sink (1985) discusses seven such measures, namely, product quality, quality of work life, innovation, profitability, effectiveness, efficiency and productivity. Out of these seven performance measures, profitability has been the measure most commonly incorporated in management's decision-making processes. The remaining six measures have not received as much attention by researchers or practitioners and consequently have not been as extensively used by management for its long- or short-term policy making. The reasons for this phenomenon have been discussed extensively by Sumanth (1984), the NASA Symposium on Productivity and Quality (1984) and others. However, it is felt that the following three managerial concerns are important: First, the modelling of these remaining six measures at the organizational level is either ambiguous or difficult or both. Second, data which support the testing of different models which estimate these measures is either not collected or if it is collected, it is not in the proper format. Third, management is not clear about the informational content of these measures and how they can be utilized to support the decision making process.

Two of the seven performance measures, namely, efficiency and productivity, are examined in this study. Specifically, there are two main objectives of this research effort. First, to empirically measure and evaluate total and partial productivity ratios for three linerboard manufacturing facilities which belong to the same firm. Second, to compare the productivity performance results with those results derived from evaluating technical efficiency measures for the same facilities. These technical efficiency measures have been previously derived by Triantis (1985). By accomplishing these objectives, two of the previously mentioned managerial concerns are addressed. First, the modelling of efficiency and productivity at the plant and process levels is undertaken for the linerboard technology, and second the informational content of both measures is examined.

In order to ensure brevity of presentation, mill productivity and efficiency measures are presented only for mill I. Also, only the productivity and efficiency performance of the pulp production and chemical recovery process for mill II is specifically discussed, since process data is accumulated by the firm only for mill II. The results for the other processes, namely, the steam generation, boiler feedwater treatment and paper production operations, are not included in this discussion. Data is obtained from the mills' monthly operating statements. The mills are studied for 25 months, i.e., from December 1980 through December 1982. During this time period no major technological changes are observed. The remainder of this presentation is organized as follows: In the next section, the efficiency and productivity concepts and models are briefly explained. In section 3, the linerboard manufacturing technology and the input/output variables are discussed. Section 4 follows with the empirical results, and section 5 concludes.

2 CONCEPTS AND MODELS OF EFFICIENCY AND PRODUCTIVITY

The decomposition of efficiency into its technical and allocative components was first introduced by Farrell (1957). Technical efficiency measures can be classified as either output increasing or input decreasing. Output increasing measures indicate the maximum amount by which output can be increased and still be producible by a given vector of input quantities. Input decreasing measures denote by how much a productive unit can reduce its input usage and still be capable of producing a constant level of output. Färe and Lovell (1978) proved that these two measures are identical only when the productive unit is experiencing a constant return to scale technology. However, the ranking of efficient performance is identical whether the output increasing or input decreasing measure is used as long as they are based on the same frontier surface. In this study, plant output increasing technical efficiency measures for pulp and linerboard facilities estimated by Triantis (1985) are further examined.

The estimation of the output increasing technical efficiency measures hinges upon the estimation of the production frontier (Schmidt (1985)). Two sets of efficiency measures are examined in this study. The first set of output increasing technical efficiency measures is obtained from a full-frontier production function specification, as formulated originally by Afriat (1972). Additionally, a log-linear production function form is specificed for each mill and production process and the corrected ordinary least squares methodology is used to estimate the production function parameters (Kopp (1981), Greene (1980)). Consequently, both total and partial input technical efficiency measures are obtained for the mill and its processes. The partial efficiency measures for each input indicate how efficient each input is, if the plant were to produce the actual level of output for that month and use the actual input levels of the remaining inputs (see, Kopp (1981)). Both the total and partial efficiency measures use the estimated production function frontier as a standard of efficient performance.

The second set of technical efficiency measures is obtained by computing a convex hull which envelops the data. Färe, Grosskopf and Lovell's (1984) linear programming formulation based on Shephard's (1970,1984) linear representation of technology is used. The weak cone output increasing technical efficiency measure[1] is used to rank efficiency performance. This measure assumes a constant returns to scale environment and uses the isoquant subset for its reference set[2].

Sumanth (1984) defines total productivity as the ratio of total output to the sum of all inputs and it reflects the joint impact of all the inputs in producing the output. Partial productivity is the ratio of output to one type of input and represents the impact of this input type in producing the output. Total and partial productivites are calculated for the mill and its processes first using the APC (1981) model for which the profitability ratio equals the productivity ratio times the price recovery ratio. The productivity ratio represents the amount of resources used to produce the plant's output. The price recovery ratio indicates how input costs behave in relation to output prices. Given the nature of the cost accounting data it was not always feasible to separate the productivity ratio from the price recovery ratio because prices or quantities were not defined for all inputs. In this case, only the values of inputs were available. Also, Gold's (1982) productivity network was obtained by computing partial productivity ratios. It was not feasible to model the cost and profitability networks given the nature of the mill data. However, the advantage of presenting productivity relationships as a network of interactions is that management can track causal relationships through the network, which may explain why certain productivity ratios have changed. The productivity measures obtained using Gold's (1982) model were consistent with the results obtained by the APC (1981) model. Consequently, in order to ensure brevity of presentation, only the APC (1981) productivity results are reported.

3 THE LINERBOARD MANUFACTURING TECHNOLOGY: INPUT/OUTPUT PRODUCTION SPECIFICATIONS

The empirical results presented in this paper pertain to linerboard manufacturing facilities. A brief description of the linerboard technology and the production function specification is given in this section. For a more detailed description, see Triantis (1985). From a papermaker's point of view, paper is an engineering structure composed of cellulose fibers. By selecting the fiber type and adjusting its properties during processing in a mill, different types of paper can be produced.

The main source of fiber is wood. Pulpwood is defined as a wood which has been cut or prepared from trees primarily for processing into pulp. The pulpwood may arrive at the mill in the form of sawdust, logs, wood chips or whole trees. Most of this wood fiber must be converted into suitable chip form in order for it to be cooked. The chips are subsequently steam cooked within a chemical solution in a pressure vessel or digester. The mills under investigation use the Kraft or alkaline pulping process which uses an alkaline solution of caustic soda and sodium sulfide. After the completion of the cooking process, the contents are discharged into a receiving tank, a process known as blowing the digester. The products of the cooking process are the pulp and the black liquor which contains the original pupling chemicals and the dissolved wood substances. Even though the ratio of pulpwood to chemicals present in the digester is four to one, the cost of chemicals is twice that of pulpwood. This necessitates the chemical recovery process.

To prepare the pulp for paper processing, refining is usually the first step undertaken. Refining is defined as the mechanical action or beating of the fiber. This beating increases fiber flexibility and, in turn, its internal bonding ability. After the refining process, and before the pulp is distributed to the paper machine, chemicals are added to the pulp mixture to alter its properties. Once the pulp has been prepared, it is distributed to a paper machine where the pulp is transformed into linerboard. Based on the above descriptions of the linerboard manufacturing technology and the availability of the appropriate cost accounting data, the following mill input/output specification was used to estimate the production frontier either using the full-frontier production function or the linear programming models. This specification also served as the basis for the APC (1981) and Gold's (1982) productivity models, i.e.,

LPR = f(CAP, LAB, NMAT,ENE). (1)

LPR represents linerboard production, CAP stands for capital, LAB is labor, NNAT stands for net raw material usage (primarily chemical and fiber usage) and ENE represents energy usage (primarily electricity, steam, fuel oil, gas and black liquor fuel usage). The input/output specification for pulp manufacturing is given by the following relationship:

PUPR = g(PUENE, PUWAT, PUCHEM, PUWOOD). (2)

PUPR represents pulp production, PUENE is the energy usage during pulp operations, PUWAT stands for water usage, PUCHEM represents the chemical usage and PUWOOD the wood usage. For more details on the definition of these variables refer to Triantis (1985).

4 EMPIRICAL FINDINGS

The efficiency measures for mill I are presented by Table 1. The full-frontier technical efficiency measures for mill I were calculated by assuming a log linear production function form for equation (1) and using the corrected ordinary least squares methodology (Kopp (1981), Greene (1980)). As a result of this estimation process, the following full-frontier production function is obtained (Triantis (1985)).

$$LPR = (0.184) * (CAP)^{.286} * (LAB)^{.340} * (NMAT)^{.126} * (ENE)^{.543}. \quad (3)$$

Also, the weak cone output increasing technical efficiency measures are given by Table 1. R_1 and R_2 indicate the ranking of each observation based on the full-frontier and weak cone output increasing technical efficiency measures respectively.

In terms of ranking, differences are observed between the full-frontier and the weak cone measure as indicated by R_1 and R_2. The fact that these disparities in ranking are observed poses the problem that the evaluation of individual observations is sensitive to the method by which the efficiency measures are obtained. However, Seaver and Triantis (1986) have analyzed the same dataset using a k-clustering methodology and multiple outlier diagnostic tests. They have found that for those observations with the largest disparity in, ranking extreme operating or weather conditions were encountered by the mill. For example, for observation 14, for which the largest discrepancy in ranking exists between the two measures, extremely bad weather conditions led to increased fiber losses and excess energy usage[3]. Given these external weather conditions it is not clear from the mill's monthly operating statement whether the plant operated efficiency or inefficiently.

The total and partial productivity measures for the mill are given by Table 2. These productivity measures represent both price and quantity changes.

They have been labeled as profitability measures within the APC (1981) modelling framework. The base year prices and quantities for these productivity measures are the average values for the prices and quantities for the year 1979. The total productivity measure isolates observations 18, 17, 25, 1 and 13 as representing the worst mill performances, whereas observations 5, 9, 2, 3, and 4 as representing the most productive months. As can be observed, there is some degree of consistency between the productivity and the efficiency results. Observations 18, 25 belong to the inefficiency group, whereas 5, 9 belong to the efficient group of observations. On the other hand, obervation 17 which represents April, 1982 has been flagged by the efficiency measures as belonging to the efficient group observations. During that month the mill was idle 80% of the time due to lack of orders. However, good energy and raw material usage was observed. This is shown emphatically by the partial energy productivity measure but not by the partial raw material productivity measure. Thus, it is not clear whether observation 17 can be classified as being overall productive. Observations 1, 13, 25 represent the month of December for the years 1980–1982. The Christmas holiday season causes the mill to be idle approximately 20% of the time. These conditions cause a non-normal operating environment during these months. Even though the mill operated more efficiently during observation 13 (which is also shown by the partial productivity measures for capital labor and energy), as opposed to observations 1 and 25, the total productivity measure was not sensitive to these differences. Last, the partial capital productivity measure isolates observations 1, 8 and 13 as being most efficient. This group was flagged as being one of the most influential subgroups of the dataset by the k-clustering methodology (See Seaver and Triantis (1986)). It is not surprising that these observations are grouped together by the k-clustering algorithm since they all exhibit a high partial capital productivity.

Table 3 gives both the technical efficiency and total productivity measures for the pulp mill.[4] The full-frontier technical efficiency measures were obtained by assuming a log linear production function form for equation (2) and by using the corrected ordinary least squares methodology the following efficiency standard was estimated.

$$PUPR = (1.86) * (PUENE)^{.134} * (PUWOOD)^{.679} * (PUWAT)^{.272} * (PUCHEM)^{-.057}. \quad (4)$$

Also, the weak cone technical efficiency measures are included in Table 3.

The total productivity measure isolates observations 20, 14, 2, 21 and 10 as representing the worst mill performances, whereas observations 6, 5, 7, 11 and 3 are flagged as representing the best mill productive performances. Again, there is some degree of consistency between the productivity and efficiency measures. Observations 2, 14 and 10 have been flagged as being inefficient, whereas observation 6 has been flagged as being efficient. However, observation 21 is isolated as non-productive by the total productivity measure and efficient by the technical efficiency measures. This comes as no surprise if the partial productivity measures for wood is examined. During observation 21 the mill recorded its highest partial wood productivity performance. Thus, it is not surprising that the technical efficiency measures classify this observation as efficient. Also, the inconsistency for observation 20 which is flagged as efficient by the weak cone measure, but isn't by the full-frontier, is partially explained by examining the partial productivity measures. Observation 20 performs poorly with respect to energy and wood usage and exhibits an average input utilization with respect to the remaining inputs. Thus, one would be reluctant to classify observation 20 as being efficient.

5 CONCLUSIONS

In order to measure and analyze productive performance at the plant and process levels, the manager or researcher can use two sets of performance criteria, i.e., efficiency and productivity measures. For this reason, technical output increasing efficiency measures and total and partial productivity measures were computed and analyzed for linerboard manufacturing facilities. The following two general conclusions can be drawn from the experience with these measures. First, there is some degree of consistency between the two approaches in evaluating past productive performance. For this reason, each type of measure can be used to clarify possible ambiguities found by implementing the other measure. For example, the partial productivity measures did clarify some of the ambiguities created by evaluating the technical efficiency measures. Second, in evaluating past productive performance, partial productivities should be evaluated with total productivities together in order to obtain a more accurate picture of past performance. This is a well documented recommendation made by most researchers in the area of productivity.

Given the usual limitations of the cost accounting data, it is not always possible to separate the price from quantity changes when measuring and evaluating productivity measures. This also holds true for the efficiency measures. If the researcher does not have access to individual input and output prices then it becomes difficult to separate technical from allocative efficiency[5] performance. Given the appropriate datasets, it would be interesting to investigate how allocative efficiency measures perform with respect to productivity measures. It is felt that the more research is conducted with respect to the implementation of the various performance measures, the more knowledge will be acquired about their informational content and the less resistent management will be with respect to their use as decision making tools.

TABLE 1 Technical Output Increasing Efficiency Measure for Mill I[6]

Obs.	Full-Frontier[a]	R_1^b	Weak Cone	R_2^c
1	.942	13	.949	16
2	.901	20	.954	15
3	.922	14	.946	18
4	.901	19	.945	19
5	.953	11	1.000	1
6	.894	22	.897	24
7	.920	15	1.000	1
8	.985	3	1.000	1
9	.960	9	1.000	1
10	.891	23	.925	22
11	.899	21	.947	17
12	.903	17	.937	20
13	.961	8	1.000	1
14	.891	24	1.000	1
15	.910	16	.929	21
16	.901	18	.983	14
17	.967	5	1.000	1
18	.881	25	.847	25
19	.986	4	1.000	1
20	.955	10	.999	11
21	.963	7	.993	12
22	.968	6	.988	13
23	1.000	1	1.000	1
24	.986	2	1.000	1
25	.953	12	.911	23
Avg.	.936		.965	

a These efficiency measures are based on the following standard:
$LPR = (.184) * (CAP)^{.286} * (LAB)^{.340} * (ENE)^{.543} * (NMAT)^{.126}$
The average values for the efficiency measures for each input are given as follows: Capital: .798, Labor: .826, Energy: .886, Raw Materials: .616.

b R_1: ranking based on the full-frontier technical efficiency measure.

c R_2: ranking is based on the weak cone technical efficiency measure.

TABLE 2 Total and Partial Productivity[a] Measures for Mill I Implementing the APC Model

Obs.	Total	Capital	Raw Mat.	Energy	Labor	R_3^b
1	.874	1.076	.782	.855	.879	21
2	1.059	.938	1.092	1.041	1.031	3
3	1.043	.914	1.156	.970	1.079	4
4	1.037	.971	1.086	.979	1.010	5
5	1.111	.911	1.246	1.025	1.146	1
6	.936	.937	1.195	.927	.494	12
7	.929	.892	.815	1.120	1.027	15
8	.994	1.317	.894	1.029	1.124	8
9	1.082	.998	1.052	1.094	1.169	2
10	.976	.914	1.068	.941	.977	9
11	1.007	.960	.987	1.015	1.028	7
12	1.024	.967	1.133	.972	1.044	6
13	.886	1.479	.799	.987	.905	20
14	.968	.889	1.625	.940	.894	10
15	.906	.851	1.208	1.001	.993	18
16	.858	.945	.895	1.077	.949	22
17	.707	.933	.707	1.146	.231	24
18	.627	.828	.740	.769	.406	25
19	.965	.899	1.452	.988	.984	11
20	.934	.913	1.173	1.048	.993	13
21	.932	.881	1.411	.962	.980	14
22	.916	.858	1.225	.976	.940	17
23	.917	.856	1.239	.999	.820	16
24	.898	.861	1.257	.973	.978	19
25	.717	.765	1.121	.811	.745	23

[a] According to the APC [9] model formulation, these measures were labeled as profitability measures, i.e., both price and quantity changes are incorporated in these values.

[b] R_3 ranking is based on the total productivity measure.

TABLE 3 Output Increasing Technical Efficiency and Total Productivity Measures for the Pulp Mill

Obs.	Full-Frontier[a]	R_4^b	Weak Cone	R_5^c	Total Prod.[d]	R_6^e
1	.865	17	.956	16	.882	18
2	.880	15	.900	22	.812	22
3	.916	7	.966	13	.885	17
4	.921	6	.994	8	1.046	5
5	.905	8	.966	14	1.116	3
6	.908	10	1.000	1	1.329	1
7	.903	11	.975	12	1.140	2
8	.923	4	.977	10	.929	15
9	.901	12	.976	11	.979	8
10	.863	18	.905	21	.879	20
11	.873	16	.933	18	1.057	4
12	.847	22	.959	15	1.046	6
13	.844	23	.852	23	.882	19
14	.841	24	.833	24	.691	23

TABLE 3 (continued)

Obs.	Full-Frontier[a]	R_4^b	Weak Cone	R_5^c	Total Prod.[d]	R_6^e
15	.922	5	.997	6	.897	16
16	.908	9	.992	9	.956	12
17	.895	13	.996	7	.931	14
18	.856	20	.919	19	.939	13
19	.852	21	.910	20	.961	10
20	.885	14	1.000	1	.676	24
21	1.000	1	1.000	1	.865	21
22	.858	19	.950	17	.964	9
23	.933	3	.999	5	.960	11
24	.937	2	1.000	1	1.013	7
Avg.	.893		.915		.952	

a Full-frontier measures are based on the following efficiency standard:
$PUPR = (1.86) * (PUENE)^{.139} * (PUWOOD)^{.679} * (PUWAT)^{.272} * (PUCHEM)^{-.057}$

b R_4: ranking is based on the full-frontier technical efficiency measure.

c R_5: ranking is based on the weak cone technical efficiency measure.

d According to the APC [9] model formulation, these measures were labeled as profitability measures, i.e., both price and quality changes are incorporated in these values.

e R_6: ranking is based on the total productivity measure.

6 FOOTNOTES

1. The weak cone output increasing technical effiency measure $W_O^k(y^i,X^i)$ represents by how much output y^i can be increased when a productive unit uses the input vector X^i in a constant returns to scale technology. X^i represents the vector of p inputs, y^i denotes the scalar output and i is an index denoting each observation(i = 1,2,...,n).

2. For detailed definitions pertaining to the isoquent subset and the weak cone technical efficiency measure see Färe, Grosskopf and Lovell (1984).

3. Information about the operating conditions is obtained from the monthly operating statements for each mill.

4. Only 24 observations are reported for the pulp mill since incomplete data concerning raw material usage was recorded for observation 25. The total productivity measure in this case represents changes in output quantity only with respect to changes of both input prices and quantities. This is because pulp prices were not recorded by the mill. Also, the partial productivity measures were not included in Table 3.

5. Allocative efficiency is defined as the extent to which the firm or plant is

successfully cost minimizing. For more information as to how to model allocative efficiency see, Färe, Grosskopf and Lovell (1984).

6. I wish to thank Sami Kabbani for assisting in this research effort by conducting the bulk of the computer work.

7 REFERENCES

Afriat, S.N., 1972. Efficiency Estimation of Production Functions. International Economic Review, 13:569-598.

Färe, R. and Lovell, C.A.K., 1978. Measuring Technical Efficiency of Production. Journal of Economic Theory, 19:150-162.

Färe R., and Grosskopf, S. and Lovell, C.A.K., 1984. Measurement of Efficiency of Production. Kluwer-Nijhoff Publishing, Boston, MA.

Farrell, M.J., 1957. The Measurement of Productive Efficiency. Journal of the Royal Statistical Society, Ser. A, 120, part 3: 253-281.

Gold, B., 1982. Practical Productivity Analysis for Management: Part I. Analytic Framework. IIE Transactions, 14:227-242.

Greene, W.H., 1980. Maximum Likelihood Estimation of Parametric Frontier Production Function. Journal of Econometrics 13:5-25.

Kopp, R., 1981. The Measurement of Productive Efficiency: A Reconsideration. Quarterly Journal of Economics, 96:476-503.

NASA, 1984. White-Collar Productivity and Quality Issues. Proceedings of the NASA Symposium on Productivity and Quality: Strategies for Improving Operations in Government and Industry. Washington, DC.

Ruch, W. A., 1981. Your Key to Planning for Profits. Productivity Brief 6, American Productivity Center, Houston, TX.

Schmidt, P., 1985. Frontier Production Functions. Econometric Reviews, 4:2.

Seaver, B.L. and Triantis, K.P., 1986, The Implications of Using Messy Data to Estimate Production Frontier Based Technical Efficiency Measures. Submitted for publication to the Journal of Business and Economic Statistics (unpublished).

Shephard, R.W., 1970. Theory of Cost and Production Functions. Princeton University Press, Princeton, NJ.

Ibid, 1984. Semi-Homogeneous Production Functions and Scaling in Production. In: Eichorn, W. (Editor), Henn, R., Opits, O. and Shephard, R.W. Production Theory. Springer Verlag, Berlin, Heidelberg, New York.

Sink, D.S. (1985). Productivity Management: Planning, Measurement and Evaluation, Control and Improvement. John Wiley and Sons, New York.

Sumanth, D.J., 1984. Productivity Engineering and Management. McGraw-Hill, New York.

Triantis, K.P., 1985. The Measurement of Technical Efficiency of Production: The Case of Linerboard Manufacturing Facilities. Submitted for publication to the Journal of Applied Economics (unpublished).

D. WHITE-COLLAR/KNOWLEDGE-WORK PRODUCTIVITY MEASUREMENT

Productivity Management Frontiers — I, edited by D.J. Sumanth
Elsevier Science Publishers B.V., Amsterdam, 1987 — Printed in The Netherlands

KNOWLEDGE WORK: A CONCEPTUAL ANALYSIS AND STRUCTURE

M.G. BERUVIDES and D.J. SUMANTH, PhD.
Department of Industrial Engineering, University of Miami, Coral Gables, Florida 33124 (USA)

ABSTRACT
Beruvides, M.G., and Sumanth, D.J., 1987. Knowledge Work: A conceptual analysis and structure.

The need to define, describe and quantify work has, and continues to be, an important factor for management. Using this information is a vital tool for any manager. What is termed blue collar labor is an area of productivity which has been well defined and analized. There is a challenge, however, to define the knowledge-work and/or white-collar work, before actually exploring ways to measure them.

Knowledge/white collar productivity has various characteristics that makes it unique from blue-collar work from a productivity point of view. Characteristics such as discretion, decision making, design, organization, ect. are areas on which various researchers have based their definitions and studies. An analysis of the implications of knowledge work such as the measure of output, long time periods for work completion, and non-repetitive nature of work cycles, are presented.

The various concepts, such as the cost of information, cost displacement, intellectual technologies, work units, and organizational structural elements, are analyzed for their merits to this field of study. Some of these concepts areused as structural blocks in the formation of a conceptual framework to define and express knowledge work.

As a major foundation, the concept of the Total Productivity Model (TPM) is used. Definitions for blue-collar, white-collar, and knowledge work are given. A distinction is made between white-collar productivity and knowledge-work productivity. Also, a conceptual mathematical model to describe the blue-collar, white-collar, and knowledge-work is presented with examples.

1 INTRODUCTION

Measuring and evaluating human resources has always been a major concern to every industry. In the past. the measure of labor productivity was restricted to what is termed 'blue collar' work. But, two major developments have shown the need to expand the scope of blue-collar labor productivity concept to what many referred to as 'white-collar' work and 'knowledge' work. These two developments are: (1) the advances in the art and science of productivity measurements(partial, total-factor, and total productivity) (Kendrick and Creamer(1965), Sumanth, (1984) and Sink, (1985); and (2) the fundamental change in the U.S. economy from a manufacturing to a service-oriented economy. As the existing manufacturing industries become more automation- and computer-oriented, there is a greater proportion of white-collar workers entering the work force. Peter Drucker (1974) puts it very boldly when he points out that "the manual worker is yesterday- and all we can fight on that front is a rear guard action. The basic capital resource, the fundamental investment, but also the cost center of a developed economy is the knowledge worker who puts to work what he has learned in systematic education, ideas, and theories, rather than the man who puts to work manual skill or muscle."

2 KNOWLEDGE WORK CHARACTERISTICS

In the literature, knowledge work and knowledge worker are not easily distinguished. Although many have a conceptual understanding of what knowledge work is, few have been able to determine or measure it specifically. John Grahn (1981) defines the knowledge worker as one who "ceates, designs, and develops things or ideas. The knowledge worker may plan, schedule, control, direct, supervise, oversee, research, analyze, and make decisions. The knowledge worker's tasks involve the exercise of judgment with an area of discretion." Also, the description of the knowledge work in the literature includes the terms: 'long cycle time,' 'communications skills,' 'leadership,' and 'creativity.'

William Ruch (1982) mentions two major dimensions that characterize knowledge work: discretion (the judgment often required in some tasks), and the degree to which a physical product is involved in the process. We agree with Grahn and Ruch that 'discretion' is an important aspect of knowledge work. The second of Ruch's dimensions (the degree to which a physical product is involved in the process) is concerned with another major characteristic of knowledge: the intangibility of the output produced. To establish a viable measure for knowledge work, many obstacles must be overcome. The perception is that outputs are generally intangible, with long time-lags for work completion, and that, often, work cycles are nonrepetitive. There is a tendency of the productivity analyst to measure activities rather than results, compounded by what Mundel (1983) recognizes as the confusion of efficiency and effectiveness with productivity. This makes the measurement of knowledge work a really challenging issue.

3 SOME RECENT APPROACHES

There has been much work done by practitioners and researchers to measure white-collar productivity. Calmes (1981), McGee and Welsh (1982), and Anthony (1984) are but a few studies reported by the practitioners. The use of traditional industrial engineering techniques such as time studies, work sampling analysis, and other traditional methods have been applied to "solving" the problem of white-collar productivity measurement. Clearly, these are inadequate, since they ignore many intangible elements of the output generated by a knowledge worker.

Edelman (1981) uses the cost of information as a method of measuring white-collar productivity. With the use of traditional 'cost displacement' methods, and identification and control of organizational 'total cost information,' productivity improvements are implemented at the 'white collar' level. Although this is distintly for heavily information-based industries, it illustrates a viable method of measuring white-collar productivity.

Marvin Mundel (1978) uses the concept of 'work units' to establish a cycle of control in which to develop a measure of productivity for service organizations. This is believed to be one of the ways to address the problem of measuring 'work units' for each specific job. Mundel (1983) describes his basic productivity equation as:

$$\text{Productivity} = \frac{\text{Sum } AO_m / RI_m}{\text{Sum } AO_b / RI_b} \qquad (1)$$

where AO = Aggregated Output; RI = Resource Input; m = measured period; and b = base period.

The measure for productivity for service and government organizations (in Mundel's method) is developed by taking sampled observations and establishing the activities that comprise the major percentage of work-time allocation. When this is done an algorithm for such activities is developed and used with the above equation, to obtain a productivity measure. The algorithm is based on a list of inputs and outputs established through extensive sampling-observation studies. Mathematical expressions for partial resources inputs (support personnel) and direct resources input (direct personnel) as well as aggregate outputs (partial and total aggregate) are presented. The inputs and outputs are weighted when necessary to eliminate external distortion, and are all calculated on a measured-year/base-year basis.

Robert Lehrer (1983) approaches White-collar Productivity through a conceptual framework based on 'structural elements that have a direct impact upon the validity of contribution in achieving organizational objectives. The structural elements he considers are: organizational structure, service interchange structure, information structure, physical resources/technology structure and human resources structure. The structural elements bring together all major areas which influence knowledge work directly and indirectly. The work unit analysis is stressed as the basis for establishing the measurement techniques.

Curley and Pyburn (1982) investigated an interesting term which they call 'intellectual' technology. They make a very good distinction between skill-based workers and knowledge workers, based on the technology used by the worker, and the requirements of that technology on the worker. The two types of technologies they define are: 'industrial' and 'intellectual.' These two technologies differ along two dimensions: the type and the range of application of the technologies. The type of learning to effectively use the technology is important. According to Curley and Pyburn, the intellectual technology requires more ongoing adaptive learning or 'Type B learning.' The range of intellectual technology is more flexible. The industrial technology is distinctively 'Type A learning' (training) and is not as flexible in its range. Although both these technologies require both types of learning, the amount of breadth is different for each of them. This technological spectrum analysis presents an excellent description that gives another perspective with which to distinguish knowledge work from the blue-collar work.

4 DEFINITIONAL PROBLEMS

Three basic terms have been commonly associated with workers: blue-collar, white-collar, and knowledge-worker. Generally, everyone has a conceptualization of what constitutes a blue-collar and white-collar worker. However, the distinction is not so clear when comparing a white-collar worker with a knowledge worker. In fact, for the most part, these terms are used synonymously. But are these terms indeed synonymous? We agree with Ruch (1980) who points out 'not all white collar workers are knowledge workers." If this is so, then, what distinguishes one from the other? What are the distinct characteristics of each of them?

There does exist one definite factor of distinction in all work; it is the degree to which output is defined. In blue-collar work, the output is clearly defined; it is a direct function of all the inputs used to produce this output. That is, a blue-collar worker's productivity when making a 'widget' on a 'widget maker' is a function of the worker's wages, the raw materials used, the machinery used, the energy expended, and to a lesser degree, all other expenses incurred. So we can reasonably argue that the output of blue-collar work is directly related to the sum of the inputs used.

For 'non-blue collar' work, on the other hand, two possibilities exist for outputs. Outputs can be tangible (clearly definable) or intangible. In all probability, most 'non-blue collar' work will be a combination of tangible and intangible outputs. When the tangibles comprise the major part of a non-blue collar worker's output, then, we could call this 'knowledge work.'

We formally define these three types of work as follows:

Blue Collar Work is all work of a manual or physical nature whose output is clearly definable or tangible, and whose inputs are clearly definable and directly influential in the output produced with no discretion (or little, if any) permitted in the job task. (Recall that 'discretion is the judgement based on knowledge, experience, and insight that are often required in some tasks).

White Collar Work is all work of any nature whose output is composed mainly of tangible factors with some intangible factors, and whose inputs are definable and directly influential in the output produced, with little or no discretion in the job task.

Knowledge Work is all work of any nature whose output is composed of some tangible factors but is mainly of an intangible nature, and whose inputs are not as clearly definable and can, but not necessarily, have to be influential in the output produced, with substantially high discretion permitted in the job task.

The characteristics defined above are summarized in Table 1.

In a true sense, then, knowledge work can produce an output with no inputs (other than salary). Note, however, that if taken to the extreme, even salary can be excluded. One can conceptualize an unemployed physicist developing a thought or concept that is of economic value. So, an idea or concept can ultimately be an output with no materials used, no energy consumed (except physical), and no capital or other expense employed. The output of knowledge work is mainly knowledge.

Once a definition of the different types of work is clarified, it makes more sense to label them. As Ruch (1982) points out, what is"sorely needed is a comprehensive logical topology of white-collar workers..." Although this is true and possible, once a definition for the different types of workers is agreed upon, the work would never be generic in nature because a worker in one company might not have the same type of work responsabilities as a worker in another company with the same job title. So, while the topology is needed, it will have to be developed at the company level in order to be realistice and meaningful.

TABLE 1
Characteristics of the different types of work.

	Type of Work		
	Blue Collar	White Collar	Knowledge
Nature of Work	manual or physical	any nature	any nature
Outputs	clearly definable (tangible)	mainly tangible with some intangible factors	some tangible factors, but mainly intangible
Inputs	clearly definable and directly influential in output	definable and directly influential in output	not as clearly definable and can be, but not necessarily are influential in output
Degree of Discretion	practically none	little or none	very high

5 PROPOSED FRAMEWORK FOR DEFINING KNOWLEDGE WORK

There exist three basic measures of productivity: partial productivity, total factor-productivity, and total productivity. The Total Productivity Model developed by Sumanth (1979, 1984) considers total productivity as the ratio of the sum of total tangible outputs to the sum of the total tangible inputs. It is expressed mathematically as

$$TP = \frac{O}{\Sigma_j \; I_j} \tag{2}$$

$$TP = \frac{O}{I_H + I_M + I_C + I_E + I_X} \tag{3}$$

where O = total tangible output; I_H = human input; I_M = material input; I_C = capital input (including both working and fixed capital); I_E = energy input; I_X = other expenses input; and j = input factor.

6 MATHEMATICAL CONCEPT

Using Sumanth's Total Productivity Equation and the definition of knowledge work given earlier, we can conceptually state the total productivity for knowledge work as:

$$TP = \frac{O}{I_H + I_T} \tag{4}$$

where $I_T = I_M + I_C + I_E + I_X$

I_T, which is the sum of the material, capital, energy, and other expense inputs, is virtually negligible when compared to the human input (salary) in the case of knowledge work. Now, we can further write the above expression (4) as:

$$TP = \left(\frac{O}{I_H}\right)(D_f)$$

where D_f is a constant multiplier called the Departmental Factor which is the form: $D_f = I_H \,/\, (I_H + I_T)$.

The departmental factor D_f represents the portion of human input in the total tangible input. It is of importance because its value, established through historical data, can be a measure of the relationship of the cost of labor to the sum of costs in a department. The fluctuations in the D_f value can alert management to take timely action in manpower planning.

This partitioning of the equation does not alter it in any way, and serves to emphasize the empirical belief that knowledge is mainly a function of human input. Thus, we can re-write the equation as follows:

$$TP = \frac{O}{I_H} \cdot \frac{I_H}{I_T} \cdot \frac{1}{\left(1 + \frac{I_H}{I_T}\right)}$$

$$TP = \frac{O}{I_H} \cdot t \cdot \frac{1}{(1 + t)} \tag{6}$$

$$\text{where} \quad t = \frac{I_H}{I_T} \quad \text{(we can call this the 't-ratio')} \tag{7}$$

Thus, we can express equation (6) in terms of t as:

$$TP = \frac{O}{I_H} \cdot \ t\,(1 - t + t^2 - t^3 + \ldots) \quad \text{for all } 0 \le t \le 1 \tag{8}$$

Mathematically, it is important to note that t lies in the range of $0 \le t \le 1$. This is generally true for blue-collar work and white-collar work. If t is greater than 1.0 ($t > 1$), that is, I_H is larger than I_T, we believe that this is representative of most knowledge work. In fact, if t is greater than 1.0, the equation

(8) above results in a negative value for total productivity, which is an infeasible condition. Total productivity can be zero or positive, but never negative. The change in total productivity (ΔTP) can be negative, but not the level of productivity itself. So then, the equation for total productivity for knowledge work must be expressed differently as follows:

$$TP_k = \frac{O}{I_H} \cdot \frac{I_H}{I_H + I_T}$$

$$= \frac{O}{I_H \left(1 + \frac{I_T}{I_H}\right)}$$

This can be approximated as

$$TP_k = \frac{O}{I_H} \cdot (1 - t_k + t_k^2 - t_k^3 + \ldots) \text{ for } 0 \leq t_k \leq 1 \qquad (9)$$

$$\text{where } t_k = \frac{I_T}{I_H} \quad \text{(we can call this the '}t_k\text{-ratio')} \qquad (10)$$

We expect that generally, most blue-collar or white-collar work would have a t value less than 1.0, and most knowledge work would have a t value greater than 1.0 (so the 't_k ratio' must be used in the case of knowledge work).

The use of the t expression has a two-fold purpose. First, it is an approximation of the D_f factor. Secondly, the t-factor is a ratio that assists in defining the work that is under consideration. That is, the t-ratio or t_k-ratio gives the productivity analyst, a measurable quantity to assist in determining what type of work is being analyzed (blue-collar, white-collar, or knowledge work).

Examples: Let us take the following example for a blue-collar worker. Let O = \$4,200; I_H = \$1,500; I_M = \$400; I_C = \$1,100; I_E = \$500; and I_X = \$100.

Obtaing the total productivity in the normal way, we get:

$$TP = \frac{4,200}{1,500 + 400 + 1,100 + 500 + 100}$$

$$= 1.166 \ \$/\$$$

If the same values are used for a white-collar worker, we get:

$$\frac{O}{I_H} = \frac{4,200}{1,500} = 2.8$$

where $I_T = I_M\ I_C + I_E + I_X = 400 + 1,100 + 500 + 100 = 2,100$

Therefore, $\frac{H}{I_T} = \frac{1,500}{2,100} = .714$, which is less than 1.0.

Notice that the t-ratio satisfies the condition $0 \leq t \leq 1$. The D_f factor is

$$D_f = \frac{I_H}{I_H + I_T} = \frac{1,500}{3,600} = .417$$

Now using the appropriate expression for TP given by equation (8), we get:

$$TP = (2.8)(.714)\ 1 - (.714) + (.714)^2 - (.714)^3 + \ldots + (.714)^{10}$$
$$= 1.192\ \$/\$$$

Now, let us consider the following data for a knowledge worker. Let O = \$5,500; I_H = \$3,500; I_M = \$200; I_E = \$400; and I_X = \$200.

Here, $\frac{O}{I_H} = \frac{5,500}{3,500} = 1.571$

$$I_T = I_M + I_C + I_E + I_X$$
$$= 200 + 200 + 400 + 200$$
$$= 1,000$$

So, $\frac{I_H}{I_T} = \frac{3,500}{1,000} = 3.5$, which is greater than 1.0

Therefore, we have to use the t_k-factor $= \frac{I_T}{I_H} = \frac{1}{3.5} = .286$

The D_f factor is $D_f = \frac{I_H}{I_H + I_T} = \frac{3,500}{4,500} = .778$

Using the normal method we get:

$$TP = \frac{5,500}{3,500 + 200 + 200 + 400 + 200} = 1.222\ \$/\$$$

By the approximation method given in equation (9), we get:

$$TP = (1.571)\ \ 1 - (.286) + (.286)^2 + (.286)^3 + \ldots + (.286)^6$$
$$= 1.224\ \$/\$$$

What is important to note here is that the approximation works, and that through historical data, the D_f factor as well as the t factors can be calculated and known for any specific department. This would facilitate the ease of future calculations and help establish trends as to how these factors are fluctuating.

If one were to let $D_f = 0$ what would the values of I_H, I_T, and t be? The equation for D_f is:

$$D_f = \frac{I_H}{I_H + I_T}$$

For D_f to be equal to zero, I_H must be 0, so t = 0 too at $I_H = 0$. Now, if we let $D_f = 1.0$, then I_T must be equal zero; and if I_T is equal to 0 then $t = \infty$. In Table 2 are listed values for I_H, I_T, and t with respect to different values of D_f ranging from 0 to 1.0.

TABLE 2
Values of I_H, I_T, and t with respect to D_f

D_f	I_H	I_T	t
0	0	I_T	0
.25	I_H	$3I_H$	.33
.50	I_H	I_H	1.0
.75	I_H	$.33I_H$	3.0
.8	I_H	$.2I_H$	5.0
.91	I_H	$.1I_H$	10.0
.99	I_H	$.01I_H$	100.0
1.0	I_H	0	∞

The tabulated data are plotted in Figure 1. This graph illustrates the D_f function. From the graph, we can easily distinguish where the blue collar/white collar region ends and the knowledge work region begins. The line of demarcation is $D_f = .5$, that is, $I_H = I_T$ and $t = 1.0$ which is consistent with the definitions expressed earlier.

7 OUTPUT MEASURES FOR KNOWLEDGE WORK

The biggest problem encountered by most researchers and companies in trying to measure knowledge worker productivity lies in the output measure. The lack of physical output and the intangibles such as creativity, organization, teaching skills, communication skills, planning, intuition, and many others which we might term the 'therbligs of knowledge work,' will probably always present a problem. Through the work done by many researchers such as those mentioned in this work, the cost of information, 'intellectual' technologies, work units, and other factors not yet explored, a better measure of knowledge output can be obtained. Packer (1983) has a very interesting way to qualify subjective responses (outputs) using a complex statistical procedure. The important idea of this work is that Packer does not pursue quantifiable measures which do not really represent a true expression of productivity; for example, to try to equate how productive a computer programmer is with the amount (in pages) of computer output that is generated. Packer's method may prove to be a good way to approach the output measure problem.

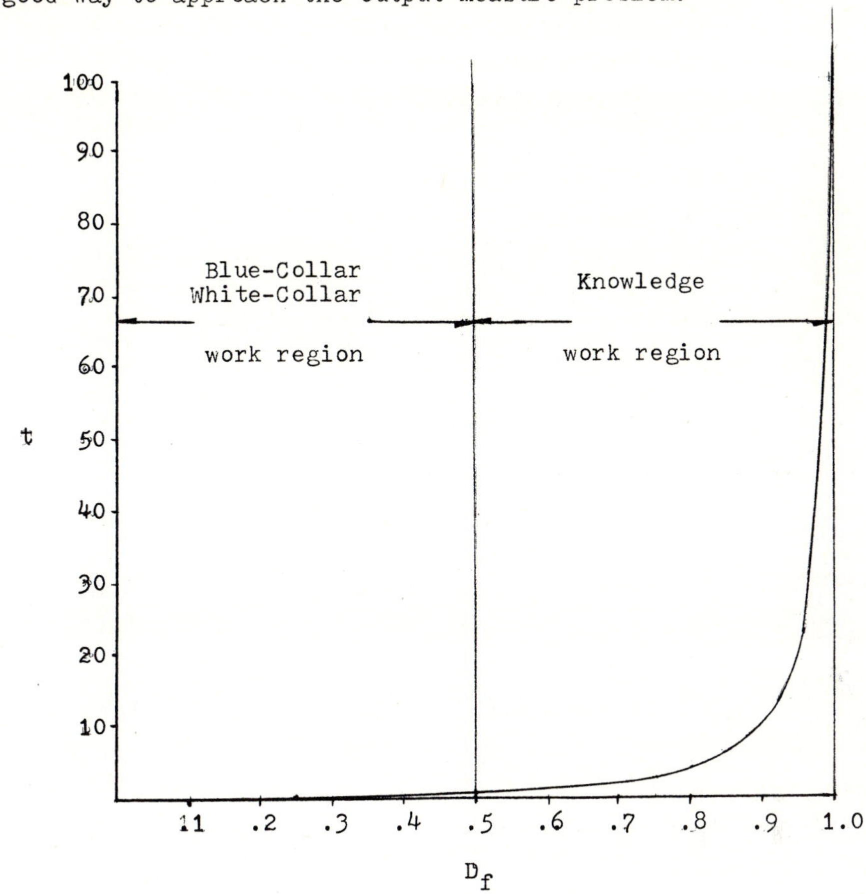

Fig. 1. Graph of tabulated data

8 CONCLUSION

This article has attempted to bring together a relevant composite of knowledge about the understanding of white-collar and knowledge-work productivity. The definitions given present an attempt to clarify some misunderstandings that exist. There are some gray areas which need to be explored further, but the definitions and equations presented give a starting point for further research. There still exists the problem of output measures for knowledge work.

The t-ratio, the t_k-ratio, and the definitions presented give a conceptual method to distinguish between blue-collar, white-collar, and knowledge work. The D_f factor could be established through historical data. Its fluctuations and trends might prove to be a measure in itself. This is an area that we will continue to study. Much more analytical thought is needed to clarify the meaning of white-collar and knowledge-work before we can really address their productivity measurement.

9 REFERENCES

Anthony, G.M.,"IEs Measure Work, Write Standards for White Collar Workers at Financial Institutions," Industrial Engineering, Vol. 16, No. 1, Jan. 1984, 77-81 pp.

Calmes, R.A., "Career Planning and Performance Appraisal Techniques for Improving Salaried Employee Productivity," IIE Fall Conference Proceedings, 1981, 29-34 pp.

Curley, K.F. and Pyburn, P.J., "'Intellectual' Technologies: The Key to Improving White Collar Productivity," Sloan Management Review, Fall 1982, 31-39 pp.

Drucker, P.F., Management: Task, Responsabilities, Practices, (New York: Harper & Row, 1974).

Edelman, F., "Managers, Computer Systems and Productivity," MIS Quarterly, Sept. 1981, 1-19 pp.

Grahn, J.L., "White Collar Productivity: Misunderstanding and Some Progress," Personnel Administrator, August 1981, 23-31 pp.

Kendrick, J.W. and Creamer, D., Measuring Company Productivity: Handbook with Case Studies. Studies in Business Economics, No. 89. (National Industrial Conference Board, New York, N.Y., 1965).

Lehrer, R.N., White Collar Productivity, (New York: McGraw-Hill, 1983).

McGee, J.P. and Welsh, R.L., "The Need to Redirect Industrial Engineering toward Office Productivity," IIE Annual Conference, 1982, 361-367 pp.

Mundel, M.E. Measuring+and Enhancing the Productivity of Service and Government Organizations, Hong Kong: Asian Productivity Organization, 1978).

Mundel, M.E. "Wanted: A System that Provides 'Box Scores For Productivity,'" Industrial Engineering, Oct. 1983, 74-81 pp.

Packer, M.B., "Analysing Productivity in R & D Organization," Research Management, Jan-Feb, 1983, 13-20 pp.

Ruch, W.A., "Knowledge Worker Productivity," Manufacturing Productivity Frontiers, August, 1980, 11-14 pp.

Ruch, W.A., "The Measurement of White Collar Productivity," National Productivity Review, Autumn 1982, 416-426 pp.

Sink, D.S., Productivity Management: Productivity Measurement, Evaluation, Control and Improvement, (New York: John Wiley, 1985).

Sumanth D.J., Productivity Engineering and Management, (New York: McGraw-Hill, 1984).

Sumanth, D.J., Productivity Measurement and Evaluation Models for Manufacturing Companies, Ph.D Dissertation, IIT, 1979 (Published by Univ. Microfilms International, #80-03,665).

Productivity Management Frontiers — I, edited by D.J. Sumanth
Elsevier Science Publishers B.V., Amsterdam, 1987 — Printed in The Netherlands

A MODEL FOR MEASURING R & D PRODUCTIVITY

TOMISLAV MANDAKOVIC
Associate Professor of Decision Sciences, Florida International University, Miami, Florida, 33199, and Research Associate, Technology Management Studies Group, the University of Pittsburgh.

WM. E. SOUDER
Professor of Industrial Engineering and Director of the Technology Management Studies Group, the University of Pittsburgh, Pittsburgh, PA 15228.

ABSTRACT

Numerous models are available for measuring productivity in production environments. However, these models are not so easily applied in R&D environments, where the outputs are more difficult to quantify. We propose a method to measure R&D productivity. This method examines the actual performance of each project relative to its expected performance and the resources allocated to it. The method uses a model formulated as a 0-1 integer program, which takes into account hierarchies within the R&D organization. The productivity of each hierarchy is measured by an aggregation process over n planning periods. A decomposable interactive approach is used to solve the model. Good results are obtained using a modification of a well-known heuristic.

INTRODUCTION

Productivity improvement is a pressing issue for modern organizations (5). Productivity improvement may be viewed as a repetitive cycle of four activities (6). As illustrated in Figure 1, these four activities are work definition, productivity measurement, productivity evaluation and productivity improvement (5). The last two activities are associated with the day-to-day operation of an organization.

In organizations where the outputs and inputs are tangible, the measurement of productivity can be designed on the basis of ratios, i.e., output/input. However, there are many organizations whose outputs are not readily quantified, (i.e., service organizations) or whose inputs are not directly related to specific outputs (i.e., hierarchial organizations). Here, the measurement of productivity can become a cumbersome if not infeasible task (5). Research and development (R&D) activities are an example. The outputs are often intangibles like knowledge or know-how. Moreover, R&D departments often consist of many different personnel skills and varieties of equipment that are not exclusively dedicated to any one product or output. R&D departments typically serve and/or report to several other entities within the total company (3).

FIGURE 1: Productivity Cycle

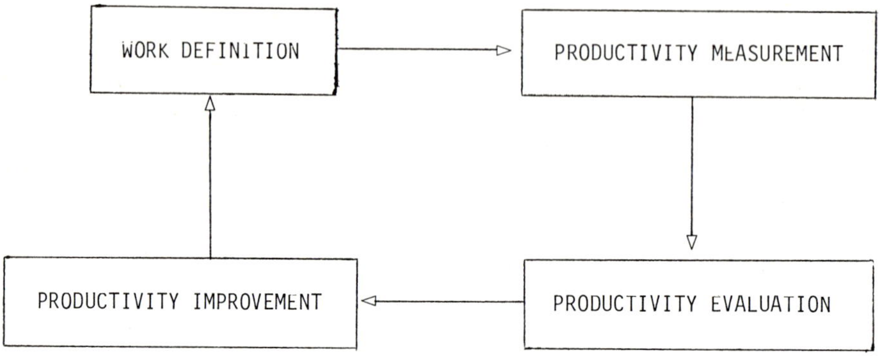

The approach discussed here begins with the annual budget exercise. It makes use of an interactive process, modeled through an integer program, to allocate resources to each new or on-going project. As part of the resources allocation process, information about each project's expected accomplishment is recorded. Before the next period's allocation, the actual outputs from each project can be compared with the expected results, which were defined during the previous resource allocation process. A set of productivity indices are developed as a result of this exercise.

THE PRODUCTIVITY MODEL

The Allocation Process

The objective of this process is to allocate resources for the next budget period among on-going and new projects. It is equivalent to a budget exercise or to a project selection process, where each project can have different funding levels. The participants in the process are the R&D managers (the resource allocators) and the project managers (the resource requirers). In a typical environment, project managers would submit project proposals and their corresponding resource requirements. Then, R&D managers would allocate the available resources in the "best" possible manner.

In our model, each project has mutually exclusive alternative versions. Thus, a different version of each project can be presented at each stage of the allocation process. We define a stage as the decision to accept or reject a project version and to make the corresponding update of the available resources.

For each version, a level of accomplishment must be defined. This is the expected output of that version, e.g.:

$$\text{Output} = \overline{\alpha} \cdot \overline{A} \qquad (1)$$

where $0 \leq \alpha \leq 1$, and

$\overline{A}$ = vector of expected accomplishments, assuming full funding.

If the expectations are to be expressed for specific time periods, then $\overline{A}$ becomes a matrix with as many rows as there are periods. Also, each version should indicate the resource usage, for every type of resource. Thus the inputs will be defined as:

$$\text{Inputs} = R\ (I,J,K,L) \tag{2}$$

where R (I,J,K,L) represents the amount of resource K required by version J of project I during time period L.

The goals of the allocation process are to:

- find an efficient resource allocation for the next budget period,
- achieve emotional commitment by the scientists by including them as participants in the allocation process, and
- obtain productivity standards for each project.

In order to achieve the above objectives, the process uses a mathematical program with 0-1 variables. Here, X(I,J) represents version J of project I. If X(I,J) = 0 the version is not selected in the portfolio. If X(I,J) = 1, then version J of project I is selected for the portfolio. The completed formulation of the integer model is included in a separate section in this paper.

The procedure which allows one to meet the three goals of the allocation process is as follows:

1. R&D management issues the resource availability vector;
2. each project manager selects the version of project that has the "best" chance of being accepted in the portfolio, given the availability of resources;
3. among the candidates (one version for each project) R&D management selects the one which has the "best" overall benefit for the organization;
4. R&D management issues a new availability vector (return to step 1) and this process reiterates until the resources are exhausted.

The process is illustrated in Figure 2. A more detailed description of the process and its mathematics are found in Mandakovic (2) and in Mandakovic and Souder (4).

The Productivity Measure

As a result of the above process, there is a quantified input in R (I,J,K,L) and an output given by A. Then, the following partial and total productivities may be calculated:

FIGURE 2: Allocating Process

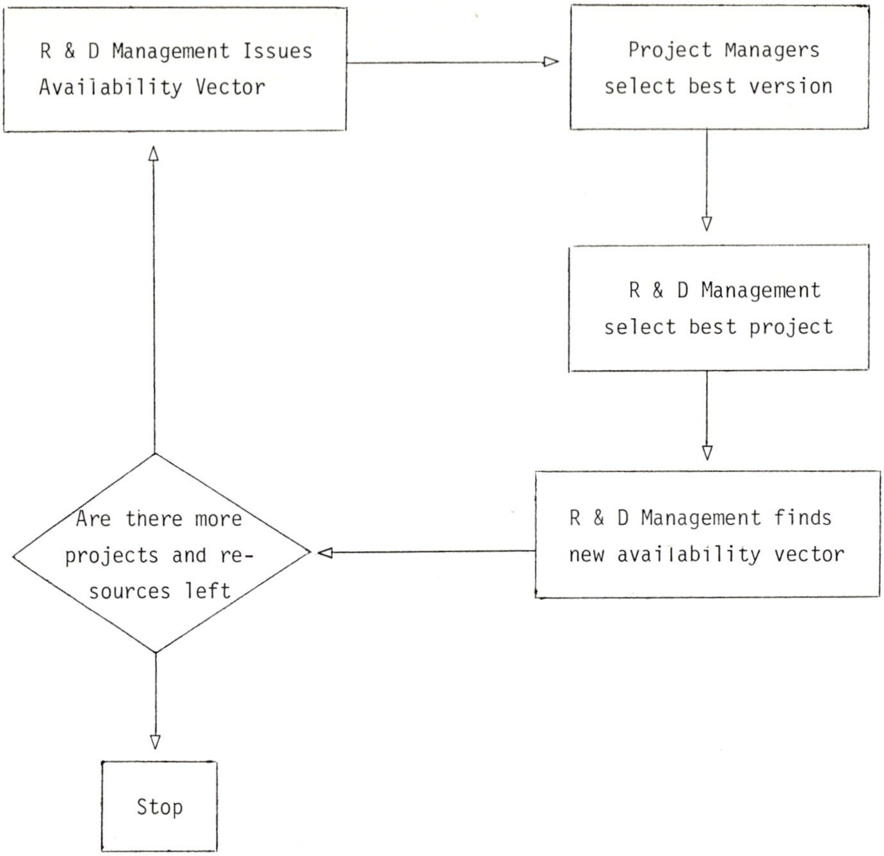

- the partial productivity of a specific accomplishment M within version J of project I, with respect to resource K during time period L, that is, PP (I,J,K,L,M). Here

$$PP\ (I,J,K,L,M) = \frac{\text{Effective M/Expected M}}{\text{Resource K}} \quad (3)$$

- the partial productivity of a specific project with respect to each resource, PP (I,J,K,L). Here

$$PP\ (I,J,K,L) = \frac{\text{Aggregated benefit for version J of project I}}{\text{Resource K}}$$

or

$$PP\ (I,J,K,L) = \frac{\text{Effective benefit/Expected benefit}}{\text{Resource K}} \quad (4)$$

- the partial productivity of the hierarchy with respect to a given resource during period L, PP (K,L). Here

$$PP\ (K,L,) = \frac{\text{Effective benefit of Portfolio/Expected benefit}}{\text{Resource K}} \quad (5)$$

- the total productivity of a given project I during period L, TP (I,J,L). Here

$$TP\ (I,J,L) = \frac{\text{Benefit of project}}{\text{Total resources utilized}},$$

  and J is the selected version of Project I
- the total productivity of the portfolio during period L, TP(L). Here

$$TP\ (L) = \frac{\text{Aggregated benefit of portfolio}}{\text{Total resources utilized}}$$

The measurement process may be performed at any time during the budget period. For example, periodic measurements of the partial productivities with respect to a resource K may detect some on-going problems within a given project:

- if the partial productivity index is considerably higher for one of the resources, the possibility of having a low use of that resource should be investigated.
- if the partial productivity index is significantly lower for a specific resource, then the possibility of an ineffective use of the resource should be investigated.

The Integer Program

The detailed formulation of the model is given in Mandakovic (2) and in Mandakovic and Souder (4). In general, the model is formulated as

$$\text{Maximize} \sum_{I} \sum_{J} V(I,J) * X(I,J) \quad (6)$$

subject to

$$\sum_{I} \sum_{J} A(I,J,K,L) * X(I,J) \leq B(K,L) \qquad (7)$$

for K = 1,2,3,..., AR resource types and L = 1,2,3,..., NP time periods,

$$S(X(I,J),M) \leq Q(M) \qquad (8)$$

for M = 1,2,3,..., NR organizational constraints (as defined below)

$$\sum_{J} X(I,J) \leq 1 \qquad (9)$$

for I = 1,2,3,..., N projects, for J = 1,2,3,..., NI versions of project I, and

$$X(I,J) = 0 \text{ or } 1$$

Here, V(I,J) represents the value or reward for undertaking and successfully completing version J of the Ith project, and X(I,J) represents the decision to either select or reject version J of the Ith project. When X(I,J) = 0 the version is rejected; otherwise it is selected. Here A(I,J,K,L) represents the amount of resource K required by version J of project I during time period L, and B(K,L) represents the availability of resource K during period L. Here, S is a function that relates the project versions to an organizational variable M, and Q is a dimensional equivalent of S in terms of M. Note that three types of constraints are defined: budgetary constraints, organizational constraints and exclusivity of versions. The first type, represented by equation 7 above, is the resource scarcity or budget constraint. The second type, represented by equation 8 above are organizational constraints involving dependencies among projects, environmental restraints, etc. The third type, represented by equation 9 above takes care of the infeasibility of having two versions of the same project in the portfolio.

This model has the structure of a decomposable mathematical program (1,2,4). Because of the constraint on the integer value for the decision variables, the dual pricing scheme of linear programming is not applicable (2). However, a heuristic proposed by Toyoda (7) and modified by Mandakovic (2) can be used to obtain a near-optimal optimal solution.

The heuristic is based on the penalization of the use of scarce resources by the projects. It selects one version of one project at a time. The project selected is the one with the highest benefit, where the benefit for each project has been penalized according to its usage of available resources.

The selection of a project is made at the highest hierarchical level. Once a version of a project is selected, new versions of the project are not considered. Also, after the selection of each project, the availability of each resource is corrected. This is done by subtracting the allocation which was made to the selected project. The new availability vector is then transformed into a penalty vector. Each penalty vector component is inversely proportional to the corresponding component of the availabilty vector. This information is sent to the lowest level of the hierarchy, i.e., project managers. At this level a selection of the best version for each project is made by the project manager. The higher level then selects the best project among the ones sent by the project managers. The process concludes with the exhaustion of resources or of candidates. The whole procedure is illustrated in Figure 2.

SUMMARY AND CONCLUSIONS

An R & D productivity measurement model has been described. The model describes a hierarchical project selection decision making process. It takes into account the typical hierarchical structure of technology-based organizations. The measure of productivity is the familiar ratio of outputs/inputs but these dimensions are multidimensionally defined. Partial productivities are computed as the ratio between particular or aggregated outputs and a single resource. The aggregation of output and/or inputs allows one to calculate total productivities for a project or for a portfolio.

The advantage of using the project selection process is the quantification of each project's expected output. Moreover, the relation between the expected output for different levels of resource allocation (input) is obtained through the process. The quantification of outputs and inputs is made by each project manager, in an interactive bidding process. In this way, communications between administrative and scientific-technical groups have a common language. Control decisions are made by the administrators on the basis of information provided by the scientists themselves. Another advantage of the proposed method is that productivities may be compared among projects, among laboratories, etc. This comparison is made possible by the uniformity of the information collection process across all levels. It should be noted that productivity comparisons are not made only at total level. The comparison of partial productivities with respect to a given resource among projects can help to establish efficiency patterns.

REFERENCES

(1) Baumol, W.J., and T. Fabian, "Decomposition, Pricing for Decentralization and External Economies," Management Science, Vol 11, No. 1, September 1964, pp. 1-32.

(2) Mandakovic, T. "An Interactive Model for R&D Project Selection Decision Making in Hierarchical Organizations," Ph.D. Dissertation, Department of Industrial Engineering, University of Pittsburgh, Pittsburgh, PA., 15261, 1978.

(3) Mandakovic, T., and W.E. Souder, "A Flexible Hierarchical Model for Project Selection," R&D Management, Vol 15, No. 1, January 1985, pp. 23-29 .

(4) Mandakovic, T., and W.E. Souder, "An Interactive Decomposable Heuristic for Project Selection," Management Science, Vol 31, 10, pp. 1257-1771, October 1985.

(5) Mandakovic, T., and W.E. Souder, "Measuring Productivity in Hierarchical Organizations," to appear in International Journal of Production Research.

(6) Summanth, D.J., Productivity Engineering Management, McGraw-Hill, 1984.

(7) Toyoda, Y., "A Simplified Algorithm for Obtaining Approximate Solutions to Zero-One Programming Problems," Management Science, Vol 21, No. 12, August 1975, pp. 1417-1427.

Productivity Management Frontiers — I, edited by D.J. Sumanth
Elsevier Science Publishers B.V., Amsterdam, 1987 — Printed in The Netherlands

MEASURING KNOWLEDGE WORK -- "HOW TO"

JAMES A. CUSHNIE
Stevenson Kellogg Ernst & Whinney, 2300 Yonge Street, Toronto, Ontario, M4P 1G2 (Canada)

ABSTRACT

The economic downturn of the early 1980's has forced many companies to successfully downsize their white collar staff while maintaining or increasing business levels. Many of these companies are now looking for ways of counter-balancing the traditional forces to hire, to prevent a reoccurrence of this downsizing.

One approach is to implement a work measurement program for their knowledge workers. A work measurement program will help identify the current workload and staff levels appropriate for that workload. A new approach to work measurement called Structured Estimating is proving successful for groups of knowledge workers performing the same function, such as design engineers, computer programmers and tax auditors.

INTRODUCTION

One of the tools to effectively manage knowledge workers is WORK MEASUREMENT. Work measurement played an important role in North America's early dominance in manufacturing and it will play an equally significant role in the management of knowledge workers.

The time is right for the introduction of work measurement as a tool to manage and motivate knowledge workers. A study conducted on behalf of Steelcase found that a majority (55%) of office workers favoured "having how much work they produce on the job measured." A full 75% of the executives polled felt that it was "likely that their company will increase efforts to measure office worker productivity." This is not surprising. Psychologists and management scientists have been telling us for a number of years to:

- Set clear goals.
- Tell employees what is expected of them.
- Provide employees objective feedback on performance.
- Plan your work.

Introducing work measurement will help provide these things for the knowledge worker. Other benefits which are often overlooked when considering the introduction of a work measurement plan include:

- Ability to objectively determine appropriate staff levels.
- Making visible, efficiency improvements.
- Assessment of current workload distribution.
- Ability to simulate the impact on workload of new products/services.
- Comprehensive methods review.
- Detailed documentation of methods for training purposes.

You might ask, considering all of the benefits, why doesn't everyone have work measurement programs? The answer, I believe, is simply it is hard work. It is hard work to set up an effective work measurement system, particularly if you have had little experience in work measurement. The Steelcase survey found that 49% of executives who had attempted measuring the productivity of office workers said that it was either very difficult or impossible. Difficult yes, but not impossible. The competitive advantages that can be achieved by a properly installed system are well worth the effort.

Much work has been done in work measurement in the manufacturing environment, but it needs to be rethought for application to knowledge workers. Knowledge work is different than manufacturing work in that it tends to:

- Be more variable in nature.
- Be longer in duration.
- Have less distinguishable outputs.
- Have a higher percentage of think time.

These factors make it more difficult and more costly to develop accurate standards. However, these factors also make the accuracy of the standards less important because the tasks are not performed as frequently.

Structured Estimating is a modification of classical work measurement techniques for measuring variable long cycle work. Structured Estimating is based on the benchmark and slotting approach used in Universal Maintenance Standards (UMS). The UMS approach was developed specifically for variable, long cycle work found in equipment maintenance functions (Wilkinson, 1970).

KEY CONCEPTS

Structured Estimating uses key concepts from UMS, specifically the use of time ranges and spreadsheets. By using time ranges, we allow a task time to be represented by a time range rather than a specifically calculated time. For instance, we might consider the time to make a telephone call at 2-5 minutes versus a calculated average of 2.86 minutes. The use of time ranges recognizes the fact that tasks are variable and that high precision is not appropriate in the knowledge worker environment. Other advantages of using a time range include:

- Simplified application of standards.
- Reduced standard development costs.
- Knowledge workers are more comfortable with a time range.

Each time range is represented by the midpoint of the range. This allows us to use a single value for all tasks that fall within the range. We can use these midpoint values to set targets and measure performance. Over a period of time, variation of actual task times will average out to the midpoint of the range. This averaging (balancing) out will occur provided the task times are evenly spread across the time range (slot). An illustration of a time range or slot is shown in Exhibit 1.

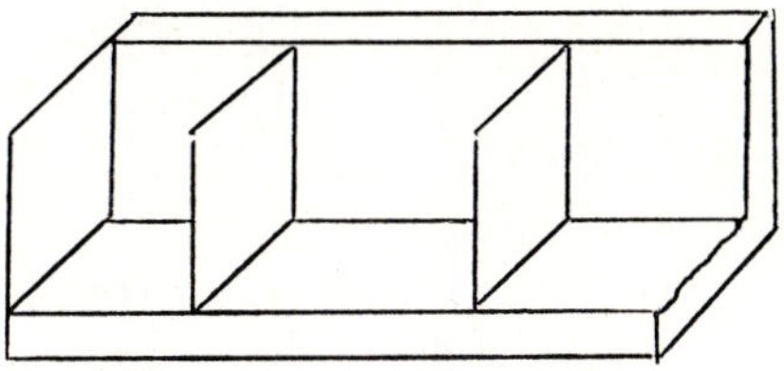

Slot		A		B		C
Slot Range (hrs)	0		.5		3.0	
Slot Midpoint		.25		1.5		4.0

EXHIBIT 1. Time range illustration

The size of the slots can be calculated and the error introduced by using the slots can be determined for a given reporting period. The key factors that affect the size of the slot are the accuracy of the measure desired (usually 5% or 10%) and the reporting period (usually a month for knowledge workers). An important consideration is the ease of use of the time ranges. Therefore, while the slot sizes can be calculated very precisely, they are usually rounded off. This sacrifices some accuracy but makes them easier to use.

The progression of time ranges is developed by the use of a formula. The formula takes into account two things:

- Subsequent time slots become larger while maintaining the same relative deviation within the range.
- Larger relative deviations are allowed on the smaller time slots.

The formula used to develop the progression of the ranges is:

$U_n = n(n+1)c$ where U_n = upper bound of slot n

n = slot number

c = midpoint of first slot

The second key concept is the use of spreadsheets. A spreadsheet is a format used for displaying the task variables and time ranges. An example of a spreadsheet is shown in Exhibit 2. Spreadsheets enable the estimator to make quick time estimates for each

Thorne Stevenson & Kellogg
Management Consultants

SPREAD SHEET
Time in Hours

Position: Level 1
Location:

Function: Routine Order — Task: 8.0 SITE TIME - 1 MAN OPERATION — April 30, 1985 (Page 1 of 2)

Time scale: 0 — .5 — 1.5 — 3 — 5 — 7.5

Description	Group A 0	Group B .25	Group C 1.0	Group D 2.25	Group E 4.0	Group F 6.25
		Plans are sufficiently detailed as to only require verification				
AERIAL						
NO STRUCTURE						
Non-urban						
Car access		0-350m	351-1150m	1151-2375m	2376-4050m	
Foot access						
No blazing		0-125m	126-450m	451-875m	876-1425m	1426-2000m
Blazing required		0m	1-175m	176-375m	376-675m	676-1000m
Urban						
Car access		0-150m	151-525m	526-1125m	1126-1900m	
Foot access		0-75m	76-300m	301-600m	601-975m	976-1400m
EXISTING STRUCTURE						
Non-urban						
Car access		0-400m	401-1350m	1351-2775m	2776-4725m	
Foot access		0-150m	151-475m	476-1000m	1001-1700m	1701-2575m
Urban						
Car access		0-175m	176-625m	626-1275m	1276-2200m	2201-3325m
Foot access		0-100m	101-325m	326-700m	701-1200m	1201-1825m
BURIED/CONDUIT						
Non-urban						
Car access		0-250m	251-850m	851-1750m	1751-2975m	
Foot access		0-200m	201-650m	651-1375m	1376-2325m	2326-3525m
Urban						
Foot access		0-125m	126-425m	426-875m	876-1475m	1476-2250m
UNDERGROUND	no field work required		inspect 1 manhole	inspect 2-3 manholes	inspect 4 manholes	inspect 5-6 manholes
APPARATUS CASES - AER/BUR.		1-2 cases	3-6 cases	7-13 cases	14-21 cases	
RETERMINATION IN C.O.			1-5 cables	6-11 cables	12-19 cables	

Function: Routine Order — Task: 8.0 SITE TIME - 1 MAN OPERATION

EXHIBIT 2. Spreadsheet example

task while knowing only the basics of the task (approximate magnitude of key variables). The task times can then be added to arrive at a total project estimate.

Often people who use estimates are surprised when their estimates turn out to be off the mark. They are usually off the mark because they have forgotten to take into account some key variable. We developed Structured Estimating to do three things:

- Increase estimate accuracy.
- Improve consistency of estimates.
- Allow standards to be changed easily.

Structured Estimating works because it breaks the job down into bite-size pieces that can be estimated using consistent data. By breaking the project into tasks and estimating the tasks individually, it is easier to deal with the variables and hence the variations in time to complete a project.

For example, if you were trying to estimate the time to travel from your office to some other city by plane, you might be hard pressed to come up with an accurate estimate. However, if you break the trip into its various components (tasks) it will become much easier to estimate. You might choose to break the trip up into the following tasks: taxi ride to the airport; wait at airport; flight time; retrieve luggage; and taxi to the hotel. It becomes easier to estimate each individual task because you have isolated one variable for each. You can then consider each task or variable separately rather than all the variables at once. It also limits the impact of error. A significant error in any one task is minimized because it is only part of the overall time.

The example spreadsheet shown in Exhibit 2 was developed for the task, site time, which was performed by Outside Plant Design Engineers for Bell Canada. Their function is to design and maintain the physical telephone network (cable) to provide service to their customers. The site time task involves assessing the proposed physical location of the new plant and making all the necessary measurements to be able to issue a work order so that the construction department can build it.

This task has two key variables: the length of plant (cable in meters) and the type of plant being placed (aerial, buried or underground). By knowing the type of plant and approximate length of cable, the Engineers can estimate the time to complete this task. For example, if the Engineer was estimating the task, site time, for 500 meters of buried cable in an urban setting, they would use 2.25 hours (Group D on the spreadsheet). The other tasks (steps) involved in the project are estimated in a similar fashion. It is a very quick procedure. In fact, we find that most projects can be estimated in less than five minutes. Even projects requirirg hundreds of man-hours to complete.

SETTING THE STANDARDS

Developing the standards is a key part of the measurement system. I suggest using knowledge worker estimates as the basis for determining standard task times. However, it is only one of many approaches that can be used. I have used predetermined motion time standard (PMTS) techniques such as MTM and MOST to develop the standards. However, I recommend knowledge worker estimates in most cases because:

- It provides direct knowledge worker input.
- It is substantially cheaper to develop than the use of PMTS.
- It provides comparable accuracy to PMTS for work that is variable in nature.
- Knowledge workers feel more comforable if the standards are based on their experience.

Setting the standards by using estimates is more involved than just asking knowledge workers how long it takes to do their job. A structured approach developing full documentation is necessary to ensure both accuracy and consistency. Without accurate and consistent standards, it is difficult to establish creditability.

At Bell, we used a four-step approach to standards development. The steps are:

1. Conduct preliminary interviews to determine tasks.
2. Conduct detailed task interviews to collect task time estimates.
3. Develop spreadsheets and time ranges.
4. Test and modify the spreadsheets.

We conducted preliminary interviews to identify the countable outputs and identify the tasks required to design and issue a work order. We found that the bulk of the Engineers' work consisted of issuing work orders and that there were two classifications of work orders: routine and specific estimates. The difference between them being the capital cost. Specific estimates having a higher capital cost required additional financial and planning controls. The routine order tasks we identified were:

Initial investigation
Site time
Travel time
Dealing with outside parties
Splicing design
Physical design
Order review

In the next step we conducted detailed interviews on each task to identify the sub-tasks and their associated times. We interviewed a minimum of three different Engineers for each task to help minimize biases. Surprisingly, little variation existed between the time estimates we collected from the different Engineers. In fact, any differences we did find were largely due to problems in defining the tasks.

We went to great lengths to document fully what sub-tasks were included in each task and how much time was allowed. This documentation, while time consuming, was critical in ensuring consistency and accuracy in the development of the standards and provided credibility and focus during the implementation.

The next step was to design the spreadsheets and time slots. We went through many iterations in this design phase until we had a spreadsheet that was simple and quick to use.

The final step in the development of standard times involved testing and making modifications. We tested the standards first within the project team, which included a member who had extensive experience in the Outside Plant Engineering function. We tested the standards by checking the reasonableness of total order times. We compared hypothetical jobs at both extremes, in terms of time required, with data we collected in the initial interviews. We also picked a sample of completed orders and reviewed estimated standard hours with the Engineers who completed the order. These tests resulted in a number of modifications and adjustments to both the format and standard times used.

The second stage of testing was to introduce and use the standards for a one-month period in a pilot location. This test resulted in a number of further modifications and provided standards which held up well in the subsequent full implementation.

The structured approach I have outlined is important because the standards will be questioned right from the beginning. The best way to get through this period is if the standards are correct and documentation exists showing exactly how the standards were developed. Good documentation usually shifts the focus of any standards disagreement from an impasse situation -- "that's not enough time" -- to consultative situation -- "is this task included"? Answering the question, "is this included?" is easy with good documentation, and builds credibility. This question often leads to another -- "should it be included"? This question can then be resolved and, if required, added to the standard. It often results in an agreement that the task in question adds no value, should not be included in the standard and more importantly should no longer be carried out.

THE BELL SYSTEM

The work measurement system at Bell has important benefits for all levels of management, knowledge workers, middle management and senior management.

From the point of view of the knowledge workers, they have better tools to plan their work and to monitor their plans. They can now estimate the time to complete each order before they actually do the work. Some Engineers were doing this prior to the introduction of the system; however, even for these people, the increased accuracy and consistency obtained by using spreadhseets is beneficial.

Middle management benefits from the system because they now receive monthly reports which quantify and summarize the work their Engineers are doing. This information is used to plan for productivity improvements. Emphasis is placed on identifying potential improvement areas and forecasting the impact of improvements on current operations. Taking positive action to achieve the planned improvements is stressed because improvements don't come just by wishing them. The monthly performance report provides information to be used to plan productivity improvement and to help keep attention on the efficient and effective use of a key resource -- knowledge workers.

Senior management benefit from haivng more productive staff but also now have a tool for objectively determining staff levels across the 44 Outside Plant Engineering locations. The staff level assessment is based on a bottom-up process where workload is forecast and man-hours required to complete the workload are estimated using spreadsheets. These man-hours are then adjusted to account for projected efficiency, vacations, absence and special projects.

STRUCTURED ESTIMATING APPLICATIONS

Bell has installed a measurement system covering some 300 knowledge workers that will:

- Measure productivity gains.
- Provide management with the tools and incentive to look for future productivity improvements.
- Provide senior management with an objective tool to allocate staff.

This system is only the beginning and is only one direction Bell is pursuing. As a tool to identify potential and quantify impacts, it is the first major step to improved productivity and profitability in the Outside Plant Engineering function.

Structured Estimating techniques have been applied in other knowledge worker jobs such as tax auditors at Revenue Canada, dairy inspectors at Agriculture Canada, and computer programmers at Royal Insurance of Canada. Productivity increased by 20% in the first six months at the insurance company.

Work measurement is not the only answer for knowledge worker productivity, but it is an effective tool in many situations.

REFERENCES

Gates, R.E., 1972, "Comparative Estimating for Indirect Work Evaluation", Work Study and Management Services, February 1972

Steelcase, 1980, The Steelcase National Study of Office Environments, No. II, "Comfort and Productivity in the Office of the 80's", Conducted by Louis Harris & Associates, Inc., 87p

Wilkinson, J.J., 1970, "MP&E Report on Universal Maintenance Standards", Modern Power and Engineering, March 1970

Productivity Management Frontiers — I, edited by D.J. Sumanth
Elsevier Science Publishers B.V., Amsterdam, 1987 — Printed in The Netherlands

MEASURING AND IMPROVING "WHITE COLLAR" PRODUCTIVITY: A NASA CASE STUDY

D. SCOTT SINK and SANCHOY K. DAS
Virginia Productivity Center, Virginia Tech
Blacksburg, VA 24060, U.S.A.

THOMAS C. TUTTLE
Maryland Center for Productivity & QWL, University of Maryland
College Park, MD 20742, U.S.A.

ABSTRACT
Sink, D.S., Das, S.K., and Tuttle, T.C., 1987, Measuring and Improving "White Collar" Productivity: A NASA Case Study.

This paper presents a description of a "white collar" productivity management process that is being "pilot-tested" in the Engineering Directorate at Goddard Space Flight Center. The "pilot test" is a very small project, by NASA standards, however, it represents a potentially significant "white-collar" productivity improvement process. The paper presents a brief background for the study, describes the specific approach and techniques employed, and discuss results to date. The project was nearing completion at the time of writing, so results are still unclear at this point. Early indications are that the process tested, has merit in a technical and professional setting.

1 INTRODUCTION

A major key to improving America's productivity dilemmas is in the office and not on the factory floor (Thurow, 1984). The character of the "white collar" worker and his work is substantially different from that of the factory worker. In studying performance it is critical to differentiate between productivity in technical and professional settings. Knowing what to do, having consensus about what should be done, having commitment to do what should be done by those who have to do it - these are the really critical issues in the "white collar" setting (Sink, 1985a). Ineffective delegation, lack of leadership, poor planning process, lack of team-work, poor situational leadership skills, poor communication and coordination, poor attention to detail, lack of knowledge about measurement and evaluation, etc.- these are the critical factors/roadblocks to productivity in most American organizations. These are the "white collar" performance issues that must be dealt with in the 80's and 90's in order for American industry to survive and compete. This paper presents a description of a "white collar" productivity management process that is being pilot-tested in the Engineering Directorate at NASA Goddard Space Flight Center.

2 BACKGROUND

NASA has, over the past several years, undertaken numerous productivity and quality improvement projects. Many of these initiatives have been supported by the Director for NASA Productivity Programs. Each NASA installation has developed a formal productivity and quality improvement plan. NASA contractors have made productivity improvement an explicit objective. NASA held a National Symposium on Quality and Productivity in 1984, which was a major success. A second biennial symposium will be held on building a quality ethic, measurement and gainsharing, revitalizing organizations, and participative management. There are numerous pilot projects being executed within NASA in an attempt to develop new and effective ways to improve productivity and quality. These initiatives range from technological transfer, capital investment, automation, facilities upgrading to procedures streamlining, quality circles, improved training incentives and gainsharing, and measurement (NASA report 1985). It is not apparent whether a consensus plan for the productivity management effort exists at NASA headquarters or within the centers themselves. However, there is considerable activity and NASA is committed to accomplish excellence in this area as in others. NASA headquarters had a strong desire to do something in the area of "white collar" productivity measurement. This pilot study, then, is one of many ongoing within NASA.

The intended focus of study was on "white collar" productivity measurement. Goddard Space Flight Center (GSFC) was selected as the host site and the Engineering Directorate was the specific pilot study system. The purpose of the effort was to have a professional management research organization test and assess two specific strategies and techniques for improving productivity at GSFC. The principal investigators (Sink & Tuttle) met with management of the Engineering Directorate several times to discuss the approach, and to obtain approval to proceed. Neither the pilot study itself nor the approach were forced on the Directorate. In fact, the approach adopted was modified significantly as a result of discussions between the PI's and Directorate management. The ultimate decision was to come at productivity measurement indirectly, through the "back door" if you will. The project was designed to be completed in approximately six months.

3 APPROACH

In meetings between the PI's and the Engineering Directorate management it was strongly felt that, there would be a resistance to an "imposed" project focusing on productivity measurement. Alternative approaches were discussed. The need for educational interventions and strategic planning for the productivity improvement effort were discussed. GSFC management felt that engineers, scientists and other professionals:

1. Will resist direct attempt to measure productivity.
2. Misunderstand productivity.
3. Will respond more favorably to efforts to improve productivity/ remove roadblocks.

Our experience coincides with these beliefs. Most managers are impatient and are often resistant to measurement efforts (Tuttle and Sink, 1985). Most managers want to see results, they want to act, put our fires, prevent fires. It was thus decided to focus first on a participative process and evolve towards measurement. The strategy and techniques for productivity improvement and measurement in this "white collar" setting were therefore developed and selected based upon client/user perception of what would ultimately be most successful. Table 1 depicts the approach finally developed and adopted. It was decided to pilot test two specific processes:

- Roadblock Identification, Analysis, and Removal Technique (RIART)
- The Process Flow Analysis Technique (PFA)

TABLE 1

Project calendar and eight basic steps/tasks.

		Month
1.	Initial inquiry and contact	0
2.	Preliminary planning discussions	1-2
3.	Contract Approval	4
4.	Top management orientation and performance improvement planning sessions	6
5.	Develop implementation plan and select branches and sections	7
6.	Test RIART	8-13
7.	Test PFA	8-13
8.	Final report/evaluation	13

The techniques were pilot tested in two different branches within the Directorate. The calendar for the testing process is shown in table 2. The project was initiated in a participative manner, to provide visibility and a feeling of ownership. Once contract approval was accomplished, a top management orientation workshop was scheduled. The entire Engineering Directorate management team was assembled and involved in the planning workshop for the effort. Approximately 45 section heads, branch chiefs, and staff were in attendance. The manager of the Engineering Directorate opened the workshop with a brief presentation on why they were in attendance and gave a review of mission statements, goals and objectives. This provided an excellent backdrop for the workshop and

TABLE 2

Calendar for process testing

	Month
Test RIART	
1. Roadblock Identification & I/O Analysis	8
2. Action Team Proposal Development	8-10
3. PAT-PAG Briefings	8
4. PAT-Mgmt. Review Team session	11
5. Management Review Decisions	12
6. Implementation	12-13
7. Measurement & Evaluation	12-13
8. Final Report	13
Test PFA	
1. Identify Key Processes	8
2. For Process Flow Analysis Team	8
3. Define Process Steps	9-10
4. Validate Process Description	11
5. Identify Key Variances	11
6. Identify Process Improvements	12
7. Define Process Criteria	12
8. Change and Track Results	13

we believe instilled a sense of genuine commitment to the effort. The objective of this top management orientation and performance improvement planning session was to:

- Create awareness
- Expose the top management group to "Productivity/Performance" basics
- Expose the management group to structured group process (NGT)
- Expose management to RIART and PFA.
- Involve management in a strategic performance improvement planning process
- Identify volunteer branches and sections for the pilot studies.

3.1 "Back Door" Approach to Performance Management

The essential components of a performance management effort, and how they fit together, is shown in figure 1. All the components must be present for the effort to be complete (Sink, 1985). As a performance management program evolves, from stage to stage, each of the components are sequentially incorporated into the process. The first stage involves the creation of a sense of awareness, and an understanding of performance basics. The next, or concurrent stage, is a strategic planning process. This should provide a crystallized vision of where the system should/must head in the next 2-5 years, and it must effectively link these goals to action planning and implementation. During the next stages measurement, evaluation, control and improvement processes are executed in that order. The final stage involves refinement and maintaining of excellence.

In the GSFC case it was felt, the prevalent attitude of the employees was to resist measurement efforts. This attitude would thwart the productivity management effort, and there was a need to modify the approach shown in figure 1. This would help ensure the success of the effort. It was thus decided to take a "back door" approach and focus first on a participative effort at improvement, and then follow up with measurement (figure 2). Kanter (1983) states, "employees can be energized - engaged in problem solving and mobilized for change - by their involvement in a participative structure that permits them to venture beyond their normal work roles to tackle meaningful issues." Similarly, we thought, a participative improvement effort would serve as an "unfreezing stage". This stage would create a motivation to change and improve performance. Following this process, we expect, there would be little resistance to measuring the improvements and performance in general.

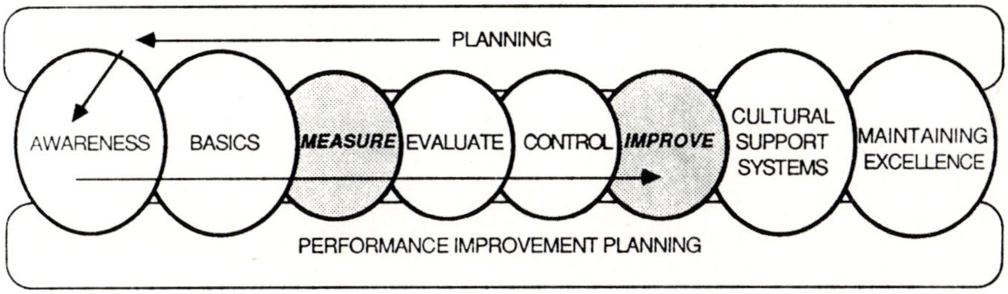

Fig. 1. Performance/productivity management process/elements.

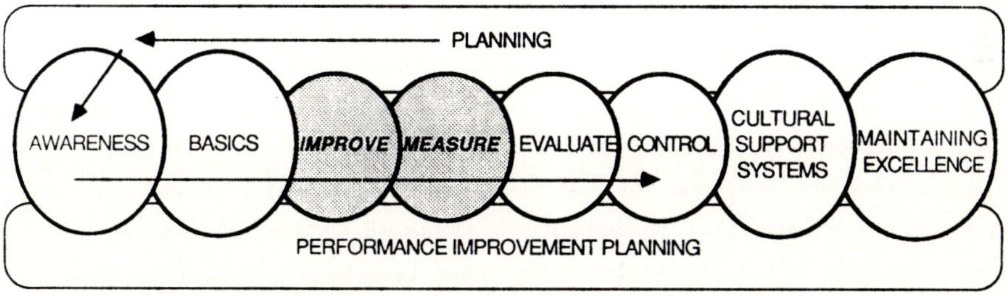

Fig.2. The "back door" approach.

3.2 Assumptions and Reality

In developing a "white collar" productivity management program, there is a tendency to make assumptions that are not valid. For instance it is assumed the organization has a well thought through, participatively developed, and communicated, comprehensive and integrated, and commonly accepted productivity management strategy. Management, staff and employees believes that productivity needs to improve. Measurement is a salient issue for all managers, and "white collar" productivity is the right issue.

In reality these assumptions are not valid in the majority of cases. Most (maybe all) organizations do not have (or know how to develop) a well thought through, participatively developed, well communicated, comprehensive and integrated, and commonly accepted productivity management strategy (at any level). Most American managers, staff and employees believe that maintaining the "status quo" is sufficient for success, survival and competitiveness. Measurement is not a particularly salient issue/concern for most managers. Productivity is not the most critical performance criteria in many "white collar" settings. The true state and attitude of the organization must be considered when designing a program. Inability to do this will result in failure. Success can be ensured if the productivity management effort is participatively developed. The two techniques tested here are of this nature.

4 ROADBLOCK IDENTIFICATION, ANALYSIS, AND REMOVAL TECHNIQUE

The Roadblock, Identification, Analysis, and Removal Technique (RIART) focuses upon identifying roadblocks to performance factors and/or problems that are causing us to be less productive than we could or should be. RIART is an action oriented, pragmatic approach to getting professionals proactively involved in productivity improvement and eventually measurement. In this process the group identifies these roadblocks. An analysis of each roadblock is then performed to ascertain the who, what, how, and when of removing the roadblock. Forms and specific techniques are utilized to accomplish these objectives. The RIART technique evolved from the Performance Action Team Process research and development described by Sink (1985) and Sink and Kiser (1985). The steps in the technique are fairly simple and easy to execute. They are:

1. Selection of the strategic business unit, branch, department, or focal system.
2. Orientation, awareness, and basics background. Training and development session.
3. Input/Output analysis (Internal customer analysis, Process activity analysis)
4. Nominal Group Technique session to identify and prioritize performance roadblocks (Sink, 1983).

5. Formation of action teams to analyze barriers roadblocks.
6. Development of scoping proposals to remove roadblocks. Review, evaluation, and approval of these proposals.
7. Implementation, project management, accountability systems.
8. Development of scoreboards to track and evaluate success of implementation.
9. Integration with performance improvement goals and objectives from upper management performance improvement planning efforts.
10. Visibility room development.
11. Evaluation, restructure rewards systems, etc.

As is the case with most interventions of this type, the skill with which these steps are executed is critical to the ultimate success of the technique. In our view, the most critical element aside from leadership qualities is the accountability system. Being able to assign the right people towards eliminating the "right" roadblock and supporting them with necessary resources will ensure improvement. The keys are discipline, accountability, leadership, persistence, patience and consistence. These are the necessary ingredients for a commitment to excellence. They are not easy to attain, but are representative of the quality of an organization.

The RIART technique was implemented at the Semiconductor Development section. Each action team developed a proposal focussed at removal of an identified roadblock. Further, at least one measure that could be used to evaluate the success of the roadblock removal process, was developed. We found the participants to be interested in the process and motivated to work on proposals. but contrary to expectations, we observe that participants had "rusty" problem solving skills. This is probably due to lack of experience in this type of activity rather than lack of ability.

5 PROCESS FLOW ANALYSIS

The PFA technique was implemented at the Environmental Test section. The PFA was addressed at analyzing and improving the process by which cost and schedule estimates for required tests of hardware are made. The PFA methodology followed was developed at the Maryland Center for Productivity and Quality of Work Life. The steps are:

1. Identify critical/key process(es).
2. Form Process Flow Analysis team.
3. Define process steps.
4. Validate process description.
5. Identify key variances.
6. Identify process improvements

7. Define process criteria.
8. Implement changes and track results.

A key process is one that consumes a high level of resources, or is critical in terms of impact on organizational performance. The key process tested in this study was the Test Action Request (TAR) process. A team of knowledgeable people were selected to serve as the PFA team. This group included representatives from all organizational units that are affected by TAR. Over a series of meeting members of the PFA team defined the process steps. This step involved a consideration of the context in which the key process is carried out, and the overall budget process which impacts the TAR. The definition of TAR steps was validated by interviews with other people who are conversant with the process. The next step involved the development of variances between "what happens" and "what is desired". A list of critical variances was developed by the PFA team. Using the key variances as a database, needed process improvements were identified. The focus was on identifying changes in the TAR process that would help control the key variances. The final step in PFA involves implementing the process improvements identified. By tracking results over time and, data permitting, making comparisons with other measures, it is possible to appraise the impact of the improvement process.

6 CONCLUSIONS/SUMMARY

The project has recently been completed and the final report is under preparation. It is expected that NASA will implement the changes identified, and will track results. Several immediate conclusions have already been drawn and are presented here:

1. Performance Action Team process has good potential as a "White Collar" productivity management strategy and approach.
2. RIART is a very pragmatic and fairly simple technique that appears to be suitable for highly technical settings with scientists, engineers, and other professionals.
3. PFA appears to be best suited as a very detailed diagnostic technique that would be used for certain roadblocks.
4. Productivity measurement appears to be best approached as an outcome of improvement efforts (indirectly) in "white collar" settings.
5. Productivity measurement and improvement efforts can be implemented in the absence of an overall plan and in the absence of overall top management

commitment through involvement. But, we doubt that our effort will have long term impact without this.

6. The study confirmed our belief in the need to continue to aggressively experiment with participative management approaches.
7. The PAT process must be implemented top down.
8. An internal change agent/process manager is essential.
9. People in "white collar" settings are ready for participative management but are skeptical and often cynical.
10. Top management, in many (if not most or all) cases do not understand productivity, do not understand the nature of changes taking place, are often not good leaders, do not realize that the most important thing they can do is to manage culture and the process.

The issue of management's commitment to the process is critical to the success of any productivity management program. We believe the participants in this study were skeptical of management's commitment to the process, and the probability of process continuity. This had a negative impact on participants commitment to the process. This behavior is common to most American organizations and is not unique to NASA or the Federal Government.

"White Collar" employees are ready and should participate in planning, problem solving and decision making focussed at improving performance. Management processes, which provide a mechanism for motivating this activity, are desperately needed by American organizations. The RIART and PFA techniques are first steps towards achieving this mechanism.

7 REFERENCES

Kanter, R.M., (1983). The Change Masters: Innovation & Entrepreneurship In The American Organization, Simon & Schuster, New York.

Sink, D.S., (1983). Using the Nominal Group Technique Effectively, National Productivity Review, Spring.

Sink, D.S., (1985). Productivity Management: Planning, Measurement & Evaluation, Control & Improvement, John Wiley & Sons, New York.

Sink, D.S., (1985a). Performance and Productivity Management in "White-Collar" Settings, Productivity Management, Vol. 4(2), Spring.

Sink, D.S. and Kiser, K.J.,(1985). Performance Action Teams: Update on Continuing Developments, Procs. Annual Int. IIE Conf., Institute of Industrial Engineers, Norcross, Georgia.

Thurow, L., (1984). Revitalizing American Industry: Managing in Competitive World Economy, Calif. Mgmt. review, No. 1, Fall.

Tuttle, T.C. and D.S., Sink (1984). Taking the Threat Out of Productivity Measurement, National Productivity Review, Winter.

Productivity Management Frontiers — I, edited by D.J. Sumanth
Elsevier Science Publishers B.V., Amsterdam, 1987 — Printed in The Netherlands

AN UPDATE ON THE STUDY OF PRODUCTIVITY MEASUREMENT AND INCENTIVE METHODOLOGY

D. SCOTT SINK and PAUL E. ROSSLER
Virginia Productivity Center, Virginia Tech
567 Whittemore Hall, Blacksburg, VA 24061 U.S.A.

A.K. (Shoni) DHIR
LTV Aerospace and Defense Co., Vought Aero Products Division
P.O. Box 225907, Dallas, TX 75265 U.S.A.

ABSTRACT
Sink, D.S.; Rossler, P.E.; and, Dhir, A.K., 1987, "An Update on the Study of Productivity Measurement and Incentive Methodology."

This paper describes Phase III of a five-phase research project attempting to investigate and develop useful productivity measurement and evaluation techniques in the defense community which integrate with government-to-contractor incentives. The Phase III research involved a "paper-test" of three measurement models and techniques. A major outcome of Phase III is a comprehensive and integrated productivity measurement methodology for defense contractors and subcontractors.

1 INTRODUCTION

In previous articles and literature, (Sink, 1985; Sink, Tuttle and DeVries, 1984; Sink, DeVries and Tuttle, 1984; Sink, Dhir, and Roberts, 1985) we have reported on research investigating state-of-the-art approaches for measuring and evaluating productivity in the defense subcontractor community. This article will update you on that continuing research. We will report on Phase III of a V-phase project which is attempting to investigate and develop useful productivity measurement and evaluation techniques for the defense contractor community.

Phase III was completed on January 31, 1986, and our final report has been submitted. We will briefly review the five phase project goals and objectives and highlight results from Phases I and II. Next, we will provide a detailed description of Phase III complete with results and conclusions. One of the major outcomes of Phase III is the development of a generic productivity management methodology for defense contractors and subcontractors. This methodology will be presented and discussed.

2 OVERALL STUDY DESCRIPTION

The overall goal of the five-phase study is to identify and develop productivity measurement and evaluation methodologies and models which will effectively integrate with government to contractor incentive methodologies.

The government (client/customer) understandably wants/needs to improve the performance of defense-related systems (i.e. reduce costs, increase quality, improve responsiveness, etc.) for acquisition purposes. The contractor understandably wants/needs to improve its performance (i.e. increase profits, reduce costs, increase quality, improve productivity, spark innovation, etc.) so it will be competitive, grow, and survive in both the short-and long-term. Government to contractor incentive/gain sharing methodologies such as the IMIP are viewed as a way to satisfy the goals of both the government and defense contractors.

A program of the scope and character of incentive methodologies such as IMIP is complex. For the intended desired outcomes to be achieved, the many elements of an overall program must be compatible. Critical elements involved are those of measurement and evaluation. Why? First, one cannot manage what cannot be measured (Sink, 1985). Second, the government cannot share benefits unless these benefits can be verified (Virginia Productivity Center/VPI&SU, Final Report, March, 1986). Third, properly designed measurement and evaluation systems motivate, promote, and encourage proactive productivity improvements (Sink, 1985). Last, measurement and evaluation are needed to control improvement interventions and ensure effective and efficient execution of productivity improvement interventions (Sink, 1985).

Phase I of the overall study identified and described current productivity measurement practices in the defense contractor community. Phase II focussed on the development of a taxonomy of productivity measurement theories and techniques. Phase III involved a "paper test" (evaluated, on paper, the applicability, ease of use, etc.) of three productivity measurement approaches identified in the Phase II contract.

Phase IV and V have been combined into a single eighteen month project which would:

1. Resolve specific developmental needs of the three models by a limited field test.
2. Complete the development of an integrated productivity management methodology.
3. Review and evaluate the models and the integrated methodology with defense contractors in a workshop setting.
4. Develop an Implementation Guide for the integrated methodology.

3 RESULTS OF PHASES I AND II

Phase of the overall study was executed by the Army Procurement Research Office (APRO). The purpose of Phase I was to 1) identify and describe current productivity measurement practices in the defense contractor community and 2) develop specific definitions of contractor productivity appropriate to the defense industry. A summary of the results reported in the 1984 (June) APRO

Final Report is presented:

1. Contractors ranked profitability most important on a list of organizational performance evaluation factors. Productivity was usually ranked fifth, after profitability, effectiveness, quality, and efficiency.
2. Problems encountered by contractors measuring their productivity were usually due to the complexities of quantifying and relating various input and output factors involved.
3. There was no evidence of a total factor productivity measurement system implemented by contractors although some attempts were being made to develop such.
4. Production cost visibility varied widely among contractors, but all could provide direct labor and material costs through work center tracking. Indirect cost visibility is becoming increasingly important as cost drivers shift from direct to indirect areas
5. Tracking the impact of an investment for productivity improvement in indirect areas gets obscure, and these areas frequently increase with a decrease in direct costs.
6. Investments were mostly for competitive and technological reasons rather than simply for cost reduction on the current contract.

Phase II, executed by the Oklahoma Productivity Center and the Maryland Center for Productivity and Quality of Working Life, focussed on identifying and describing currently available productivity measurement theories and techniques (see Sink, Tuttle, and DeVries, 1984 and Sink, Tuttle, DeVries, and Swaim, 1983). Techniques identified included both productivity and surrogate measurement systems. Surrogate or substitute measures are those which measure variables related to productivity (e.g. scrap reduction, cost reduction), but do not measure productivity (output/input) directly. Productivity improvement efforts and accomplishments can be, and are being, measured without the aid of productivity measurement and evaluation techniques.

One productivity measurement technique, the MFPMM, and two surrogate techniques, Cost Benefit Analysis/Cost Benefit Tracking (CBA/T) and the the DCF/SSA used for IMIP, were identified as having most promise to link productivity measurement to incentive methodology. The MFPMM is an intermediate, plant level measurement technique used to analyze changes in profits and productivity and simulate the impact of productivity improvement interventions (see Sink, 1985; Sink. Dhir, and Roberts, 1985; and Sink, Tuttle, and DeVries, 1984). A CBA/T technique is Price Waterhouse's Automated Cost Baseline Generator (ACBG). ACBG is the software tool which accompanies Price Waterhouse's Cost Definition Methodology CDEF (see Price Waterhouse CDEF Briefing Package, 1985). CDEF utilizes a top-down analysis technique which facilitates the 1) identification of performance and cost measurement criteria; 2) selection of improvement opportunities; and 3) economic justification of identified investments. The DCF/SSA is an engineering economic analysis of a program or project's cash flows and potential shared savings (see DOD Guide on Improving Productivity in Defense Contracting, 1985 Edition).

The Phase II Study concluded that productivity measurement technology has the capacity to provide accurate productivity data to managers. There are several reasons why industry, in general, is not taking full advantage of state-of-the-art techniques:

1. Knowledge of specific productivity measurement techniques is not widespread.
2. State-of-the-art techniques are less complex than they appear; however, they do require substantial effort to actually implement.
3. Some of the macro-measurement and surrogate techniques may be adequate for an individual manager's needs.

4 THE STUDY OF PRODUCTIVITY MEASUREMENT AND INCENTIVE METHODOLOGY

The goal of Phase III, "The Study of Productivity Measurement and Incentive Methodology" was to evaluate the productivity measurement/evaluation techniques identified in the Phase II contract. A fourth approach or methodology investigated was the LTV Vought Aero Products Division (LTV/VAPD), integrated productivity measurement system. This system is under development by LTV/VAPD and incorporates elements of the three primary models listed above. The specific objectives of the study were:

1. "Evaluate the ease of measuring and/or evaluating productivity using these three models in paper test fashion."
2. "Develop a comprehensive description of inputs and outputs for each model as applied in paper test fashion."
3. "Attempt to compare results of the paper test from the three models."
4. "Identify and describe in detail the data required to 'drive' each model. Compare and contrast data requirements for each model."
5. "Describe the level (unit of analysis) for which productivity was measured and evaluated in the field paper test. Describe the most appropriate unit of analysis(es) for each model."
6. "Evaluate the abilities of the models, in paper test application, to satisfy the overall project goal."
7. "Describe incentive/reward system(s), if any, used by the field test site participating in this evaluation."

8. "Recommend (not develop) modifications, to any or all of the three models that would make them easier to:
   (a) to use
   (b) to control
   (c) to administer
   (d) to obtain information
   (e) to use in incentive/rewards applications"
9. "Recommend whether or not to proceed with the Phase IV study."

(Virginia Productivity Center/VPI&SU, Final Report, March, 1986).

4.1 Approach

The Virginia Productivity Center (VPC) was contracted to execute the Phase III contract. The VPC paper tested the MFPMM, CDEF, and DCF/SSA as well as LTV/VAPD integrated productivity management methodology. Four subcontractors were involved: the Maryland Center of Productivity and Quality of Working Life provided overall project support; Price Waterhouse paper tested their CDEF methodology; Westinghouse paper tested the DCF/SSA; LTV/VAPD paper tested their integrated methodology and independently paper tested the MFPMM, CDEF, and DCF/SSA. Each subcontractor was responsible for the coordination and execution of their paper test.

The research team spent considerable time discussing the difference between a "paper test" and a "field test". The conclusion was that a paper test should evaluate the models on paper addressing specific questions relating to the study objectives. Development or extensive data analysis with the models was not viewed as within the scope of Phase III.

Before presenting Phase III findings, we feel it is important to state implicit assumptions and study constraints. The intent is to recognize the limitations of our ability to extrapolate the study findings to the entire defense contractor community. We do not want to mislead readers as to the extent these findings are widely applicable in all contractor settings and applications. Although, only one field test site was used, the research team has broad and extensive experience with other defense contractors.

We are confident the models themselves can, and do, work in the defense contractor environment. Each perform the tasks they were designed to accomplish. We are unsure of the extent to which these models will be endorsed by the contractor community and developed into a management system (i.e., a productivity management methodology).

We do not assume these three models represent the universe of models available to accomplish the desired objectives of the government. They certainly do not represent necessary or sufficient tools a government contractor must use to succeed in business. However, they do represent three state-of-the-art approaches to measurement and evaluation of performance in this environment. LTV/VAPD has shown that, when combined into an integrated system, they represent a sophisticated and useful management system.

4.2 Major Findings and Conclusions:

1. Individually, "None of the three models tested will accomplish all of the objectives desired by the government or by contractors."
2. "A methodology which incorporates the use of a variety of measurement and evaluation models is required if all the desired objectives of both the government and contractors are to be satisfied."
3. "Each of the three models tested has "soft spots" or current developmental problems that need to be, and are being worked on. All of these models are relatively new developments that have excellent potential."
4. "Variances in operating systems, management styles, pressures and priorities, perceived problems and opportunities, and skilled/ competent productivity management personnel will very likely make it difficult to translate and transfer models and methodologies from one company to the next." The issue/problem of translation, transfer, and effective implementation needs to be studied.
5. "Each model tested was initially designed to accomplish certain objectives. More than one model or approach are likely available to accomplish these same objectives. The challenges, then facing management in the defense industry are:
   (a) acceptance and commitment to productivity management, specifically measurement and evaluation;
   (b) design of systematic productivity management efforts that combine a set of measurement and evaluation models into an effective system."

   These models can be combined into an effective productivity management methodology as described later in this paper. Also, a case example of a defense contractor, LTV/VAPD, that has developed a productivity management methodology supports this particular finding.
6. "These three models when viewed together constitute a potentially satisfactory methodology which can accomplish what the government and contractors want to do." Each model, used independently of other models and systems, is not sufficient to accomplish the overall goals of the government and contractors. Each of the three models was designed to accomplish an important part of the overall goal DoD and contractors have established in IMIP-type programs. Of the three models tested, only the MFPMM actually measures total input-output productivity.

DCF/SSA model is primarily an analysis and decision making tool for planning and forecasting purposes. Price Waterhouse's CDEF is an approach developed to prepare performance and cost baseline data in support of commercial factory modernization or Department of Defense IMIP's.

7. The DCF/SSA Model performs well against the criteria for which it was designed. However, there are deficiencies in the software developed by Logistics Management Institute (LMI) to implement the DCF/SSA Model.
8. "The MFPMM must be modified to function in the defense industry environment." LTV/VAPD has made this conversion and has found the model useful as an integral component of their productivity management methodology.
9. "Price Waterhouse's CDEF/ACBG performs well against the criteria for which it was designed." The top down activity structure required for data analysis may differ from a company's organizational structure. Therefore, a node-tree structure must be developed. The effort required to execute this step will depend on the complexity of the company's processes or activities.
10. "Each model tested will work in the defense contractor environment." They were designed to accomplish specific objectives and are useful for those purposes. However, to develop a comprehensive management effort, a combination of measurement and evaluation techniques are required.

(Virginia Productivity Center/VPI&SU, Final Report, March, 1986).

5 PRODUCTIVITY MANAGEMENT METHODOLOGY FOR DEFENSE CONTRACTORS

During the course of this study (Phase III) the research team struggled with the distinction between a measurement and evaluation model and a productivity management methodology. We concluded that measurement and evaluation models are integral components of a productivity management methodology. We knew what the models we were testing were designed to accomplish and our findings suggest that each performs its function(s) effectively. Our conclusion is that each of the three models (MFPMM, CDEF/ACBG, DCF/SSA) can and has been used successfully in the defense contractors environment. However, the ability of a model to be applied successfully does not ensure that it will be applied successfully. Models such as these are best viewed as decision support tools. Their successfully application cannot be ensured unless the systems designer successfully builds an infrastructure for their use in the organizational system. Figure 1 depicts critical elements of the management system that these tools are a part of. This figure depicts several things. First, it depicts the measurement, evaluation, control and

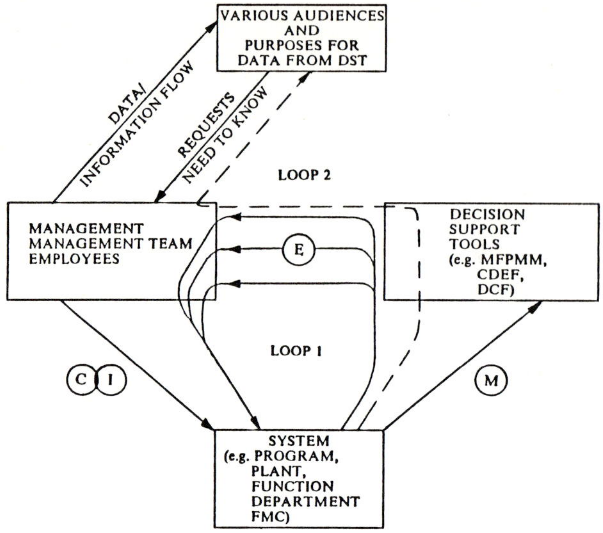

Figure 1. A management system

improvement process. Data is collected from the system(s) being managed (the M for measurement). Data is input into specific measurement and evaluation systems, models, tools. These models, such as the MFPMM, CDEF/ACBG, and DCF/SSA approaches, use specific data from the system, analyze/manipulate the data and present modified data to the manager and/or management team for the system. The modified data may also be used to satisfy the needs of other audiences. The designed intent of these models is to support decision making and to support control and improvement of performance of the system(s).

Notice that two measurement and evaluation "loops" are identified. The first loop (Loop 1) focusses upon the measurement and evaluation and control and improvements needs/demands of the manager, management team and employees. The focus of this loop is on providing the information necessary to control and improve in a timely and effective fashion the performance of the system(s) being managed. The focus of Loop 2 is on providing information to satisfy the wide variety of needs and requests of other audiences. Examples of other audiences are the government, the IRS, the comptroller, the manager's boss, the president, etc. The reason for making a distinction between these two loops of information is that different measurement systems are often required to satisfy the differing needs. System designers that fail to recognize design requirement differences for the two loops frequently fail to design successful decision support systems. They do design reports, however, they are not used effectively by the users.

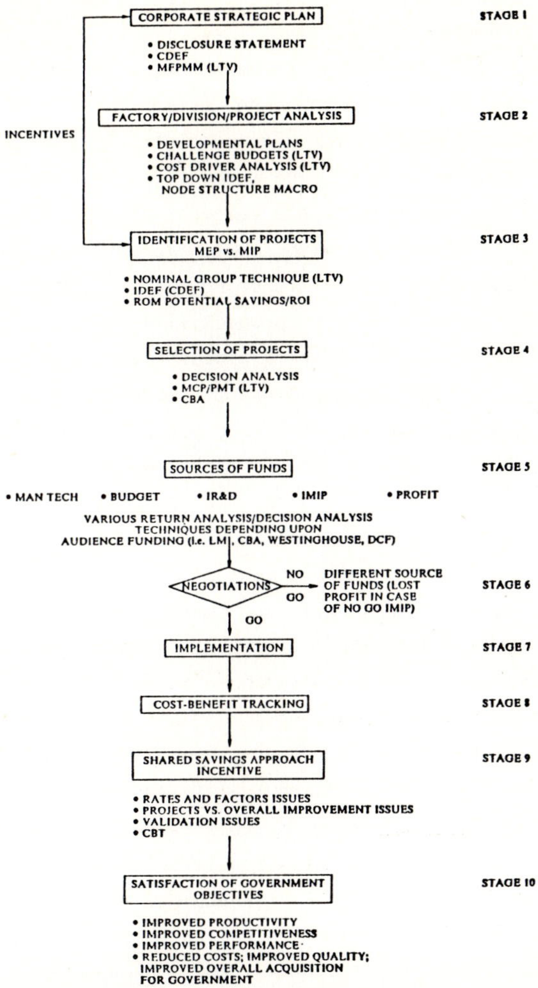

Figure 2. Generic productivity management methodology as related to the defense industry.

So, we concluded from our study that these three models tested can work in the defense contractor environment, but that they will have to be developed and implemented as an integrated management system in order to have designed impact. We took these thoughts one step further and attempted to develop a generic productivity management methodology for defense contractors. The methodology, depicted in Figure 2, has 10 stages. The methodology focusses primarily upon decisions associated with modernization investment projects (those requiring capital) versus decisions associated with what have been called modernization efficiency projects (those requiring little or no capital investment). **Stage 1** indicates the importance of driving the productivity improvement process from the results of corporate strategic planning processes. This is an attempt to ensure that our capital investment decisions for modernization are compatable and congruent with the larger scoped strategic plans. LTV/VAPD utilizes the MFPMM at this stage to support strategic improvement decisions. **Stage 2** represents the process of analyzing data for the factory, division, or project in an attempt to identify target areas for improvement. The CDEF methodology utilizes the ACBG to assist in the development of "as is" cost and performance baselines. LTV utilizes what they call a "cost driver analysis" to target improvement areas. **Stage 3** represents the process by which specific improvement projects are identified. LTV utilizes the Nominal Group Technique to generate consensus regarding improvement projects. This facilitates ease of implementation at later stages. LTV does evaluate the normatively generated projects against Stage 2 analysis to ensure quality and needed impact. **Stage 4** is the point at which actual selection of projects takes place. A variety of decision analysis techniques could be utilized to determine which projects are worthy of further development. The CDEF methodology develops and compares "as is" costs and performance in relation to "to be" costs and performance in order to select projects with the biggest potential for improvement. **Stage 5** is an obviously critical step and involves an analysis of sources of funds available to support the projects. Various decision analysis methods are required at this stage depending upon the audience/funding source. **Stage 6** is a political reality involving the negotiation process associated with obtaining support for improvement projects. Assuming that funding for the project is achieved, **Stage 7** represents the implementation phase. Many, if not most, government supported investment programs, such as IMIP, require cost-benefit tracking **Stage 8**. The ultimate goal of this overall project is to develop improved models that will enable valid evaluation of improvement projects. Did the government, the taxpayer, obtain the desired/predicted performance improvement? Improved productivity measurement techniques will play a key role in being able to answer this question. **Stage 9** is the desired outcome of the IMIP. To promote and assure productivity improvement in the defense contractor community it has been convincingly argued that there must be incentives. Stage 9 represents the point at which these incentives become a reality. The question of whether shared savings are in fact validated is, in fact, at the heart of this project. To date, shared savings are primarily based upon projected benefits rather than a systematic cost-benefit tracking process. Measurement and evaluation (validation) of improvements has been an illusive goal. **Stage** 10 is the bottom line, if you will, for the government and likely for the contractor also.

This productivity management methodology is actually just a subset of the one described by Sink, 1985. However, in the defense contractor environment this modernization investment oriented approach is certainly a key element to improved performance in the 80's and 90's.

6 REFERENCES

Contractor Productivity Measurement, APRO 83-01, Army Procurement Research Office, Ft. Lee, VA, June, 1984.

Delbecq, A.L.; Van de Ven, A.H. and Gustafson, D.H. Group Techniques for Program Planning: A Guide to Nominal Group and Delphi Processes. Glenview, IL: Scott, Forseman and Co., 1975.

DoD Guide on "Improving Productivity in Defense Contracting." 1985 Edition.

DoD Instruction 4200.15 "Manufacturing Technology Program," July 14, 1972.

DoD Instruction 5000.2, "Major Systems Acquisitions Procedures," March 19, 1980.

DoD Directive 5000.1, "Major Systems Acquisitions," March 29, 1982.

DoD Directive 5000.34, "Defense Production Management," October 21, 1977.

DoD Directive 4275.5, "Acquisition and Management of Industrial Resources," October 6, 1980.

DoD Industrial Modernization Incentives Program, Number 5000.XX, 1985 Edition.

DoD Industrial Modernization Incentives Program, Number 5000.XX-G, 1985 Edition.

Eiler, R.G., Goletz, W.K. and Keegan, D.P. "Is Your Cost Accounting Up to Date?" Harvard Business Review, July-August, 1982.

Eiler, R.G., Muir, W.T. and Michaels, L.T. "Technology Management and the Automated Factory." Material Handling Engineering, Jan., Feb., March, 1984 (3 part series).

Headquarters Naval Material Command. "Industrial Modernization Incentives Program, Circulation Comment Package." Washington, D.C., 1984.

Kaplan, R.S. "Yesterday's Accounting Undermines Production." Harvard Business Review, July-August, 1984.

Lehrer, Robert N. White Collar Productivity, Chapter 9. New York: McGraw Hill Book Co., 1983.

Price Waterhouse, Department of Defense Productivity Measurement Study: CDEF Briefing Package, August 14-15, 1985.

Sink, D.S. "Productivity Action Teams: An Alternative Involvement Strategy to Quality Circles," Annual Industrial Engineering Conference Proceedings. Atlanta, Georgia: Institute of Industrial Engineers, 1981.

Sink, D.S. "Using the Nominal Group Technique Effectively." National Productivity Review, Vol. 2, No. 2, 1983.

Sink, D.S., Tuttle, T.C. and DeVries, S.J. "Productivity Measurement and Evaluation: What is Available?" National Productivity Review, Summer, 1984.

Sink, D.S., DeVries, S.J. and Tuttle, T.C. "An In-Depth Study and Review of Existing Productivity Measurement Techniques." Proceedings. World Productivity Congress, Oslo, Norway, May, 1984.

Sink, D.S., Tuttle, T.C., DeVries, S.J. and Swaim, J. Development of a Taxonomy of Productivity Measurement Theories and Techniques. Final Report AFBRMC Contract No. F33615-83-C-5071, 30 November 1983.

Sink, D.S. Productivity Management: Planning, Measurement and Evaluation, Control and Improvement. New York: John Wiley & Sons, 1985.

Sink, D.S., Dhir, S. and Roberts, C.A. "Innovative Advances in Productivity Measurement and Evaluation in the Aerospace and Defense Industry." Annual Industrial Engineering Conference Proceedings. Chicago, IL, Institute of Industrial Engineers, 1985.

The Study of Productivity Measurement and Incentive Methodology. Virginia Productivity Center/Virginia Polytechnic Institute and State University, Blacksburg, VA, March 1986.

Tuttle, T.C. Productivity Measurement Methods: Classification, Critique, and Implications for the Air Force. Contract No. AFHRL-TR-81-9, Brooks Air Force Base, TX, September, 1981.

Westinghouse Manufacturing Systems & Technology Center. DoD Productivity Measurement Study: IMIP Briefing Package, August 14-15, 1985.

Productivity Management Frontiers — I, edited by D.J. Sumanth
Elsevier Science Publishers B.V., Amsterdam, 1987 — Printed in The Netherlands

FORMULATING QUALITY MEASURES IN SERVICE INDUSTRIES: A FIELD STUDY IN AUDITING

S.G. SUTTON, J.C. LAMPE, and E.E. ADAM, JR.
College of Business & Public Administration, University of Missouri, 312 Middlebush Hall, Columbia, MO, USA, 65211

ABSTRACT
Sutton, S.G., Lampe,J.C., and Adam, E.E.,Jr., 1987. Formulating Quality Measures in Service Industries: A Field Study in Auditing.

The major public accounting firms in the United States emphasize the quality of audit services performed to clients and to personnel performing the audit. Recent Statements on Auditing Standards (SAS) 46 and 47 and Congressional hearings (Dingell Committee) reflect the continuing concern for quality audits. Even with such emphasis, a gap remains within accounting firms between desired objective measurement and documentation regarding audit quality and actual data currently available.

This field study is directed toward improving quality measurement processes and quality measures in public accounting audits. Group techniques emphasizing audit personnel participation in quality measurement were selected as the most promising methodology.

The structured group process consisted of seven group meetings with representatives from each level of the audit team (staff through partner). The members of the audit team successfully completed each stage of the group process. The audit process was broken down into four unit operations. From over 100 deviations occurring during the unit operations, 19 were selected as key deviations. Measures were developed by the group for key quality deviations. Implementation of the quality measures is under consideration.

INTRODUCTION

The major public accounting firms in the United States are emphasizing the quality of work they perform to both personnel and clients. New personnel entering a firm are trained with an emphasis on high quality work. When pursuing new clients, as well as maintaining continuing clients, a major selling point is the high performance quality of the audit engagement.

Current trends in the public accounting profession indicate an increasing desire to attain higher levels of confidence in audit decisions. Tighter confidence levels are aimed at higher levels of quality. This trend is reflected in the evolution of the professional standards and guidelines, particularly in the recently released Statements on Auditing Standards (SAS),

46 and 47 [AICPA, 1984]. SAS 47 presents new guidelines on determining materiality limits which effectively tighten these limits and raise assurance in the audited accounts, while SAS 46 presents provisions for omitted audit procedures discovered subsequent to the report date of the audit.

The public accounting profession is currently being scrutinized more closely by the public than in past years. This is evidenced in the number of recent Congressional hearings, most prominently the hearings at Rep. John Dingell's Subcommittee on Oversight and Investigations of the House Committee on Energy and Commerce which is pressuring the profession towards higher quality of service. Other Congressional hearings resulting from the to frequent savings and loan collapses as well as the volatile state of the banking industry as a whole, have brought increased public attention to some perceived poor quality audits. Additionally, the current trend in governmental auditing towards the single audit has caused the profession to reexamine the audit process as applied to governmental and not-for-profit organizations.

CURRENT PUBLIC ACCOUNTING PRACTICE

Currently most public accounting firms generate one or both of two measures for use in audit evaluation. These measures are:

(1) budgeted hours versus actual hours spent on an engagement

(2) billing fee based on actual hours @ standard rate versus actual billed fee

These measures were validated as primary measures by the auditors participating in the group discussions conducted during this study. Because of their reliance on output quantity, the two measures are limited in scope and cannot measure the overall performance of the audit. Equally important to performance and productivity is the quality of the output. An audit team may come in under the budgeted hours for the engagement and have an 'uneasy' feeling about the audit, that is, a feeling that perhaps the audit risk is still greater than desired. On the other hand, another audit team may come in over-budget and feel exceptionally good about the audit, having very high confidence that audit risk has been minimized.

In light of recent research findings quality measurement is very important. One stream of recent research has shown that when an emphasis is placed on the measurement of the quantity produced and less on the quality of the product, the quality of the product decreases. When quality measurement is introduced in these organizations, quality tends to rise quickly [Adam, Hershauer and Ruch, 1986]. Yet another view is that both quality and production can be increased simultaneously by 'doing the job right the first time'. This later view reflects recent successes in manufacturing, especially by the Japanese [Adam and Ebert, 1986] In either view, it is crucial to measure the quality of audit work in order to trace improvement efforts and to document actual performance to assure both management and clients that a desired level of quality exists.

These research findings indicate that current audit measurement systems based on time budgets leave a firm susceptible to decreasing levels of quality. If quality measurement can be instituted and emphasized similar to the importance associated with the impact of hour and billing variances, the probability of reduced quality occurring should be decreased.

SELECTION OF A MEASUREMENT METHODOLOGY

Research into measuring productivity and quality can be approached from the viewpoint of the economist, engineer, accountant, behavior scientist, or manager. Similarly, the measurement issue varies from the macro (countries or economies) to the micro (the individual at work). Our interest is managerial and at the work group level. From the literature it is clear that measuring performance quality is much more difficult than measuring quantity (traditional productivity or effeciency measures). [Adam, et. al. Ruch (1986)].

The process selected here focuses on involvement of the audit team in developing productivity and quality measures. The method is detailed by Adam, Hershauer, Ruch [1986] and is based on work sponsored by the National Science Foundation in the late 1970's.

Participative Nominal Group Approach to Quality Measurement

Studies conducted in a wide array of service industries have strongly supported the hypothesis that there exist a large number of deviations in any given process. A deviation is any variable that has an affect on the ability to effectively and efficiently complete the process.

For the quality measurement technique selected nominal group processes are used. Nominal group techniques are designed to get the personnel who perform the tasks involved in both the determination of deviations that occur in the process and how to measure these deviations. The personnel who are directly involved in the process on a day to day basis may often be able to determine key deviations in the process that are not readily apparent to management. The procedure followed is presented in Tables 1 and 2.

Table 1

The Quality Measurement Procedure Summarized

Step	Description	Meeting/Time
1	Definition of Systems Boundaries	1A/2 hrs. *
2	Determination of Unit Operations	1B/4 hrs. **
3	Generation of Process Deviations	2B/3 hrs.
4	Determination of Key Process Deviations	---
5	Generation of Measures	3A,B/3 hrs.
6	Evaluation of Technology & System Input	4B/2 hrs.
7	Edit Measures Into Productivity Ratios	---
8	Clarify Edited Measures	5A/3 hrs.
9	Development of Questionnaire for Measures	---
10	Ranking of Measures Through Questionnaire	6A/2 hrs.
11	Completion of Productivity Ratios	---

* Group A (3)	** Group B (4)
Partner	Audit Senior (2)
Senior Manager	Audit Staff (2)
Manager	

Table 2

Defining the Steps in the Quality Measurement Procedure

1) The definition of system boundaries deals with the horizontal (where in the audit process to start and end generation of measurements) and vertical (how high in the audit supervision structure the project will generate measures) limitations of the project.

2) Determination of unit operations identifies the phases in the audit process. A phase is a process where a state change occurs in a throughout.

3) The generation of deviations begins with individual participants listing their ideas, followed by a group "brainstorming" session. Deviations are clarified and voted on to determine key deviations.

4) Determination of key deviations requires the coordinators to tabulate and review voting results on key deviations to generate a final list.

5) The group generates measures that will reflect the outcome and costs of the key deviations. Measures are proposed and clarified as in step 3.

6) The technology and sytem input analysis is designed to evaluate the environment of unit operations influencing measure and quantity of deviations. The labor intensity of the auditing process suggests inclusion of present and future audit staffing.

7) The coordinators classify measurements by deviation being measured, omitting measures not meeting project requirements. Measurements are then organized into productivity ratios and indicators.

8) The edited set of quality mesures are presented to Group A for additional clarification and feasibility evaluation of specific measure's.

9) Coordinators develope questionnaire for the set of edited measures.

10) Group A completes questionnaires, evaluating the measure's strengths in measuring the value outcome of the deviations.

11) Questionnaire results are evaluated and key measures are used to generate a preferred set of productivity ratios and indicators.

Two groups of participants, both from the office of an international public accounting firm, were used in the quality measurement procedure. The first group consisted of two seniors and two staff auditors. The second group consisted of a partner, a senior manager and a manager.

The senior-staff group (group B) was responsible for:

(1) determining how the audit process could be broken down into its component parts,

(2) determining what internal and external factors influence the audit process,

(3) determining what deviations exist in each component part of the audit process, and

(4) finding methods to measure the deviations generated.

The partner-manager group (group A) was responsible for determining the boundaries for the project, limiting the parts of the audit that measures should be generated for. Additionally the partner-manager group supplied an administrative view for the generation of measures and then evaluated the total set of measures.

The following sections of the report present an overview of the group activities completed during the process. The overview is presented in the chronological order of the steps completed and results obtained from the group meetings.

Defining System Boundaries The partner-manager group met for the initial meeting to determine the system boundaries for the project. Two sets of boundaries are required to sufficiently limit the scope of the project, a horizontal set and a vertical set of boundaries.

The horizontal set of boundaries determine where in the process to begin and end generation of measures. The goal was to specify the segment of the process where deviations affecting quality are likely to occur. The group decided that the generation of measures, within the scope of this project, should begin after the engagement has been secured and prior to the beginning of the planning for the project. The ending horizontal boundary was set at the date of issuance of the reports. This was later clarified in a subsequent meeting to include:

(1) completion of reports,

(2) billing of the client, and

(3) presentation of reports and billing to the audit committee or other appropriate person(s).

The vertical boundaries determine what personnel in the process are to be included in the measurement system. It was determined that the first level directly involved in the audit process was the staff auditors. Group discussion then led to the upper boundary being set at the partner level, a result of the group's decision that quality is affected throughout the ranks of the audit personnel and measurement should include all of these personnel.

Determination of Unit Operations. A unit operation is defined as any phase in the process in which there is a definable input entering the system, a state change occurs during the throughput of the process, and a definable output exits the process. A unit operation denotes a major change opposed to tasks which are completed but do not really change the overall throughput themselves. Tasks might include for example analytical review and confirmations. While these tasks are essential in most audits and provide specific evidence, neither task in and of itself results in a significant change in the overall audit throughput.

The senior-staff group determined that the system as defined by the given boundaries, could be broken into four unit operations: (1) engagement planning, (2) interim fieldwork, (3) year-end fieldwork, and (4) final administration (see Table 3).

Generating Process Deviations. A process deviation is any variable that has an affect on the ability to meet the objectives of the process. A deviation in the audit process would, therefore, be any variable affecting the auditor's ablitity to complete the audit effectively and efficiently.

The objective of this meeting was to have the senior-staff group determine any deviations (using a brainstorming type technique) they could think of within the given unit operations. From the total list of deviations generated, each group member selected and ranked the deviations he/she felt were key to the given unit operations. Deviations are presented in Table 4 for one of the four unit operations. The rankings by the group members were accumulated to generate key deviations for each unit operation based on the average ranking for the deviation. The key deviations will be presented in the summary section which follows.

Table 3
Unit Operations for Auditing Function

Unit Operation	Tasks	Output
Engagement Planning	1. Establish Audit Objectives 2. Logistical Arrangements 3. Risk Analysis 4. Time Budgeting 5. Draft Audit Program 6. Draft Planning Memo 7. Partner/Manager Review	Planning Memo Tentative Audit Program
Interim Fieldwork	1. Coordination with Client 2. Initial Evaluation of Internal Control 3. Analytical Review 4. Evaluation of Internal Control 5. Detailed Testing of Transactions 6. Review of Interim Work 7. Prepare Report Draft and Management Letter	Revised/Approved Year-End Procedures Reliance on Internal Control Results of Dual Test
Year-End Fieldwork	1. Coordination with Client 2. Detailed Testing of Balances 3. Subsequent Events Checklist Disclosures 4. Drafting of Statements and Disclosures 5. Analytical Review 6. Drafting of Auditor Reports 7. Senior Memo 8. Workpapers Review 9. Review for Defensibility 10. Risk Analysis 11. Issuance of Reports	Senior Memo Completed Workpapers Drafted Statements and Reports
Final Administration	1. Time Budget Analysis 2. Workpaper Filings 3. Staff Evaluations 4. Exit Conference 5. Delivery of Reports 6. Post Audit Committee Meeting	Staff Evaluations Audit Fee Billings Completed Engagement

Audit Quality Measures. A joint session with the partner-manager and the senior-staff groups was held for the purpose of generating measures for each key deviation. The senior-staff group convened again in the afternoon to help clarify the measures and discuss what types of technological tools were

Table 4
Deviations For Engagement Planning Unit Operation

1. Client availability & willingness
2. Timing of planning
3. Client understanding of audit
4. Personnel involved in planning
5. Degree of reliance on prior audit experience
6. Audit teams experience with client
7. Special requirements of engagement
8. Changes in client operations, systems and/or entity
9. Risk Assessment
10. Changes in standards and regulations
11. Client's integrity
12. Degree of conflicting job assignments
13. Clarity of writing
14. Completion of planning memo before fieldwork
15. Review of planning memo before fieldwork
16. Bundle preparation and completeness
17. Competency of client staff / system
18. First time audit
19. Staff preparedness

---

provided the auditor that may influence the outcome of the measures. The second day of meetings was with the partner-manager group with the primary objective being to take the raw measures generated in the first measurement session one and edit them into measures that were administratively usable. Table 5 presents the measures generated for one key deviation.

Ranking Measures. In the final group session, each member of the partner-manager group filled out a questionnaire designed to obtain a measure of the usefulness of each measure. The questionnaire asked each respondent to give two ratings for each quality measure listed. The first rating for each measure rated the strength of the potential measure in terms of its relevancy and meaningfulness to the given deviation. This response was made on a four point scale anchored as follows:

0 - not applicable to the given unit operation
2 - useful quality measure
1 - limited usefulness as a quality measure
3 - very strong quality measure

The second rating scale captured how much confidence each respondent had in the implementation of the potential measure. Confidence was represented as a percent ranging from 0% to 100%. Each measure was then assigned a value by

Table 5
Measures for Key Deviation 'Changes in Client'

1	# of systems changed / total number of systems
2	# of new people audit team deals with / total people deal with
3	# of new locations / total number of locations
4	# of new subsidiaries and-or divisions / total # of divisions
5	# of new products (services) / total products (services) offered
6	% change in gross margin percentage
7	% change in revenues
8	% change in net income
9	$ change in net income
10	use a likert type scale to rate the impact of changes on audit

multiplying the two rating factors together. For example, a measure for the key deviation of risk assessment was the 'percent of hours change in scope.' A participant may have decided that the given measure would be an useful quality measure for risk assessment if objective data for the measure could be collected, and rated it a 2. If this participant also felt fairly confident that this measure can be efficiently and objectively used in an audit engagement, then he/she may have given the measure a confidence rating of 90%. The value assigned the measure is the product of the two ratings (2 * .90) or 1.80.

SUMMARY AND DISCUSSION

The tabular presentation of unit operations, key deviations, and quantitatively rated measures of the key deviations summarizes the results of the study (see Table 6). In Table 6 we note there were four unit operations nineteen key quality deviations, and thirty-eight measures (the top two measures for each deviation).

The remaining step in this study is to obtain a weighting of the relative importance of each key deviation. For example, the deviation 'change in client' within the planning operation of an engagement may be relatively more or less important than the 'client competence' deviation within the fieldwork operations. If 'change in client' is significantly less important to total audit quality than 'client competence', then it should be given significantly less weight when all of the measures are combined into a single quality index.

Table 6

Highly Rated Measures For Key Deviations
(Unit Operation / Key Deviation / Quality Measure)

	Score
Engagement Planning:	
Changes in Client:	
% change in net income	2.57
# of new subsidiaries and-or divisions / total number of divisions	2.53
Risk Assessment:	
% change in hours resulting from risk assessment scope changes	2.25
did compliance test meet expected results?	2.00
Timing of Planning:	
hours of planning / hours of fieldwork	1.77
% of plan completed prior to first day of fieldwork	1.68
Fieldwork:	
Completion of Interim Fieldwork Before Year-End Fieldwork:	
% of interim work completed prior to beginning of year-end work	2.00
% of seniors and managers review of interim completed before year-end	2.00
Understanding of Client Systems:	
use a Likert scale for degree of understanding of client systems	1.53
actual hours of review / budget hours for individual systems review	1.30
Client Competence:	
number of college graduates / total number of accounting staff	2.00
use a Likert type scale for degree of understanding of client systems	1.95
Programs / Planning Memos:	
did manager approve prior to beginning of interim work?	1.98
number of post-planning changes	1.95
Audit Team Expertise:	
number of months on client / number of audit team members	2.32
number of audit team changes / total personnel during audit engagement	2.30
Client Rapport:	
use a Likert type scale to rate overall rapport with client	2.30
dollar amount of special work billed / total dollar amount billed	2.15
Thoroughness of Review Process:	
dollar value of adjustments discovered / net income	2.42
% of work reviewed by manager and senior prior to opinion date	2.40
Client Preparedness:	
number of schedules completed / number of schedules requested	2.52
dollar value of post-closing adjustments / net income	2.18
Completion of Working Papers:	
number of memos documenting issues / number of key issues	2.20
working papers approved by manager and senior prior to opinion date	2.08
Availability of Resources:	
% of tax hours performed by staff with prior experience with client	2.00
% of MCS hours performed by staff with prior experience with client	1.85
Unusual Items:	
use a Likert scale to rate effect of unusual items on audit quality	2.00
days from resolution of unusual items to sign-off of engagement	1.97
Timing Constraints:	
days between closing (or year-end) and opinion date	2.70
number of hours after opinion date / total actual hours	2.08
Final Administration:	
Client Reaction:	
dollar amount of special work billed / total dollar amount billed	2.00
use a Likert type scale for favorability of client's reaction to audit	1.67
Time Budget Analysis:	
are there explanations for budget overruns?	1.83
number of hours of unexplained variations / total hours of variation	1.73
Billing of Fees:	
service value adjustment / budgeted service value adjustment	2.00
service value adjustment / total anticipated billing	1.95
Timing and Quality of Staff Evaluations:	
number of days since off-budget to receiving evaluation	2.10
number of suggestions for improvement	1.98

This phase of the study is currently on-going.

The study progressed well, based upon group generated results. The objective of the group meetings was to generate a set of measures that can be used to evaluate audit quality. This objective was achieved. Quality measures that were scored in the range between useful and very strong (2.00 - 3.00) were found for over 75% of the key deviations. Two-thirds of these deviations had multiple measures in the 2.00 - 3.00 range.

A significant by-product of the study was <u>the awareness of the influences on quality that the participants acquired</u>. Several of the participants mentioned that they had benefitted from the group meetings by going through the structured process that required thought on factors influencing audit quality.

From a researchers point of view, this study was a necessary and important first step in developing additional measures for quality. The study is limited in scope, however. The primary limitations are the group size and the use of a single public accounting firm. The senior-staff group consisted of only four people, two seniors and two staff (the staff having less than one year experience). The partner-manager group consisted of three people, a manager, a senior manager and a partner. Even with the total willingness and enthusiasm of each member of the two groups, the size factor limits conclusions about the representativeness of the results. Replication at other sites and in other firms would be desirable as well.

A correlation was found between the key deviations selected and the type and amount of audit experience of the participant. With only certain levels and audit areas represented, the key deviations generated may not be representative of the key deviations that would be selected by a larger more diversified group.

Similarly, large variation was found in the questionnaire responses on the alternative quality measures. On some measures, each of the three respondents provided a different strength rating and/or a very different level of confidence. With only three participants, in some cases one rating, significantly different from all others, strongly influenced the average score for that measure.

In order to reduce the validity problems associated with the small group size used to determine key deviations, a questionnaire will be administered to a large group of seniors. The questionnaire will be designed to reduce two limitations in the current study; the representativeness of the current key deviations, and the possible omission of some deviations in the audit process. Similarly, a second questionnaire related to the quality measures generated, will be administered to a large group of managers and partners. The respondents will rate the existing, and any potential additional measures, to generate a more representative set of measures.

After replicating at several sites, a final set of quality measures and a dependent quality index should be produced. The measures must then be edited into an instrument that can be used in the audit environment to collect the data required for the measures. The data collection instument is the final component needed to produce a practical and applicable audit quality measurement system.

REFERENCES

Adam, Everett E. Jr. and Ronald J. Ebert. Production and Operations Management: Concepts, Models, and Behavior, 3rd Ed., Prentice-Hall, 1986.

Adam, Everett E. Jr., James C. Hershauer, and William A. Ruch. Productivity and Quality: Measurement As A Basis For Improvement, 2nd Ed., (University of Missouri-Columbia, Research Center, College of Business and Public Administration, 1986.

AICPA. Codification of Statements on Auditing Standards. American Institute of Certified Public Accountants, New York, 1984.

Jiambalvo, J. "Performance Evaluation and Directed Job Effort: Model Development and Analysis In a CPA Firm Setting." Journal of Accounting Research, Autumn, 1979 (pp. 436-55).

Productivity Management Frontiers — I, edited by D.J. Sumanth
Elsevier Science Publishers B.V., Amsterdam, 1987 — Printed in The Netherlands

COMPANY LEVEL INVESTIGATION OF PERFORMANCE MEASURES AS RELATED TO TOTAL PRODUCTIVITY

F. P. YAVUZ Suave Shoe Corporation, P. O. Box 24-9104, Coral Gables, FL 33124
and D. J. SUMANTH Department of Industrial Engineering, University of Miami, Coral Gables, FL 33124

ABSTRACT
This paper attempts to develop a methodology to reduce the universe of company level performance measures to a more meaningful set through the use of statistical techniques and investigates the effect on total productivity and net income

1 INTRODUCTION

In managerial literature, many researchers have attempted to define various measures of performance which reflect the success of companies with respect to different company objectives and goals. The operation of any organization is successful only if the management can control the entire organization as an integrated unit. As the functions, and consequently the objectives of any company are interdependent and most of the time conflicting, an optimal tradeoff is required in obtaining an at-least satisfactory level of performance for each function and objective.

Through a literature survey, it can be seen that a myriad of success indicators have been defined and are currently being used by companies at different levels of management. With the advance and acceptance of operations research techniques on a wide scale, managers have been exploring what should be optimized and whether the parameters that are optimized are true indicators of success. Hence, the basic question that will be of consideration in this article will be whether all of the commonly used performance indicators are really necessary in measuring the success of a company. If there are redundant measures, can we come up with an optimal set of measures or even a single measure that will reflect organizational performance and aid managers in the process of planning and controlling organizational performance?

A review of literature suggests that companies in general concentrate on various managerial objectives such as market standing, innovation, productivity, profitability, financial and physical resource utilization, manager/worker performance, and growth. For each of these criteria, different measures have been used to evaluate the degree of success with the ultimate objective of increasing organizational productivity. Productivity improvement has become one of the major issues in the recent years; however, the word "productivity" has been used in an haphazard manner, without considering possible theoretical relationships with either profitability, or, other measures of performance.

2 OBJECTIVES

This article will mainly deal with various measures of performance considered critical in achieving a certain combinations of goals and objectives in organizations. The specific goals of the paper are:

1. To summarize some of the most commonly used performance measures as discussed in literature,
2. To develop a conceptual framework which will distinguish the most relevant measures with respect to their predictive nature toward company goals,
3. To develop a company objective evaluation methodology which should aid managers at the company level to evaluate company objectives in a formalized manner,
4. To demonstrate the feasibility of this methodology through an empirical study of some industrial organizations in the Automotive and Automotive Parts Industry.

3 PERFORMANCE AND GOALS IN ORGANIZATIONAL UNITS

Several prominent authors in managerial literature have concentrated on defining corporate goals and success; for example, Drucker (1958), Humble (1966), Christopher (1977), Forbes (1963), Ericson (1966), and Heath (1978). Their consensus on the subject matter indicates that company goals can be summarized under the following categories:

- Profitability
- Productivity
- Market Standing
- Financial and Physical Resource Utilization
- Innovation
- Growth
- Manager/Worker Performance
- Public Responsibility

Yavuz (1982) has compiled an extensive list of performance measures relevant to the above categories (as shown in Figures 1 to 6). These measures are not meant to be a complete list, but, they include a major subset of measures that are relevant to each major organizational objective.

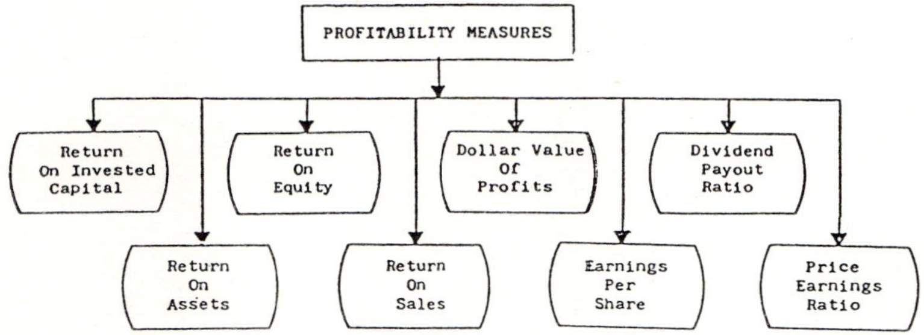

Figure 1 : Profitability Measures At the Company Level

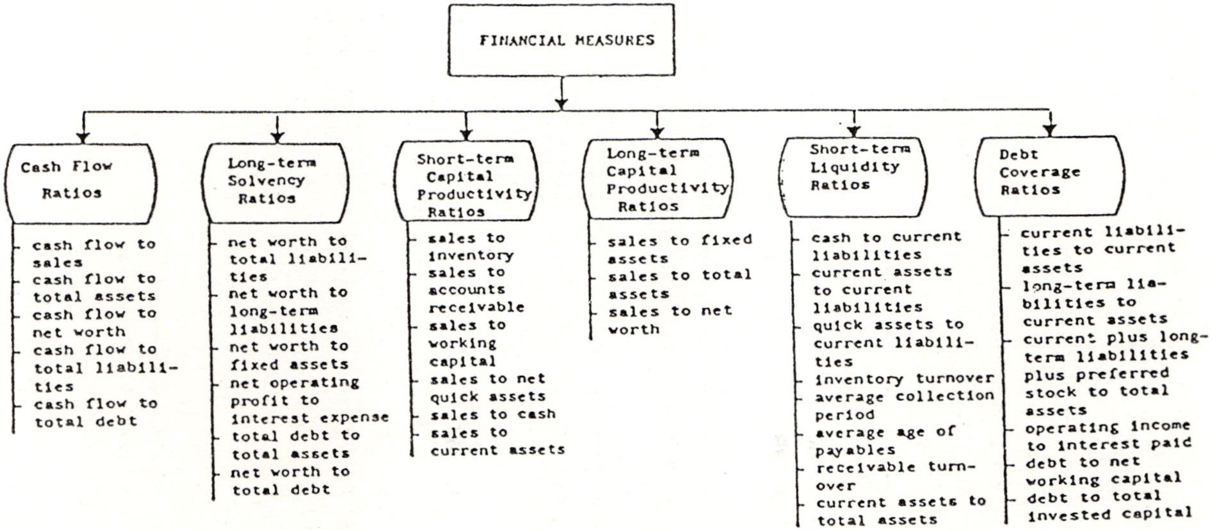

Figure 2 : Financial Measures at the Company Level

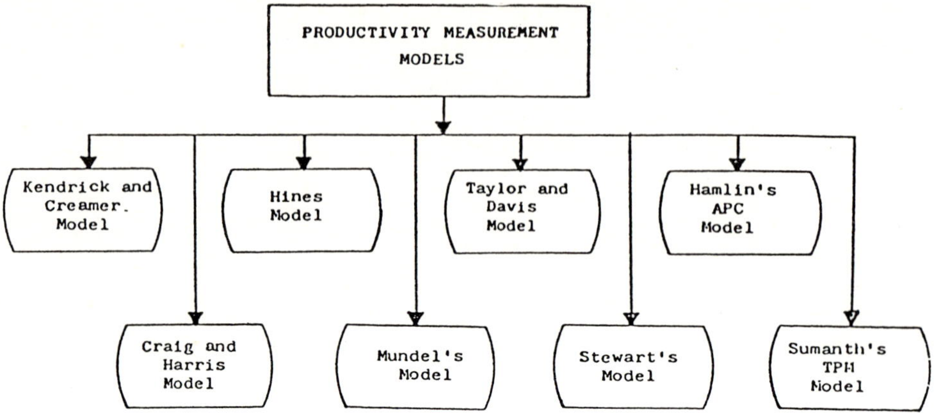

Figure 3 : Productivity Measurement Models at the Company Level

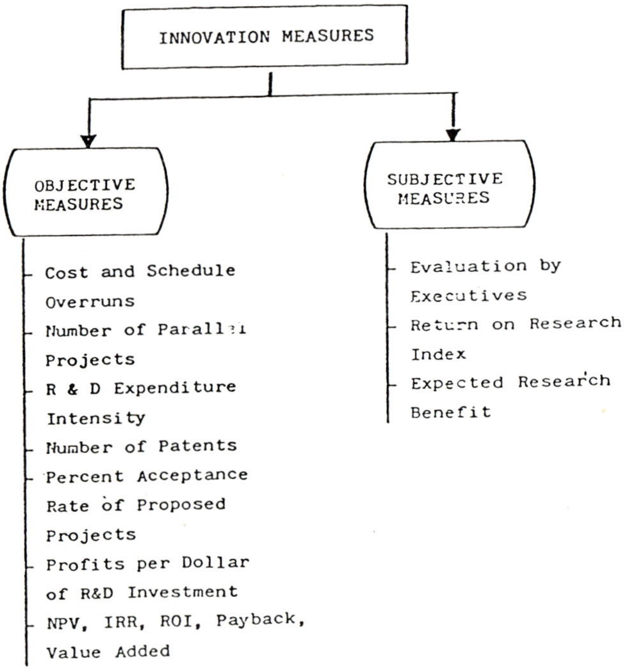

Figure 4 : Innovation Measures at the Company Level

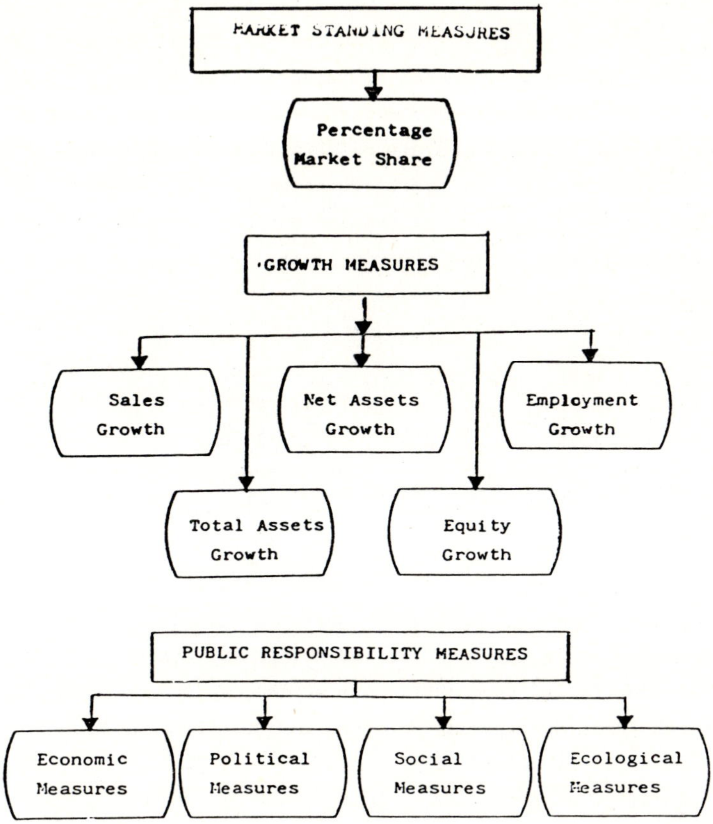

Figure 5 : Market Standing, Growth, and Public Responsibility Measures at the Company Level

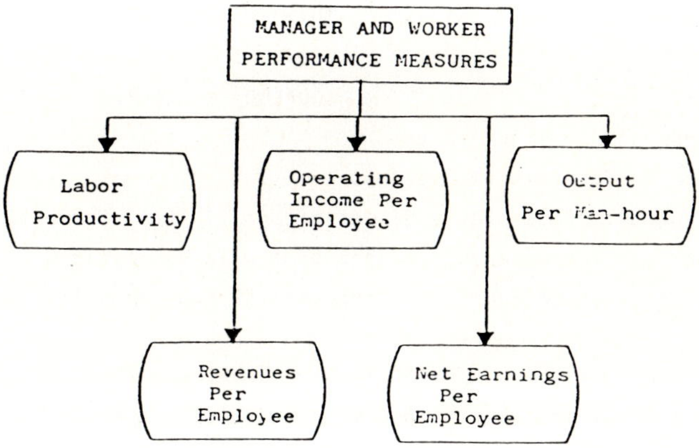

Figure 6 : Manager and Worker Performance Measures at the Company Level

4 CONCEPTUAL FRAMEWORK AND ORGANIZATIONAL OBJECTIVE EVALUATION

It is apparent that most of the performance indicators suggested at the company level can be derived from financial statements while some specific measures need to be obtained from specific reports and records which are not uniform in every company. Most companies are reluctant to display more information than required by law; thus, it becomes impossible to combine all the available measures for the purposes of drawing conclusions about companies in a specific industry. Therefore, this paper will try to reduce the set of measures to one that can be compiled based on the information in Moody's Industrial Manuals.

4.1 Notation:

TP	:	Total Productivity	S	:	Total revenues
CGS	:	Cost of Goods Sold	INT	:	Interest Paid
OP	:	Profit Before Taxes	NI	:	Net Income
CD	:	Cash Dividends	AD	:	Annual Depreciation
C	:	Cash & Equivalents	AR	:	Accounts Receivable
INV	:	Inventory	CA	:	Current Assets
TA	:	Total Assets	CL	:	Current Liabilities
NS	:	Net Shareholders Equity	EPS	:	Earnings Per Share

LTC : Long term debt/(long term debt + net worth)

EQ : Net Shareholders Equity Plus Retained Earnings(Net Worth)

LTD : Long Term Debt (LTD=TA-EQ-CL)

QA : Quick Assets (QA=C+AR)

TD : Total Debt (TD=CL+LTD)

FA : Fixed Assets (FA=TA-CA)

WC : Working Capital (WC=CA-CL)

w : Weighted Average Cost of Capital $\left(w = \frac{INT*LTC}{100*LTD} + \frac{CD*(100-LTC)}{100*NS}\right)$

IT : Income Taxes

SE : Selling and Administrative Expenses

TP : S/(S-NI-w*CA+AD) (Sumanth(1979) definition of Total Productivity)

Based on this notation, the performance measures that will be utilized in the conceptual framework are given in Figure 7. There are 42 such measures defined as R1.......R42.

Altman(1968), Deakin(1972), and Elam(1975) have developed methodologies that relate traditional ratio analysis to the prediction of corporate bankruptcy using Multiple Discriminant Analysis (MDA). MDA is a statistical technique used to classify an observation into one of several predefined groupings dependent upon the observation's individual characteristics. It is

Figure 7 : Performance Measures that can be Computed Based On the Information in Moody's Industrial Manuals

Measure Number	Definition	Measure Number	Definition
R1	CA/CL	R22	LTD/TA
R2	(CA-INV)/CL	R23	NI/S
R3	C/CL	R24	CA/S
R4	QA/CL	R25	QA/S
R5	C/CA	R26	C/S
R6	INV/CA	R27	FA/S
R7	S/(CA-INV)	R28	AR/S
R8	CA/EQ	R29	CGS/S
R9	NI/EQ	R30	WC/S
R10	TD/EQ	R31	(S-CGS)/S
R11	C/EQ	R32	(S-CGS)/INT
R12	S/EQ	R33	C/TD
R13	FA/EQ	R34	CL/TL
R14	S/INV	R35	EQ/TL
R15	CGS/AVG.INV	R36	EQ/LTD
R16	S/TA	R37	TD/WC
R17	NI/TA	R38	S/(CA-CL)
R18	TD/TA	R39	EPS
R19	OP/TA	R40	OP/NI
R20	C/TA	R41	TP
R21	CL/TA	R42	NI

primarily used to classify and/or to make predictions in problems where the dependent variable appears in qualitative from i.e., increase or decrease, bankrupt or nonbankrupt. After the dependent variable is classified into groups, MDA attempts to derive a linear combination of characteristics which "best" discriminates between these groups.

The MDA techniques will be used in this conceptual framework, to determine the most relevant set of measures that define a company objective.

4.2 Company Objective Evaluation Methodology

Given the set of "measurable" performance indicators and the statistical techniques that are available, a methodology is presented to evaluate the expected trends in various company objectives. The proposed methodology should aid managers at the company level to approach the planning process in a formalized manner, based on the general industry that a particular company belongs to. Following are the steps for this methodology:

Step 1 : Choose a company objective to be evaluated i.e., the company objective can be depicted as profitability, productivity, growth in sales, etc.

Step 2 : Gather data from financial records to calculate the values for the set of "measurable" performance indicators. Denote these indicators as R1.....Rn. Data from other companies in the same industry is relevant since this will aid in comparing company performance with industry averages.

Step 3 : If there is an increase in the objective with respect to the previous time period, create a dummy variable Y whose value equals "1". Otherwise, Y will take the value of "0".

Step 4 : Define the model as:
Y=f(R1,R2,.......Rn)
where f is a linear function.

Step 5 : Use the Multiple Discriminant Analysis (MDA) to evaluate the trends related to the objective under study. Find the coefficients of respective discriminating functions for current year's ratios, one-year-lagged ratios, up to n-year-lagged ratios.

Step 6 : Among the models developed in Step 5, choose the best discriminating model based on statistical evaluation of groups correctly classified.

Step 7 : Evaluate the objective for the coming time period to determine whether an increase or decrease is expected.

Step 8 : If an increase is expected, go to Step 11.

Step 9 : In the case of an expected decrease, identify problem areas in the management of criteria contained in the set of significant ratios determined by MDA. Rank the importance of ratios with respect to standard coefficients.

Step 10: Install improvement programs for more efficient management of the problem areas.

Step 11: Manage to maximize the increase of the accounts contained in the most significant set of ratios.

Step 12: Use the set of ratios determined by MDA in a Regression Model to determine the values of objectives for future time periods and forecasting.

This methodology is summarized as a flowchart in Figure 8.

5 EMPIRICAL STUDY

In order to illustrate the conceptual framework and the methodology discussed so far, an example will be given based on 16 Fortune 500 companies in the Automotive and Automotive Parts Industry between 1974 and 1980. The data was compiled from Moody's Industrial Manuals, and the 42 ratios (listed in Figure 7) were evaluated for each time period. These ratios constitute the

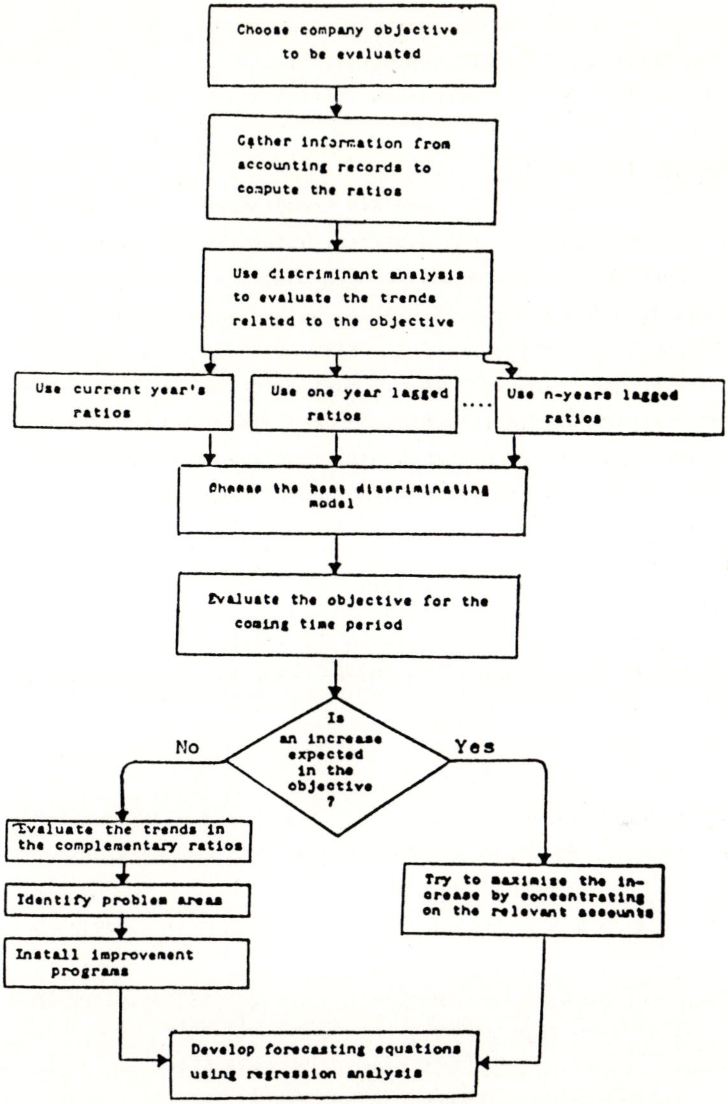

Figure 8 : Company Objective Evaluation Methodology

independent variables for the MDA analysis. The 16 companies with 7 years of data provided 112 data points for the Discriminant Analysis. Altman (1968) used 66 points, while Elam (1975) used 96 points for the same type of analysis; therefore, the sample size was considered to be large enough for this study.

Two performance objectives were selected for the purposes of this study: Productivity and Profitability. For productivity, the Total Productivity (as defined by Sumanth (1979, 1984)) was used, and Net Income was used as an indicator of profitability.

5.1 Total Productivity Analysis

The dependent variable in this case was the decrease or increase in Total Productivity. The results of the MDA analysis, using the Total Productivity measure, are shown in Figure 9. It can be seen that the one-year-lagged ratios provided the best discriminating model. Also, we see that the most important ratio in predicting total productivity is C/TA. It is interesting to note, that Deakin (1972) finds this ratio as the most important one among the one-year-lagged ratios, in the prediction of bankruptcy as well. An overview of the significant measures in the best model indicates that cash management with respect to current liabilities, current assets, and total debt are

Figure 9 : Total Productivity Trend with Respect to One Year Lagged Ratios, MDA Results

Ratio Code	Ratio	Standardized Coefficient	Rank	Unstandardized Coefficient
R3	C/CL	.35230	12	1.52886
R5	C/CA	-3.80009	2	-49.67881
R7	S/(CA-INV)	.54161	10	.20675
R15	CGS/AVG.INV	.53513	11	$.1 \times 10^{-5}$
R17	NI/TA	.68716	9	14.94940
R19	OP/TA	-.92062	5	1.70787
R20	C/TA	5.00681	1	102.16860
R27	FA/S	.90602	6	11.55284
R33	C/TD	-1.86684	3	-18.57843
R39	EPS	-.88569	7	-.24217
R41	TP	-1.00904	4	-6.82079
R42	NI	.81791	8	$.11 \times 10^{-5}$
	CONSTANT			3.52615

Percentage of Cases Correctly Classified = 83.33 %

Type-I Error = 7 %

Type-II Error = 9 %

Significance = .0000

important elements in obtaining a productivity increase for the upcoming time period.

5.2 Net Income Analysis

The dependent variable, in this case, is defined as the increase or decrease in net income. Based on the MDA results, the best model came out to be of two-year-lagged ratios. Again, C/TA was found to be the most significant ratio. The magnitude of C/TA, therefore, is a crucial parameter that should be controlled by the managers in the automotive and automotive parts industry, to optimize the overall company objectives (Figure 10).

Figure 10 : Net Income Trend with Respect to Two Years Lagged Ratios, MDA Results

Ratio Code	Ratio	Standardized Coefficient	Rank	Unstandardised Coefficient
R3	C/CL	2.68643	3	11.58491
R4	QA/CL	-1.74983	5	-3.27842
R14	S/INV	.35790	10	$.98 \times 10^{-6}$
R16	S/TA	-1.44396	6	-3.03451
R19	OP/TA	.45959	9	7.57856
R20	C/TA	4.03414	1	84.45844
R26	C/S	-2.43684	4	-87.61971
R27	FA/S	-1.28688	7	-17.24366
R33	C/TD	-3.27757	2	-31.74497
R42	NI	.50900	8	$.77 \times 10^{-6}$
	CONSTANT			10.70458

Percentage of Cases Correctly Classified = 85 %

Type-I Error = 6.3 %

Type-II Error = 8.8 %

Significance = .0000

6 CONCLUSION

We have tried, in this article, to develop an objective "Company Evaluation Methodology" that would provide a significant set of measures in depicting trends in different company objectives. The methodology presented here is aimed at reducing the available universe of performance measures to a smaller, and a more manageable, set, through the use of statistical techniques. This methodology helps managers to select a subset of measures that are most revealing and relevant to a specific company objective. Although, most of the objectives at the company level are interdependent, reduction of the set eliminates this dependency by providing distinct parameters through statistical methods.

The reduction of the set of performance indicators gives managers a clearer picture of the parameters that need to be controlled for the overall effectiveness of their organization. For example, the C/TA, having been found to be a significant measure for the companies studied here, can be used for budgeting purposes while planning the levels of cash and total assets in a company.

Data Collection for this type of analysis is not cumbersome as the information is obtained directly from the financial statements of companies. Therefore, companies in a particular industry can compare their individual performance with that of their industry. The methodology presented in this paper can be applied to other industries.

As in any area of research, there exists an unlimited scope for further research in this area of performance measurement and control in companies. We consider this research as only one of the first steps to stimulate further interest in formalized methodologies of this type.

7 REFERENCES

Altman, E.I., 1968. Financial ratios, discriminant analysis and the prediction of corporate bankruptcy. The Journal of Finance, Vol:13 (4), pp. 589-609.

Deakin, F.B., 1972. A Discriminant Analysis of Predictors of Business Failure. Journal of Accounting Research, Spring pp. 167-179.

Christopher, W.F., 1977. Achievement reporting-controlling performance against objectives. Long Range Planning, Vol:10, October, pp 14-24.

Drucker, P.F., 1958. The objectives of a business. Long Range Planning for Management.

Elam, R., 1975. The Effect of Lease Data on the Predictive Ability of Financial Ratios. The Accounting Review, January, pp. 25-43.

Ericson, F.E., 1969. Impact of cybernetic information on technology. Management Science, Vol:16, October, pp. B-48.

Forbes, 1963. Management performance: 1962, Forbes, January.

Heath, L.C., 1978. Financial reporting and the evaluation of solvency, AICPA, New York.

Humble, T.N., 1966. Standards in strategic planning and control: A conceptual case study, Bureau of Business Research, The University of Texas, Austin.

Standard and Poor's Industry Surveys, 1982. Vol:1.

Sumanth, D.J., 1979. Productivity Measurement and Evaluation Models for Manufacturing Companies, Ph.D. Dissertation, IIT, Chicago (Published by UMI, Michigan).

Sumanth, D.J., 1984. Productivity engineering and management. New York:McGraw-Hill.

Yavuz, F.P., 1982. Company-Level Investigation of Performance Measures as Related to Total Productivity, Master's Thesis, University of Miami, Dept. of Industrial Engineering.

Productivity Management Frontiers — I, edited by D.J. Sumanth
Elsevier Science Publishers B.V., Amsterdam, 1987 -- Printed in The Netherlands

DETERMINING THE PRODUCTIVITY OF OFFICE AUTOMATION AT FLORIDA POWER & LIGHT COMPANY

Richard P. Peregoy and Byron R. Humphries, Management Services Department, Florida Power & Light Company, P. O. Box 029100, Miami, Florida 33102

ABSTRACT
(A paper describing the conduct and results of an office automation pilot project at the Florida Power & Light Company)

In July 1983, the Management Services Department of the Florida Power & Light Company began a one-year pilot project aimed at determining the productivity effect of the emerging office automation technologies on the knowledge worker.

The concept of the pilot was state-of-the-art automation of one professional work group within the Company. This was accomplished through the installation of 51 IBM PC's and by providing access to central dictation, word processing and high speed photocopying to approximately 108 executives, managers, professionals and technicians. A smaller group was also provided access to a voice mail service.

The Management Services & Control Department was selected as the study group due to the diversity of its functions. Its professional staff is engaged in a broad range of manpower control, budgeting, internal auditing and internal consulting activities.

Productivity effects were evaluated for each technology through a combination of quarter-hourly activity profile time ladders and daily work activity logs. Results of the one-year study clearly indicated that personal computers running package software provided gains approximating 11.9% in professional productivity in the pilot group where approximately 26% of the work activity could be enhanced by automation. Based upon these results, the Florida Power and Light Company has proceeded with personal computer and other automation technology installations in areas with similar potential for productivity benefits.

1 PREFACE

Office Automation: one author has defined this phenomena as "the transfer, in whole or in part, of activities, such as creation, manipulation, storage retrieval, distribution, and control of data/information from knowledge workers, clerks, and secretaries to technology."

Much has been written about the loss of the United States' competitive position in the basic industries due to failure to stay up to date with the technology of those industries. Economists and corporate officials alike generally explain this failure in terms of a declining return on investment in the basic industries as the World Economy became more competitive. The ability to transfer information handling tasks to technology comes at a time when the U.S. economy is said to be shifting from an industrial base to a service and information economy. Futurists in the popular press hail office technology as the salvation of the U.S. economy. In the long run, however, the veracity of the futurists' glowing projections will be tested by the measure of return on investment; and with two thirds of the national payroll working in offices, the issues of office automation and knowledge worker productivity have a critical connotation to the economic future.

If the issues of knowledge workers office productivity are critical to the macro economy, it is apparent that these same issues are even more critical to the micro economy of the individual firm. It is the individual firm that must decide to make the substitution of capital for knowledge worker labor and it is certain that this decision will be predicated, in the long run, on competitive position and the effect on profitability. Office automation is likely to proceed only if it produces a satisfactory return on investment.

Measurement of knowledge productivity has always been a difficult and controversial subject due to the diverse nature of the work determinants involved. Nonetheless, some estimate of the impact of office automation technology upon productivity is essential if this new technology is to successfully compete for scarce capital resources. It is toward this goal of experimentation and measurement of productivity effects that Florida Power and Light undertook an "Office Automation Pilot Project" in July of 1983.

2 INTRODUCTION

2.1 Objective

The objectives of the Florida Power & Light (FPL) Office Automation Pilot were threefold: (1) to explore and measure the productivity benefits, (2) to develop a model structure for future FPL office automation, and (3) to develop the necessary interdepartmental interfaces to implement and support office automation.

2.2 Scope

The technical scope of the project encompassed personal computers, voice mail and central dictation/word processing. The physical scope involved the Management Services, Internal Auditing and Budgeting departments. Key personnel, outside of these departments, also participated in the voice mail portion of the pilot.

2.3 Background

In defining the scope of the office automation pilot for FPL seven functional objectives were identified as subject to the available automation technology:

1. Provide non-simultaneous communications.
2. Automate message and document preparation.
3. Automate filing/retrieval of documents.
4. Provide better access to information.
5. Reduce the need for co-location of functions.
6. Reduce the need for media transformation.
7. Improve data analysis.

To address these functional objectives, three technologies were selected for piloting representing the three major areas of the modern office concept; distributed data processing, non-simultaneous communication and central document processing. In February, 1983 a Microcomputer Configuration Selection Study Team was formed with the purpose of selecting the appropriate hardware and software to meet FPL's microcomputer requirements. The choices of the Team were; the IBM PC, Lotus 1-2-3, WordStar and dBASE II. The choice of software for the personal computers was predicated upon giving each professional a complete set of automated tools of his trade; i.e., writing, numerical analysis and data handling.

In order to minimize the cost of the pilot, it was decided to use a service-bureau-supplied voice mail service and asynchronous PC to PC communication to represent the non-simultaneous communication function. Central document processing would be provided by a new Word Processing Center (WPC) to be built in the Juno Beach facility housing two (2) leased Sony central dictation units, two (2) Xerox 860 word processors, and a Xerox high speed photocopier.

The study group selected to participate in the pilot consisted of engineers, financial and operational auditors, general business and financial analysts, administrators, clericals and executives performing a variety of tasks typical of Florida Power & Light (FPL) professionals. The group encompassed the Management Services Department whose function it is to measure and control labor productivity at FPL. This group was expected to be particularly alert to the impact of technology on their own productivity and uniquely qualified to objectively observe and report these effects. Another factor in the group's selection was its geographical separation with three of the participating departments split between Miami and Juno Beach.

The pilot project was initially budgeted at $283,954 for all hardware, software and outside services and was targeted to be paid back in one to three years from gains in professional productivity of from 5% to 15%.

Performance was to be measured with a combination of 15 minute time ladders and logs that reported productivity gains and losses for each Office Automation (OA) application throughout the one (1) year project. Statistical usage data supplied by the voice mail service vendor, Word Processing Center (WPC) work logs and user surveys were also used as performance measurements tools.

After consideration of the curricula of several consulting firms, training and support functions for the personal computers and voice mail were custom designed in-house through the cooperation of Management Services (MGT), Telecommunications (TEL), and Systems and Programming (SPD). An external consultant was engaged to provide training in dictation technique. The choice of internally developed training was predicated on satisfying the #3 objective and the conclusion that an internal program would be more responsive to the FPL environment.

3 DISCUSSION OF CONCLUSIONS:

3.1 Productivity Results:

Participants in the Office Automation Pilot Project were required to maintain logs recording each application of office automation technology. In this log the actual time required to complete a task was compared to the user's estimate or recollection of the time required to complete the same or similar task using the previous methods.

Office Automation Pilot participants reported 3,717 hours applied to 1,041 applications of automation technology during the study period. These applications averaged a savings of 45% over the previous methods representing 2,978 hours of professional work time saved. This magnitude of savings is considered applicable to an average of 26.5% of the professional's time which the time ladder "daily activity profile" indicated is being spent in automatable activities. The total professional productivity gain available from this configuration of office automation, therefore, is calculated to be 45% X 26.5% = 11.9%.

Productivity gains reported on personal computer tasks have averaged 44% over the course of the pilot. When that level of productivity is applied to the portion of the work day spent at PC automated activities, the available gain in personal productivity computes to be 11%.

Central dictation application averaged 48% savings over the previous methods. The actual application of dictation, however, averaged only 2% so the actual gain realized during the pilot was 48% X 2% = 1%.

The overall productivity of the Word Processing Center (WPC) could not be approached in the same manner as the other technologies because of the difficulty of deriving a "time by previous method" figure for the individual secretary method of preparation preceeding the Word Processing Center. Productivity for this element was, therefore, measured in terms of typist utilization and input/output internal to the WPC itself.

Overall, the two word processing operators produced 11,783 pages in 126 work days directly applying 34% of their available work hours to typing. This utilization factor had risen to 62% by the end of April, 1984. Details of this utilization are summarized in Exhibit I.

3.2 Impact on Daily Activity Profiles:

While the three sets of time ladders taken before and during the automation pilot showed varying work profiles, it is impossible to conclude that automation had a profound effect on the proportions of professionals' daily activity. These variances could as well be attributable to changes in work content as to automation. The data from the ladders is better applied, as above, in determining the average proportion of the professionals' day that is subject to automation.

Some general conclusions can be drawn from empirical observation about the impact of automation on daily work activity and methods:

- Most of the professionals equipped with personal computers became dependent upon them and their machines showed very high rates of utilization.

- Secretarial workload, outside the WPC, dropped precipitously with most documents being drafted to final on the PC or finalized in the WPC.

- Personal computers created a new requirement for clerical staff trained on the PC to perform data entry activities.

- Automation failed to appreciably reduce the need for hardcopy and limited printing capacity was a definite bottleneck in the system.

- Failures of both dictation and voice mail to achieve wide acceptance are partially attributable to established work methods. In the case of dictation, most authors edit during drafting and are uncomfortable without hardcopy from which to work. In the case of voice mail, most users reported preferring the assurance of a responsible person that messages would be delivered.

3.3 Impact on Work Quality:

While impossible to quantify, the impact of office automation on work quality is considered to have been profound. The personal computer in particular allowed for greater depth of analyses in every number-intensive application than would have been previously possible. Notable cases of quality improvement were observed during 1984 budget preparation and review and during the recent rate case hearing before the Florida Public Service Commission (FPSC).

The economic benefits resulting from improved work quality are as difficult to quantify as is the degree of quality improvement itself. It is apparent, however, that the ability to more finely hone and examine data is likely to manifest a favorable economic impact through improved decision-making and earlier problem recognition.

One potential pitfall in the area of quality is created by the technology of office automation. Due to the relative ease of data manipulation and retrieval, the level of detail generated can easily exceed the requirements of the project adding unnecessary time or, worse, masking the essential elements of an issue in a flood of data. Managers and professionals may guard against this pitfall by limiting the volume of data and analysis requested or supplied to that which is necessary and appropriate.

3.4 Financial Impact:

The financial analysis of this pilot project is detailed in Exhibit II along with the assumptions supporting it. In summary, the financial results were very favorable:

Net Present Value, @ 11.5%, 8 years	=	$1,250,000
Payback (years)	=	.94

3.5 Human Factors and Ergonomics:

Worker acceptance of the technology is generally cited as the critical human factor in office automation. The keys to acceptance in this pilot appeared to be (1) training, and (2) adaptability of the technology to existing work methods. In the case of the personal computer, the training was extensive and the computer fit in well with the existing work methods of the technically oriented staff. Dictation and voice mail both suffered from less intensive training programs and the necessity of changing basic work methods to accomodate the technology. Executives and workers alike preferred the assurance of a personal contact in delivering a message. Authors preferred the editorial flexibility of hardcopy drafts and did not feel confident in their ability to dictate word-critical text.

The training developed internally by SPD and TEL for the personal computer, its software and peripherals was considered excellent by the majority of participants. It succeeded in indoctrinating even the least computer literate into the use of the new tool. Some weaknesses were noted in the early versions of the WordStar and dBASE courses and these were subsequently corrected. Another weakness was in conducting training in all three software packages on three consecutive days. This problem, too, has been corrected and courses in each package are now designed to stand alone.

Problems with office automation ergonomics were relatively minor during the pilot, but raised some issues that may be significant in the future as automation proliferates. One employee refused to use any Video-Display Terminal (VDT) during pregnancy, citing the widely discussed, but unsubtantiated, connection between VDT's and problems during pregnancy. Overly warm and cramped offices also were an ergonomic result of installation of the PC's where office space was limited and air conditioning inadequate to dissipate the additional heat. Screen glare and the lack of computer-compatible furniture created some minor problems that were dealt with on an individual case basis.

Minor cases of eyestrain were the most serious medical problems attributed to the Office Automation (OA) technology. It should be noted, however, that the professional environment lends itself to frequent interuptions of VDT sessions that may minimize the development of physical problems.

3.6 <u>Voice Mail:</u>

According to user surveys and vendor provided data, 76% of the participants in the Office Automation (OA) pilot used the system "rarely" or "not at all," characterizing this test of the technology as a failure.

Three factors were the primary reasons for the low acceptance of voice mail. One reason was that the test group was too small. In addition, the group tended to be a peer group, rather than a supervisor/subordinate group (in which messages tend to go vertically more often than horizontally). When conveying information, one transaction is preferred rather than calling some people, leaving a voice message for others, and sending a written note to others. The more people on a single system, the better the chance of accomplishing the single transaction goal. Thus, a supervisor could leave a single message with a group address to all his employees.

A second reason for the low acceptance is that little or no "market research" was done to determine the demand for such a service.

The third primary reason for the failure was the lack of adequate useful features. In a real world voice mail system, many other features are available that were not available in this pilot test.

From a feature standpoint, users noted that the system did not verify delivery of messages, nor was there a means to alert the recipient that there was a message waiting. It is noted that the system would alert the sender that his message had not been heard after 24 hours had elapsed if the sender had ended the message by dialing '55' rather than just a '5'.

Finally, the service was considered to be too expensive at $50 per month per user.

4 DISCUSSION OF RECOMMENDATIONS

The concept of Office Automation (OA) requires that a working unit be provided with a set of technological tools to adapt to their particular tasks and work style. With this concept no effort is made to analyze and cost justify each application in advance; instead, the assumption is taken, as it is with a desk and a telephone, that this new technology is a basic tool of the trade to be used as appropriate. The productivity results obtained in this pilot appear to justify the concept on the basis of overall cost effectiveness. While it is not quantifiable, the synergistic effect of having the entire work unit automated simultaneously is believed to account for a large measure of the benefit. With this implementation, the technology was able to exert all of its leverage without encountering the resistance and obstacles it might have in a piecemeal implementation. No one was left without the technological means to participate, so work methods were able to change immediately and utilize the efficiencies offered by automation. Personal resistance was also lowered as the entire unit worked up the learning curve together.

In summary the "Office Automation" implementation concept is recommended as the most efficient and cost effective approach to integrating technology into the office where functions are similar to those of the pilot group. The principal elements in executing this implementation are:

> Select the general technologies (PC's, dictation, Word Processing Center, etc.) applicable to the operating unit's overall task.

- Select a standard configuration of software and hardware to ensure compatibility within and outside the operating unit.

- Implement throughout the work unit simultaneously.

- Modify internal procedures to accomodate automation and do not encourage or permit redundant manual "backups."

- Preserve and control the configuration to prevent incompatible "specialty" software or hardware from diluting the synergy effect.

- Monitor the effectiveness of automation and respond to gaps in hardware, software or training requirements.

While the Office Automation (OA) implementation detailed above is recommended for maximum cost effectiveness from the available technology, the "Technology Applications" (TA) implementation is a viable alternative requiring less initial investment. "Office Automation" is aimed at applying the most appropriate technology to virtually all of the office functions within an operating unit. The Technology Applications implementation, however, is aimed at automating only certain, predesignated tasks.

The TA implementation requires some modifications to the approach taken in the OA approach:

- The task(s) to be automated must be carefully defined and its growth projected over the life of the project.

- The choice of the appropriate technology, hardware and software, will be critical to success.

- Since the entire cost of the technology must be justified by savings on one task, the "before" costs of the task must be accurately determined and a method of measuring savings must be defined.

- Personnel selection and training are also critical factors in the success of a TA implementation. Training need not be as broad as in the OA implementation, but should be more in depth in the area of the application.

4.1 Dictation:

Central dictation, while productive when applied, suffered from a low rate of utilization. Less than 1% of the available dictation machine work day hours were used.

Seventy percent (70%) of the 36 respondents to a dictation users survey taken in May, indicated that the word critical and number intensive nature of their correspondence inhibited use of dictation. Only eight (8%) of respondents cited the Word Processing Center's service level or their level of dictation training as major barriers to dictation use.

5 DISCUSSION OF ISSUES

The results of this pilot project suggest that some extension of the technologies of office automation would be operationally beneficial and cost effective for FPL. This premise raises a number of issues for executive consideration and decision:

5.1 Implementation Strategy:

The first issue to consider for the future of office technology at FPL is an overall strategy for implementation. The more conventional "Technical Applications" (TA) approach would continue to require individual departmental study, justification and approval for each purchase. While this method ensures maximum control and budgetary restraint, it threatens to dilute the beneficial impact by eliminating the synergistic effects of a fully automated work unit. It also incurs redundant costs of study and preparation of justifications. Another potential problem with this implementation is that, as new functions are automated incrementally within a work unit, technologies will advance leaving the last increment obsolete and, perhaps, incompatible with the next increment.

An alternative to the TA implementation is the "Office Automation" (OA) implementation. This concept has some advantages over the TA approach for work units whose work functions resemble those of the Pilot group. It would bypass the study and justification process by assuming productivity benefits similar to those of the Pilot. Besides eliminating the cost of the study, the OA implementation promises early synergistic benefits of a fully interactive, automated work unit. This approach also avoids the pitfall of incompatibility by ensuring that a work unit's automation is concurrent with the same point in the state of the art.

5.2 Points of Impact

The issues inherent in the selection of an Office Automation/Technology Applications implementation strategy relate to the nature of the work and the staff composition at the point of impact. Results from the pilot indicate that maximum benefit occurs in different areas for different technologies. Analysts and other professionals constructing complex spreadsheets or drafting word critical documents benefit most from the personal computer. Supervisors use the PC to review and edit their subordinates' work, but benefit most from communications media such as dictation, voice mail and PC to PC communication. The Word Processing Center was most valuable to those who didn't have direct access to secretarial support, such as field auditors. These variables should be considered in the selection of the technology and strategy to implement and in the prioritizing of automation schedules by organizational unit.

5.3 Control/Standards

Centralized control of the selection and acquisition of Personal Computing hardware, software, and peripherals is extremely critical to ensure compatability among departments and applications. Centralized control allows for consistent and adequate support of the personal computing resources allowing the support departments to provide ongoing training and continuing technical support to users. Standards are an integral part of the control issue, and relate to the ongoing need for the establishment of corporate standards for hardware, software, and peripheral equipment. Without standards, compatability is lost, resulting in the inability to transfer data between functional groups and potentially even within the same group. Example: During the recent rate case, standards proved to be extremely valuable, enabling the transfer of data among all of the departments involved in the rate case proceedings, with this data transfer and modification occurring among multiple locations as well. Training and support will be much more complex, and potentially impossible if standards are not followed. Standards allow for control of the costs of the various components of the personal computers, enabling the company to take advantage of volume discounts on a variety of products. In this rapidly changing technology, standards allow for an orderly and planned evolution/conversion to new products. The establishment of standards prevents the proliferation of large numbers of various software and hardware products which are non-compatible, for which no training is available, and for which support will potentially be minimal. In order to function in a semi-controlled environment, it must be understood that the organization will not be able to always recommend the newest technological advances. Rather, "leap frogging" to new technologies will occur, maintaining organized upgrades ensuring compatability. It is recommended that the existing controls and procedures for the procurement of hardware and software remain in place within Corporate Services. Overall, the establishment of realistic standards is felt to be one of the most important issues and recommendations resulting from this study.

EXHIBIT I

OFFICE AUTOMATION

SUMMARY OF WORD PROCESSING CENTER PERFORMANCE

Summary for the period July 1, 1983 through April 30, 1984:

Pages Typed		4,953
Pages of Revisions Completed		6,830
. Output Pages From Dictation	431	
. Output Pages From Written Draft	11,352	
Total Pages of Output		**11,783**
Pages of Written Input	5,203	
Minutes of Dictation	1,489 (mins.)	
Work Hours From Written Draft		934
Work Hours From Dictation		66
Total Work Hours Reported		**1,000** *
Work hours available		2,948
% Utilization		34%

* Low utilization is partially attributable to under-reporting of work hours during 1983.

Summary for the period January 1, 1984 through April 30, 1984:

Total Pages of Output	7,248
Pages of Written Input	2,512
Minutes of Dictation	860
Work Hours Reported	717
Work Hours Available	1,161
% Utilization	62%

EXHIBIT II

OFFICE AUTOMATION

- FINANCIAL EVALUATION -

AUTOMATABLE ACTIVITY	AVERAGE % OF WORK DAY
. Writing	8.4
. Calculating	4.0
. Analysis*	8.0
. Editing	3.8
. Proofing	2.3
TOTAL	**26.5**

* 1/2 of "Analysis" is assumed to be automatable activity.

. Average productivity gain reported per application of automation.	45%
. Average overall productivity gain (45% x 26.5%).	11.9%
. Estimated annual payroll for 51 professionals.	$2,550K
. Annual benefit of an 11.9% productivity gain.	$ 303K
. Estimated cost of Office Automation for 51 professionals.	$ 284K
. Payback	.94 years
. Net Present Value (NPV) of $303K @ 11.5% for 8 years	$1,248K

E. PRODUCTIVITY PLANNING

Productivity Management Frontiers — I, edited by D.J. Sumanth
Elsevier Science Publishers B.V., Amsterdam, 1987 — Printed in The Netherlands

PLANNING FOR OPTIMUM PRODUCTIVITY IMPROVEMENT UNDER IMPOSED CRITERIA

D. C. BRAUER
Jersey Central Power & Light, Madison Avenue at Punch Bowl Road, Morristown, New Jersey 07960 (USA)

ABSTRACT

One means of improving productivity performance is to minimize the total resources or inputs required to produce total output. Given a set of potential productivity improvement programs, under constrained resources, the best mix of those programs must be selected to pursue. An integer programming model based on the concept of total factor productivity is presented to choose those productivity improvement programs considering limited labor, material, and capital, in addition to product selection criteria, imposed partial measures, and minimum program requirements. Once the aggregate input resources are determined, pricing can be established with respect to existing pricing levels and the productivity gains.

1 INTRODUCTION

There have been significant work and research done in measuring productivity performance and in equating higher profitability to increases in productivity. The underlying technique associated with this subject principally appears to be total factor productivity (TFP) or firm productivity measurement. By quantifying the aggregate inputs and outputs, to some base period, it is possible to analyze discrete changes in order to identify causes of productivity increase or decrease. This, in turn, can provide information with which to formulate productivity improvement and enable the undertaking of planned programs.

Knowing a firm's existing level of productivity, defined as the ratio of total output to total input, alternatives can be identified to improve productivity. However, additional investment and resources of one or more inputs may be required to reduce other inputs without having to solely increase price. Implementing those identified productivity improvement programs may exceed the amount of available resources. This would typically be determined, most ideally, in parallel with a firm's annual budgeting process. During the budgetary allocation and approval process, productivity gains, trade-offs and requirements can be directly appraised and considered. The decision that must be made in considering a set of productivity improvement programs is which ones shall be undertaken for optimum effect.

This paper presents a mathematical decision model to select those programs which will minimize the total input resources based on availability limits, and provide for further profitability, pricing and productivity performance goals. An example is presented that illustrates an application using the methodology.

2 BACKGROUND

Total factor productivity (TFP) has been applied to measure the extent of how effectively resources are used to produce a given output. These resources are typically considered to categorically be labor, material, capital and other miscellaneous expenses. Output is usually valued as the total sales revenue of products or services produced in a given period. Paramount to these inputs and output is the requirement that all quantities be expressed in terms of a constant dollar base. This will assure that inflationary and other escalation effects are removed which could distort the results. In order to arrive at the level of productivity from one period to another, a ratio of the TFP measures are taken. The TFP expression is simply defined as

$$TFP = \frac{\text{Output (Revenue)}}{\text{Labor + Material + Capital + Other.}}$$

If TFP_0 represents the initial or base measure, then the productivity performance for the current period is TFP_1/TFP_0, where the subscript "1" denotes the current period.

Craig and Harris, 1973 discussed in detail the valuation and adjustments that should be performed in quantifying TFP measurement. They additionally brought out points relating the magnitude of productivity performance and return to capital. Chaudry, 1982 described the use of TFP for corporate planning and budgetary purposes. A further model (NIPA) is also discussed by Chaudry that was formulated to explain the growth of net income in terms of TFP and other resource inputs. Thor, 1982 stresses that the key elements of productivity performance measurement are profitability, productivity and price recovery. He also discusses how these relate to one another and their importance to financial and strategic planning.

Since productivity improvement can mean higher profits beyond price increase and manipulation, a planned or proactive approach for such improvement is desirable. TFP measurement attempts to quantify the past in order to identify and understand the elemental components that lead to changes in productivity. Instead of just purely tracking historical performance, the concept of TFP measurement can be used to plan for future improvement. Obviously, the level of current performance would be necessary as a starting point for any planning effort and ultimate objective.

The planning phase can start by analyzing the firm's product or service portfolio to identify what feasible improvements can be made, and what resources will be required to enact these improvements. Typically, the major considerations will be the effects on labor, material and capital. Just consideration must also be given in identifying all affected products or services which may benefit from the infusion of a capital investment in equipment or a methods change with little or no investment. Once all the feasible improvement programs have been devised, resource requirements to support implementation will have to be assessed. More often than not, the required resources will exceed the available resources, analogous to the classical capital rationing problem. In this situation, one must choose which programs to undertake within resource limitations. The objective then becomes that of maximizing the productivity performance for the upcoming period by minimizing the aggregate input resources through the best mix of programs. One approach to choosing those programs is through the use of mathematical programming. The development of a decision model shall be subsequently presented to select those programs which will, in turn, allow for productivity improvement and profitability.

3 MATHEMATICAL MODEL

The resource minimization model can be stated as

$$\text{Minimize } Z = \sum_{k}\sum_{i\in k} \lambda_{ki}X_i(-q'_{kil} - q'_{kim}) + \sum_{k} f_k y_k + \sum_{k} w_k y_k$$

subject to

$$\sum_{i} X_i q_{oil} - \sum_{k}\sum_{i\in k} \lambda_{ki}X_i q'_{kil} \leqslant b_l \tag{1}$$

$$\sum_{i} X_i q_{oim} - \sum_{k}\sum_{i\in k} {}_{ki}X_i q'_{kim} \leqslant b_m \tag{2}$$

$$\sum_{k} a_k y_k \leqslant b_c \tag{3}$$

$$\sum_{k} \lambda_{ki} \leqslant 1 \quad ; \text{for } i = 1,2, \ldots \tag{4}$$

$$\sum_{i\in k} \lambda_{ki} \leqslant r_k y_k \quad ; \text{for } k = 1,2, \ldots \tag{5}$$

$$r_k = \{i\in k\} \; ; \text{for } k = 1,2, \ldots \tag{6}$$

$$\frac{\sum_{i} X_i q_{oil} - \sum_{k}\sum_{i\in k} \lambda_{ki}X_i q'_{kil}}{\sum_{i} X_i} \leqslant h_l \tag{7}$$

$$\sum_{k} y_k \geqslant d \tag{8}$$

$$\lambda_{ki} = 0,1 \quad ; \text{for all } k \text{ and } i \tag{9}$$

$$y_k = 0,1 \quad ; \text{for all } k$$

where,

λ_{ki} = binary decision variable for program k and product i

X_i = volume of product i

q'_{kil} = incremental labor unit cost differential for program k and product i

q'_{kim} = incremental material unit cost differential for program k and product i

f_k = periodic valued capital resource input for program k

y_k = binary decision variable for program k

q_{oil} = base period labor unit cost for product i

b_l = total labor cost limitation for upcoming period

q_{oim} = base period material unit cost for product i

b_m = total material cost limitation for upcoming period

a_k = capital requirement for program k

b_c = total capital limitation for upcoming period

r_k = set of i products contained in program k

h_l = partial labor productivity measure of the weighted average unit labor cost

d = minimum number of programs required to undertake

w_k = other miscellaneous input expenses for program k.

The objective function of the model minimizes the total input resources required to achieve all or part of the k programs for the upcoming period. This formulation represents the incremental input costs expressed in terms of base period costs. The input resources included are labor, material, capital and other expenses. Capital input, f_k is valued on a periodic basis such as the equivalent annual worth over the investment life. This should be consistent with the base period methodology in valuing capital input. The upcoming period requirements in terms of units is represented by X_i. It is assumed that this value for each product or service is provided through a forecast or known demand. The cost associated with each program will typically be represented in terms of capital and miscellaneous expense inputs of the model, while the incremental benefits will be realized predominantly through labor and material inputs, i.e. $-q'_{kil}$ and $-q'_{kim}$. Thus, the objective function attempts to select those programs that will result in the most benefit with the least cost to minimize total resources.

The first and second constraints of the model limit the total available labor and material cost for the upcoming period. This is expressed by the difference between the existing unit costs, q_{oil} and q_{oim}, and the incremental unit cost savings, q'_{kil} and q'_{kim}. Both unit costs should be all in terms of the base period. Constraint three, like the first two, limits the available capital for the upcoming period. Each program will require an amount a_k of capital to undertake. This amount is related to the objective function through f_k, the equivalent valued capital input for program k. Constraint four defines program and product uniqueness. Overlapping of programs to the same product is assumed to be infeasible. Therefore, the assignment to each product must be specific. The fifth and sixth constraints define the set of products in each program, limited by the total r_k, in conjunction with the y_k decision variable to undertake program k. If y_k equals zero then program k is not selected. Alternatively, if y_k equals one, one or up to r_k products in program k shall be affected. Constraint seven is an arbitrary constraint added to illustrate the case in which a partial measurement may have to be considered. This particular formulation requires that the weighted average unit labor cost of all products does not exceed some imposed limit, h_1. Any other desired partial measures would have to be developed and added as necessary. Constraint eight simply sets the minimum number of programs that must be undertaken. The ninth constraint defines the decision variables as 0,1 non-negative integers.

Once the productivity improvement programs are selected for the upcoming period, a further evaluation can be performed on profitability and pricing. Using the output/input concept as defined in total factor productivity, with the results of the resource minimization problem, a model can be constructed to set pricing and productivity goals. This approach then considers pricing concepts and marketing strategies, complemented through productivity gains, which can enable a firm to effectively lower selling price at or above existing margins. Alternatively, a small escalation in price, say due to inflation, or general market conditions can result in productivity and profitability gains achieved through the improvement programs and incremental pricing increase.

One such formulation relating the above factors is

$$\text{Maximize } V = \sum_i p_i x_i$$

subject to

$$p_i \leqslant u_i \quad ; \text{ for } i=1,2,\ldots$$

$$\frac{1}{TFP_o}\left[\frac{\sum_i p_i x_i + \alpha}{\beta}\right] = g$$

$$p_i \geqslant 0 \quad ; \text{ for all } i$$

where,

P_i = unit price for product i

x_i = volume of product i

u_i = upper bound on selling price for product i

TFP_o = base period productivity (output/input)

α = other sources of income

β = total input resources required, based on productivity improvement programs

g = productivity goal.

The objective function seeks to maximize total sales revenue based on pricing effects applied to given product volumes. Constraint one imposes an upper limit on individual product price setting. The second constraint requires a productivity improvement level to be attained over the base period taking into account all required resources and forms of revenue. Constraint three simply sets non-negativity for the decision variable p_i. An alternative approach would be to structure the formulation in terms of a goal programming problem with respect to g, the productivity goal.

4 NUMERICAL EXAMPLE

Suppose a particular firm has identified five various productivity improvement programs that affect eight different products. An analysis performed on these eight products has identified the incremental material and labor unit cost savings, as well as the investment requirement to accomplish each program. Table 1 displays the cost components associated with each program and product. In this particular example, each program requires a certain amount of capital investment to enact the material and labor effects. However, depending upon the specific situation, it is very possible that a program may require no capital, and only a one time miscellaneous expense.

Table 2 displays the unit volume requirements for each of the eight products, the base unit material cost and the base unit labor cost. Let us also assume that capital is valued on an equivalent capital return basis of 12% over the expected life of the investment. This condition is shown on Table 2 directly below the product volume and cost data. The budgetary ceiling on material, labor and capital for the upcoming period, expressed in terms of the base period has also determined to be \$8,048,750, \$17,934,000 and \$400,000, respectively. Last of all, the partial unit labor cost measurement has been targeted not to exceed \$137.43.

Applying the given mathematical model with the above numerical conditions and constraints result in the following program and product selection.

Program (k)	Products (i)	Capital (a_k)
1	6	\$ 75,000
2	1,2,3,7	150,000
3	5	50,000
5	4,8	95,000
		\$370,000

TABLE 1

NUMERICAL EXAMPLE

		Unit Cost Avoidance - Product (i)															
	Capital	1		2		3		4		5		6		7		8	
Program (k)	Invest. (a_k)	Mat.	Labor	Mat.	Labor	Mat.	Labor	Mat.	Labor	Mat.	Labor	Mat.	Labor	Mat.	Labor	Mat.	Labor
1	$ 75,000		3.2				1.6		2.5				3.8				
2	150,000	1.5	2.0	1.9	2.4	0.7	1.5			1.0	2.2			2.1	2.5		
3	50,000							0.5	1.8	1.2	2.0	0.8	1.5	1.8	2.3		
4	100,000				2.8										3.5		4.0
5	95,000	1.0	1.5					1.6	2.0							2.1	2.5

TABLE 2

NUMERICAL EXAMPLE

Product (i)	Units (X_i)	Base Unit Costs ($/Unit) Material ($q_{oim}$)	Labor (q_{oil})
1	20,000	$ 40.0	$150.0
2	7,500	95.5	180.0
3	15,000	25.0	90.0
4	30,000	75.5	125.0
5	10,000	25.5	100.0
6	8,000	125.0	185.5
7	15,000	100.0	200.0
8	25,000	45.5	120.0
	130,500		

Program (k)	Capital Investment (a_k)	Capital Valued (1) Input (f_k)	Investment Life (Yrs.)
1	$ 75,000	$13,274	10
2	150,000	22,024	15
3	50,000	10,956	7
4	100,000	17,698	10
5	95,000	13,948	15
	$470,000	$77,900	

(1) 12% pretax capital return rate

The value of the objective function associated with the solution is -$429,450. This represents that net decrease in input resources in undertaking the selected programs and products. The partial unit labor cost measurement is $135.20, within the constrained value of $137.43.

Given the incremental decrease of resources required for the upcoming period, total input resource requirements can be determined using the existing base period. At this point, pricing can then be examined in conjunction with this productivity gain. Specifically, analyses can be performed utilizing the previously given pricing model to investigate productivity, pricing and profitability effects.

5 CONCLUSION

The models in this paper show one way in which productivity improvement planning can be approached. This approach emphasizes resource minimization, based on TFP input. Pricing and profitability can be directly assessed with the effects of planned improvements integrated into the upcoming periods TFP. This enables investigation of various pricing and marketing strategies to be fully explored and examined.

6 REFERENCES

Canada, J. R. and White, J. A., 1980. Capital Investment Decision Analysis for Management and Engineering. Prentice-Hall, Englewood Cliffs.

Chaudry, M. A., 1982. Projecting Productivity to the Bottom Line. American Productivity Center, Productivity Brief 18. Report, 8 pp.

Craig, C. E. and Harris, R. C., 1973. Total Productivity Measurement at the Firm Level. Sloan Management Review, 14: 13-28.

Davis, H. S., 1955. Productivity Accounting. University of Pennsylvania Press, Philadelphia.

Taha, H. A., 1982. Operations Research, 3. Macmillan, New York.

Thor, C. G., 1982. A Complete Productivity Measurement System. American Productivity Center. Report (unpubl.), 3 pp.

Productivity Management Frontiers — I, edited by D.J. Sumanth
Elsevier Science Publishers B.V., Amsterdam, 1987 — Printed in The Netherlands

PRODUCTIVITY PLANNING METHOD USING INPUTS COSTS PROJECTIONS

ROGER J. MEIMBAN
Department of Wood Science and Technology, College of Forestry, University of the Philippines at Los Baños, College, Laguna 3720 Philippines

ABSTRACT
Meimban, R.J. 1987. Productivity planning method using inputs costs projections.

The productivity planning method presented considers the relationship of the unit price and unit cost of an input to its productivity. Target factor productivity levels are established using forecasts of unit prices. The heuristic approach ensures that planned productivity levels are realistic while keeping the resulting total unit cost under control even if there is an increase in the prices of inputs. This is achieved by identifying the inputs that strongly influence the total unit cost. Allocation of productivity improvement efforts is then carried out to relieve the pressure on factors where improvement is extremely difficult to attain.

1 INTRODUCTION

Productivity planning, as defined by Sumanth (1984), is concerned with the setting up of target levels of total and/or partial productivities so that these levels can be used for comparison purposes in evaluating productivity performance as well as for mapping strategies of improving productivity. With today's competitive environment, productivity planning then becomes a necessity for business survival.

Among the recommended approaches to productivity planning is the use of forecasting techniques where the estimate of productivity in the future is based on past performance (Sumanth, 1984). In this paper another method is presented by considering the relationship of the unit price and the unit cost of an input to its factor productivity. The method is based on the premise that "if the increase in price of a factor input is not matched with a corresponding increase in its productivity, an increase in the unit cost results". In such case, a forecast on the unit price of the input can reasonably lead to the productivity conditions under which the unit cost can be kept under control. As it seems to be, improvement in productivity performance is gauged by how well it controls the resulting unit cost. At any rate, the ultimate goal in productivity improvement is the reduction in the total unit cost.

2 BASIC CONCEPT

The expressions used in this paper are built on the maxim that "the unit cost* (C) of an input multiplied by its factor productivity (Π) equals the price (P)". This arise from the definition of factor productivity which is the ratio of the physical output (V) to the factor input (Q), and as derived by Eilon (1984):

$$\Pi = \frac{V}{Q} \tag{1a}$$

$$= \frac{P}{C} \tag{1b}$$

The maxim can then be directly presented as

$$\Pi C = P \tag{2}$$

3 EXPRESSION FOR THE CHANGE IN Π

The expression for the relative change of a factor productivity can be derived by differentiating over equation (2) and dividing it by the initial value of Π; or

$$\frac{\delta P}{\Pi} = \frac{C\delta\Pi}{\Pi} + \frac{\Pi\delta C}{\Pi} + \frac{\delta\Pi\delta C}{\Pi} \tag{3}$$

By defining

$X^* = \frac{\Delta X}{X}$ or the ratio of any change in X to its original value

expression (3) can be simplified (recall that $\Pi = P/C$) as

$$P^* = \Pi^* + C^* + \Pi^* C^* \tag{4}$$

From the above equation, the expression for the relative change in factor productivity then takes the form

$$\Pi^* = \frac{P^* - C^*}{1 + C^*} \tag{5}$$

Likewise;

$$C^* = \frac{P^* - \Pi^*}{1 + \Pi^*} \tag{6}$$

Thus it is verified that if $\Pi^* > P^*$, the unit cost will decrease ($C^* < 0$). The different interesting circumstances that may arise are summarized as (Meimban, 1986):

$$\Pi^* \gtrless 0 \quad \text{if} \quad P^* \gtrless C^* \tag{7}$$

---

*The unit cost, as used consistently in this paper is defined as the cost of an input per unit of output.

and

$$C^* \gtrless 0 \quad \text{if} \quad P^* \gtrless \Pi^* \tag{8}$$

The last condition simply shows that an improvement in productivity may not reduce the unit cost unless such improvement compensates the associated increase in the unit price. This condition is the basis in the development of a productivity planning approach that aims to ensure the reduction or control of the unit cost by establishing the target productivity levels using the estimated changes in the unit price.

4 THE TARGET LEVEL OF Π

To make use of the preceding conditions, let

τ = planning period (forecast period)

t = base period

$\Pi_{t+\tau}$ = the desired productivity level at future period, $t + \tau$

$P_{t+\tau}$ = forecast value of the unit price made at period t

$C_{t+\tau}$ = resulting unit cost at period $t + \tau$

$\Delta\Pi = \Pi_{t+\tau} - \Pi_t$

$\Delta P = P_{t+\tau}$ p P_t

$\Delta C = C_{t+\tau} - C_t$

It is assumed that the values of C_t, P_t and Π_t are already established with certainty and $P_{t+\tau}$ can be estimated using available forecasting techniques. The discussion of the appropriate forecasting techniques is beyond the scope of this paper but a number of them is presented by Johnson and Montgomery (1974).

Using the notations, equation (6) can be expanded as:

$$\frac{\Delta C}{C_t} = \frac{\dfrac{(P_{t+\tau} - P_t)}{P_t} - \dfrac{(\Pi_{t+\tau} - \Pi_t)}{\Pi_t}}{1 + \dfrac{(\Pi_{t+\tau} - \Pi_t)}{\Pi_t}} \tag{9}$$

$$= \frac{(\Pi_t P_{t+\tau} - P_t \Pi_{t+\tau})}{P_t(\Pi_{t+\tau})}$$

Thus the following conditions can be generated:

$$C^* \lesseqgtr 0 \quad \text{if} \quad \begin{cases} \Pi_t P_{t+\tau} \lesseqgtr P_t \Pi_{t+\tau} & \text{(10a)} \\ \text{or;} \\ \dfrac{\Pi_t P_{t+\tau}}{P_t} \lesseqgtr \Pi_{t+\tau} & \text{(10b)} \\ \text{or;} \\ \dfrac{P_{t+\tau}}{C_t} \lesseqgtr \Pi_{t+\tau} & \text{(10c)} \end{cases}$$

The last condition (10c) establishes the target level ($\Pi_{t+\tau}$) based on the forecasted value, $P_{t+\tau}$, and the initial value of the unit cost, C_t. It is important to point out that for a particular factor input, it is not always necessary to force its factor unit cost to decline, Depending on the desired level of change of a factor input unit cost (as discussed elsewhere), a corresponding level of factor productivity can be computed from equation (9), that is;

$$\Pi_{t+\tau} = \frac{\Pi_t P_{t+}}{\Delta C\Pi_t + P_t} \qquad \text{(11a)}$$

$$= \frac{P_{t+\tau}}{C_{t+\tau}} \qquad \text{(11b)}$$

5 COST IMPROVEMENT ALLOCATION

The analytical procedure developed cannot be readily used for total productivity planning. Among the reasons are:

1. For each of the factor input i individual target ΔC^i must first be established and consequently that of the total unit cost, ΔC^T.
2. For the given system of productivity measurement (1a and 1b), there exists a maximum attainable level of $\Delta\Pi^i$ for a given planning period. Thus it may not always be possible to reduce a factor input unit cost even with an improvement in the factor productivity. And for the ultimate goal of forcing the total unit cost to decline, the increase in the unit cost of one input must be properly compensated by the decrease in one or more of the other factor inputs unit cost.

For this cost allocation phase of the planning approach, one of the methods suggested is derived from Eilon's relative unit cost increment model (1983, 1984). In this model, the relationship of the total unit cost (C) to the total cost (K) and volume (V) is given by

$$C^* = \frac{K^* - V^*}{1 + V^*} \qquad \text{(12)}$$

To illustrate its application in productivity planning, let us assume the factor inputs cost model used by the same author. This model is of the form

$$K = M + L + E \tag{13}$$

and the relative total cost change is given by

$$K^* = f_M M^* + f_L L^* + f_E E^* \tag{14}$$

where M = material cost

L = labor cost

E = cost of other inputs

f_i = cost proportion of factor input i

Thus equation (12) can be expanded to

$$C^* = \frac{1}{1 + V^*} (f_M M^* + f_L L^* + f_E E^* - V^*) \tag{15}$$

If one of the factor inputs costs is allowed to change proportionately with volume (for convenience, it is assumed that $V^*>0$), e.g., materials, the condition for the total unit cost to decline ($c^*<0$) is if

$$L^* - V^* < - (\frac{f_E}{f_L} (E^* - V^*) \tag{16}$$

i.e., given that $M^* \simeq V^*$.

Incidentally, the system of expression (16) can be used to allocate factor unit cost changes, and consequently the factor productivities. This is done by dividing both sides of the inequality by $(1 + V^*)$,

$$\frac{L^* - V^*}{1 + V^*} < - \frac{f_E}{f_L} (\frac{E^* - V^*}{1 + V^*}) \tag{17}$$

Furthermore, it can be shown that the factor unit cost changes can be expressed as

$$C^*_L = \frac{L^* - V^*}{1 + V^*}$$

and

$$C^*_E = \frac{E^* - V^*}{1 + V^*}$$

Thus expression (17) is equivalent to the form

$$C^*_L < - \frac{f_E}{f_L} C^*_E \tag{18}$$

Relationship (18) then, gives the proper combination of the factor inputs unit costs changes. At this point, the desired levels of the factor productivities, given the proper combination of the unit costs changes can be established. To this end, the following heuristic is suggested:

1. Make forecasts of the individual unit price (p^i) for period $t + \tau$ and establish the corresponding $\Pi^i_{t+\tau}$ using equation (10c). Based on past and current performance, check whether the resulting factor productivities are attainable.
2. For the factor productivity levels which are extremely difficult to meet, make estimates of the $C^i_{t+\tau}$ by dividing the $p^i_{t+\tau}$ with the corresponding initial value of Π^i_t. This tests the effect of operating at the existing productivity levels for period $t + \tau$ on the unit cost.
3. Compute the total unit cost resulting from the different factor productivities; consider the attainable level of $\Pi^i_{t+\tau}$ (from Step 1), if there is any. The total unit cost, (C^T), is computed as

$$(C^T = \Sigma C^i)_{t+\tau} \qquad (19)$$

4. If no reduction in total unit cost results, compute the individual unit cost change $(C^i)^*$, i.e., $\Delta C^i/C_t$, using the unit cost estimates from Step 2. Select the factor input associated with the min $\{C^*\}$ and/or the factor with the readily attainable $\Pi^i_{t+\tau}$ (from Step 1). This is tantamount to selecting the factor input whose planned productivity level can absorb the associated factor unit price increase. Notice that as $C^* \rightarrow 0$, then $\Pi^* \rightarrow P^*$.
5. Set the selected factor input to vary proportionately with the relative change in output (V^*). This allows the formulation of conditions similar to (18). Any discrepancy in the assumed approximation will be adjusted later.
6. Formulate the conditions similar to (18), and determine tow the factor unit costs must be related to attain a decline in the total unit cost. The suggested starting factor for immediate adjustment is the factor input with the highest cost proportion, i.e., it should be decreased to an attainable level that will minimize its strong influence in the total unit cost.
7. For the preliminary unit costs combination, compute the associated $\Pi^i_{t+\tau}$ and check whether these target factor productivities are attainable. If a factor productivity is not attainable, make further adjustment in the cost combination until implementable level of $\Pi^i_{t+\tau}$ can be achieved.

8. For every attainable factor productivity combination, compute the resulting total unit cost (by equation 19) and determine if the resulting improvement is desirable enough. Recall that the ultimate objective is for the combination with min $\{C^T\}$.

6 CONCLUSION

The productivity planning method presented establishes target factor productivity levels based on estimated unit price and unit cost of inputs. It's effectiveness relies more on the accuracy of the cost estimates and price forecasts. Initially, the method is intended for a one-period-ahead plan (i.e., τ=1) but if forecasts accuracy permits, then planning could be extended to several time horizons. At this point, it must be emphasized that the estimates and forecasts of parameters must be tempered with judgement before they are considered in planning.

The approach provides relationships on how factor productivity improvement must vary. The productivity combinations consider how one or more factor productivity improvement relieves the pressure on the factors where improvement is extremely difficult to attain.

The analysis presented also allows a firm to prioritize productivity improvement on factors that critically affect the total unit cost. And once the productivity benchmarks are established, the firm can then review the means (e.g., materials, methods study, etc.) to achieve the desired levels of unit cost and productivity. Thus, the approach is beneficial to a firm seeking ways to recover cost increases through productivity improvement.

7 REFERENCES

Eilon, S. 1983. Effect of productivity on unit cost. Paper presented at the 7th International Conference on Production Research, Ontario, 1983 and published in Int. J. Prod. Res., 1984, vol. 22, No. 5, pp 831-839.

Eilon, S. 1984. The art of reckoning: analysis of performance criteria. Academic Press, London. 507 pp.

Johnson, L.A. and Montgomery, D.C. 1974. Operations research in production planning, scheduling and inventory control. John Wiley and Sons, Inc., N.Y. 525 pp.

Meimban, R.J. 1986. Analysis of factor inputs costs and productivity relationships in lumber production. (Unpublished M.S. thesis, UP College of Forestry, College, Laguna, Phil.) 72 pp.

Sumanth, D.J. 1984. Productivity engineering and management. McGraw-Hill Book Co. 547 pp.

Productivity Management Frontiers — I, edited by D.J. Sumanth
Elsevier Science Publishers B.V., Amsterdam, 1987 — Printed in The Netherlands

A METHODOLOGY FOR COMPREHENSIVE PRODUCTIVITY PLANNING

JOHNSON AIMIE EDOSOMWAN
Dept. 146, Bldg. 012, International Business Machines Corporation, P.O. Box 950, Poughkeepsie, New York 12602, USA

ABSTRACT
Edosomwan, J. A. 1987. "A Methodology for Comprehensive Productivity Planning."

The productivity crisis in the United States and the rest of the world has been well documented. In order for organizations to successfully compete in the world market and encourage gains in the growth rate of productivity, it is important for them to establish a comprehensive productivity planning program (CPPP). This paper presents a conceptual framework for productivity planning in manufacturing and service organizations. The importance of a CPPP is highlighted. A formalized approach for planning productivity is offered in a four-stage process described as the Comprehensive Productivity Planning Cycle (CPPC). Productivity planning relationships to other components of the productivity management process is discussed. Measurement instruments and techniques are suggested for each of the four stages of the productivity planning cycle. An approach and checklist for setting up a formal productivity planning program (PPP) is offered. The common problems encountered in setting up a PPP are discussed, as well as ways to counter them.

INTRODUCTION

Planning efforts exist in most organizations in a global sense. However, most of the plans are partial and based on extrapolation technique. Most firms extend the annual operating plan to a two-, five-, or ten-year horizon. The most common source of projection is based on financial forecasts and business volume demand. In some situations, a linear extrapolation is made to project future spending and other resources that will be needed. Although several publications are available on business planning, there are few if any on productivity planning. Cotton (1976), Sink (1985), Andrews (1971), Ackoff (1981), Jones (1974), Ohmae (1982), Quinn (1980), Anthony (1965), and Steiner (1979) offer useful and general approaches to overall business planning. The available productivity planning tools and methodology suffer from the following shortcomings:

1. They offer no logical framework for consistent planning within the organization.
2. They assume no significant changes in the regulatory environment.

3. They assume no significant changes in factors such as inflation rate, resources, competition and operating policies.
4. They offer no methodology for assessing the organization's strengths and weaknesses. There are few if any tools for a detailed overview of the firm, its industry, and economic climate.
5. They offer no methodology for on-going assessment of the threats and opportunities in the operating environment.
6. They offer no framework where problems are easily identified and pin pointed to a base planning process. In some situations, objectives and goals arise by default and firms have to deal with inadequate planning programs through crisis management.

DEFINITION AND BENEFITS OF COMPREHENSIVE PRODUCTIVITY PLANNING

Edosomwan (1986) defined comprehensive productivity planning (CPP) as a process by which all factors affecting an organization are considered in formulating its goals and objectives, assessing its capabilities and capacities, designing alternative courses of action for the purpose of achieving these goals and objectives, initiating necessary actions for their implementation and evaluating the effectiveness of the plan.

CPP provides the following benefits:

1. It minimizes resistance to change, reduces fear of the future and stimulates creative thinking.
2. It permits the use of objectives to analyze and prepare realistic actions.
3. It serves as a formal means of coordination and integration in establishing a productivity program.
4. It sets the framework for identifying future productivity improvement possibilities, and prepare for their possible adoption.
5. It creates a basis for effective supervision of necessary actions, and improves decision making by providing a better understanding of the future so that better decisions can be made.
6. It provides a reference framework for measuring actions and initiating corrective measures.
7. It provides a basis for evaluating operational feasibility of projects, new ventures and policy changes.
8. It provides a basis for identifying the strengths, weaknesses, threats and opportunities of the organization.
9. It provides a data base for auditing the effectiveness of staff programs and the restructuring of the organization to meet plan objectives.
10. Policies are refined and objectives set based on the diagnosis of what is really going on in the organization.

COMPREHENSIVE PRODUCTIVITY PLANNING AS A COMPONENT IN THE PRODUCTIVITY MANAGEMENT TRIANGLE (PMT)

CPP is one of the components of the productivity management triangle PMT shown in Figure 1. The productivity management triangle encompasses an information system that provides input of information relevant to the planning and analysis process, performance and measurement evaluation and control process, and improvement monitoring process for the implementation of corrective actions and techniques that improve productivity. The productivity values and indices derived in the measurement phase are used for both short-term and-long term productivity planning. Forecasting techniques are used to determine productivity values and indices for future time periods.

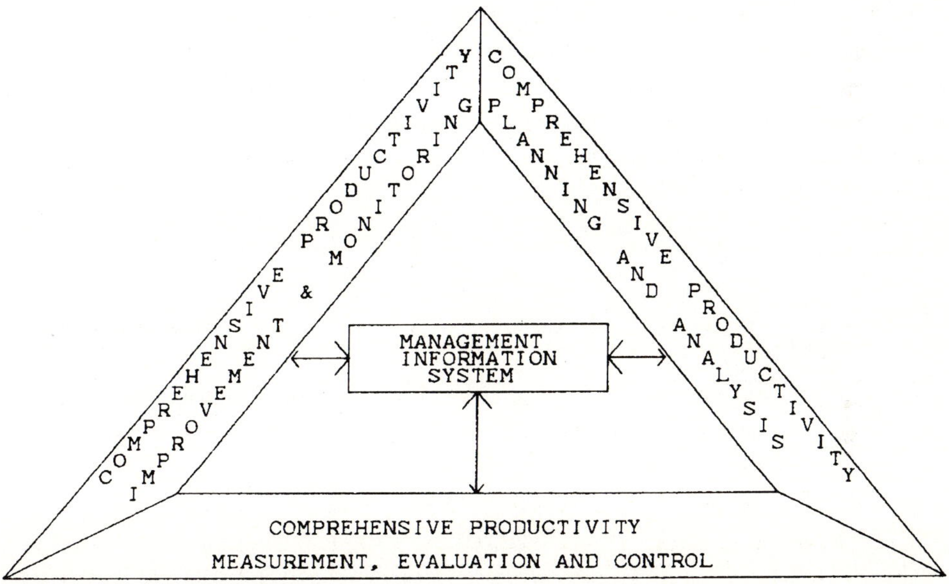

Figure 1. The Components of PMT
Source: Edosomwan, J. A., 1987. Reprinted with permission of Marcel Dekker, Inc., New York.

CHARACTERISTICS OF COMPREHENSIVE PRODUCTIVITY PLANNING

The seven major features of CPP are:

1. It must be based on viable assumptions.
2. It must have stated achievable goals and objectives.
3. It creates awareness within an organization. This can be done by explaining and defining goals and objectives to management, union, and employees.

4. CPP can have dimensions such as: time (short range, medium range, long range); organizational identification (such as certain parts of the organization - production planning, personal planning, etc.); orientation (internal, external) and scope of the plan (all factors considered - comprehensive plan, few factors considered - partial plan).
5. It requires elements of flexibility to be built into the design and it implies standards which become the basis for controlling the program.
6. It requires an input of forecasted variables since it deals with the future to make complex and interdependent decisions.
7. It requires suggestions from both management, employees, and unions in designing alternative courses of actions.

COMPREHENSIVE PRODUCTIVITY PLANNING STAGES

CPP is not a one time action, but a continuous process consisting of four interrelated stages: Productivity Planning Appraisal (PPA), Strategic Productivity Planning (SPP), Tactical Productivity Planning (TPP), and Operational Productivity Planning (OPP). These four stages are the components of the CPPC shown schematically in Figure 2.

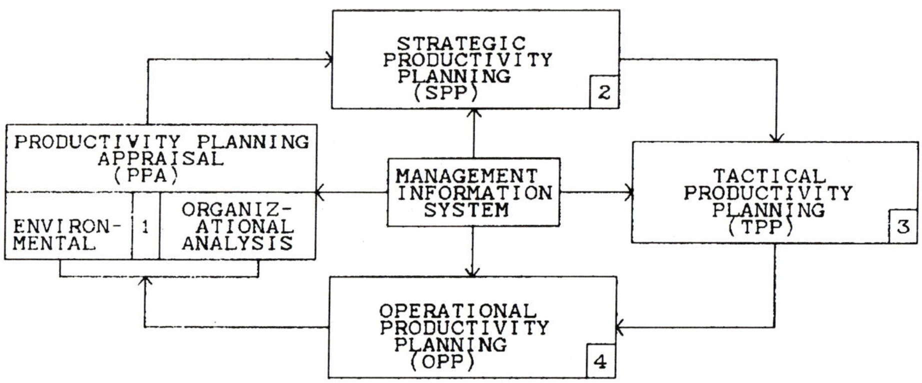

Figure 2. Components of CPPC
Source: Edosomwan, J.A. (1986 P.68)

For an organization starting a CPPP for the first time, a PPA is the first stage in the CPPC. Once the factors affecting the organization internally and externally have been assessed, a set of strategies is now formulated. Based on these strategies, short range objectives are formulated. Operational techniques are then used to implement both short and long range objectives. In order to assess the extent to which the objectives were successfully implemented, an organizational analysis is performed again. The cycle thus continues for as long as the CPPP exists in the organization.

IMPLEMENTATION METHODOLOGY

In order to implement a CPPP, major factors affecting the organization in each stage of the CPPC must be understood. Following this, an assessment of the economic impact of the plan is done. In Figure 3, the implementation steps for CPPC are presented.

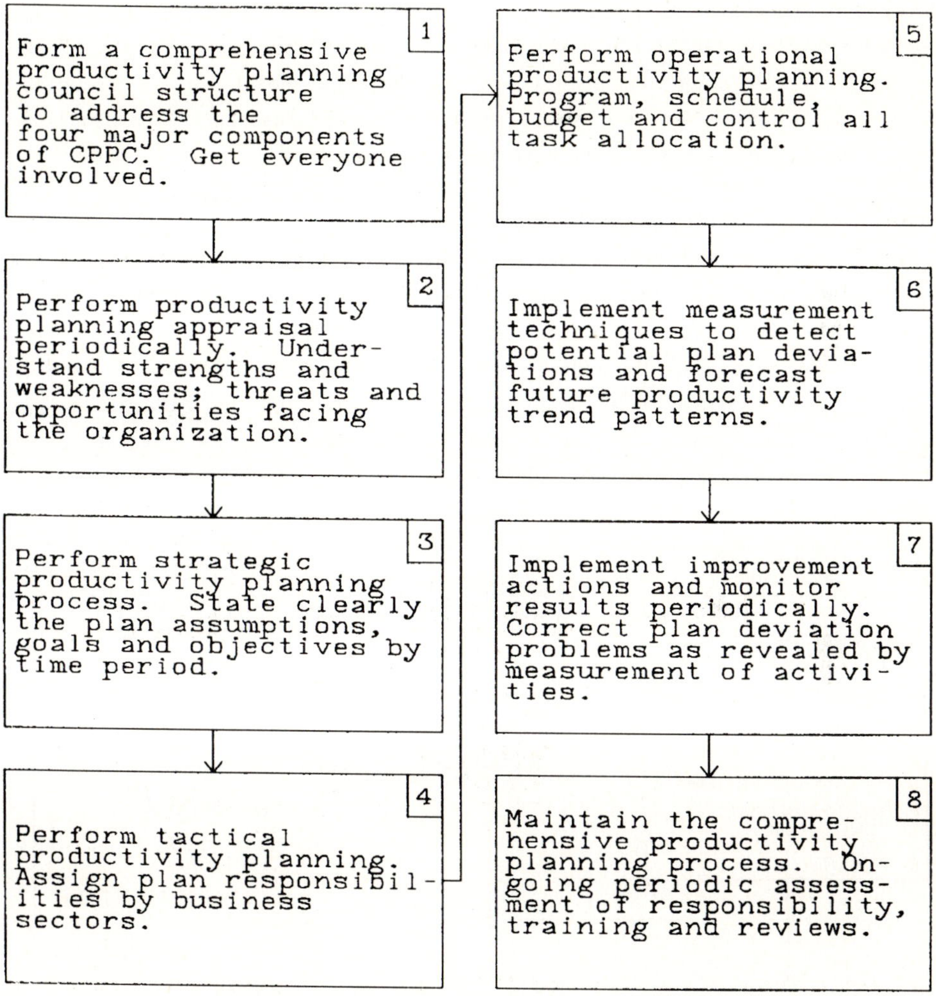

Figure 3. Implementation steps for a CPPP
Source: Edosomwan, J. A., 1987. Reprinted with permission of Marcel Dekker, Inc., New York.

STEP 1: ESTABLISHMENT OF A COMPREHENSIVE PRODUCTIVITY PROGRAM

In order to get started with a formal CPPP a planning council should be established. The council will develop basic implementation strategy including training requirements, focus on key areas of priority and structure that accommodate the four components of CPPC. The role of council will also provide technical leadership to the planning teams, facilitate total people involvement (management and employees), review plan assumptions, goals, objectives and accomplishments, and facilitate clear communication up and downwards within the organization. The planning council acts as the focal point for the organization on all planning issues. It is recommended that the council be comprised of members from all key functional areas of the business.

STEP 2: PRODUCTIVITY PLANNING APPRAISAL PROCESS (PPAP)

The purpose of the PPAP is to understand the threats and opportunities in the environment in which the organization operates, and the strengths and weaknesses of the organizations internal operating characteristics. A Productivity Plan is then developed to address both the threats and weaknesses. Action plans that will continue to maximize the opportunities and strengths of the organization are enforced. The major factors and elements to be considered in the PPAP are shown schematically in Figure 4.

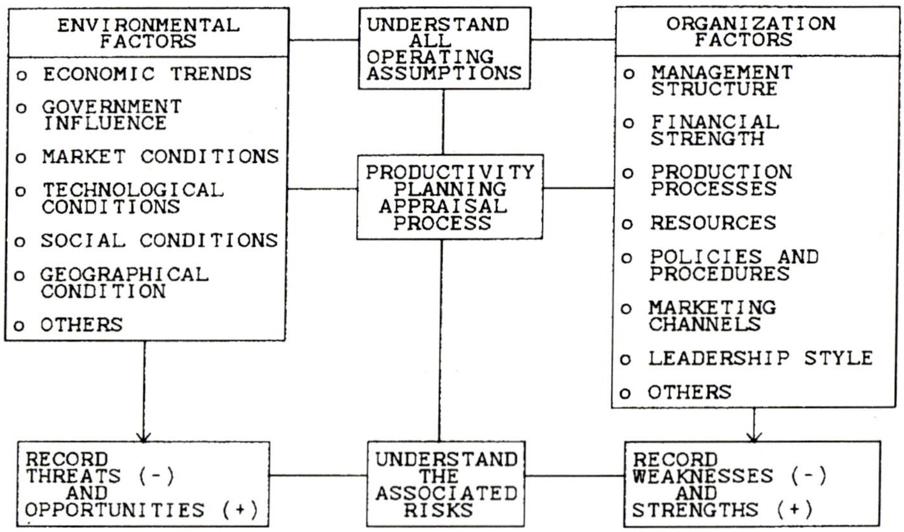

Figure 4. Factors and Elements in the PPAP.

STEP 3: STRATEGIC PRODUCTIVITY PLANNING PROCESS (SPPP)

In this step, long-range goals and objectives are formulated based on the appraisal performed in Step 2. The elements involved in the process are shown schematically in Figure 5. Additional elements involved in the strategic planning process are offered by Sink (1985) and Edosomwan (1986,1987).

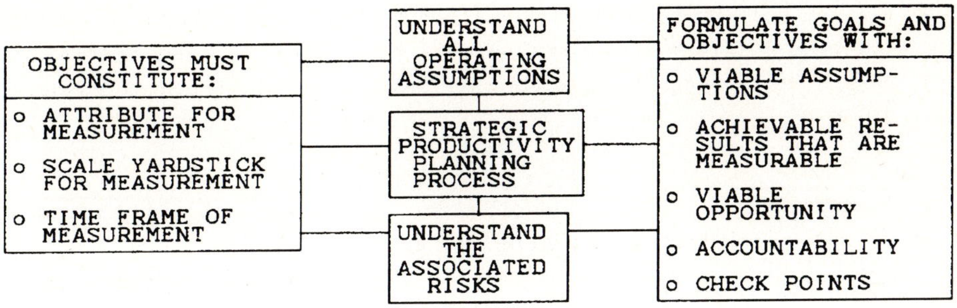

Figure 5. Elements in the SPPP

STEP 4: TACTICAL PRODUCTIVITY PLANNING PROCESS (TPPP)

The TPPP concerns itself with the means of implementing the goals and objectives specified in the strategic stage. All working assumptions and resources supporting the plans are streamlined to suit short-range business demand. The plans are evaluated and redistributed by key business sectors. The key elements involved in this process are shown in Figure 6.

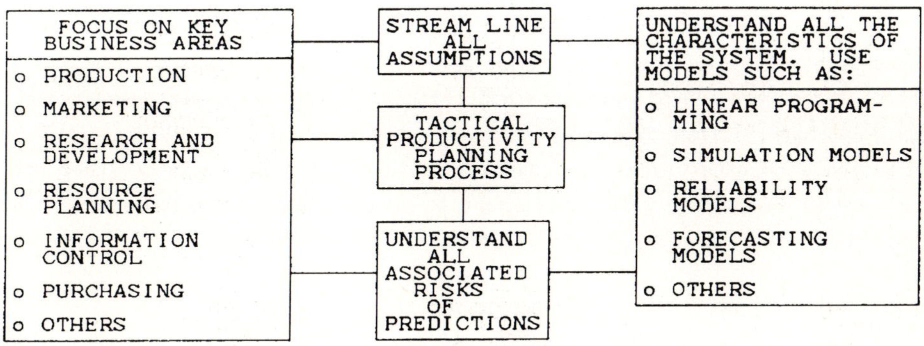

Figure 6. Elements in the TPPP

STEP 5: OPERATIONAL PRODUCTIVITY PLANNING PROCESS (OPPP)

The OPPP concerns itself with detailed implementation of the plans. It focuses on determining and delegating implementation actions and provides the means to control such actions. In this step, specific tasks are assigned to specific people with complete instructions, and progress is reviewed periodically. A system is also put in place to monitor and correct potential plan deviations. CPP implementation priorities are set in this step. This is usually based on problem areas, business demand, cost consideration, and other major factors. The elements involved in this process are shown schematically in Figure 7.

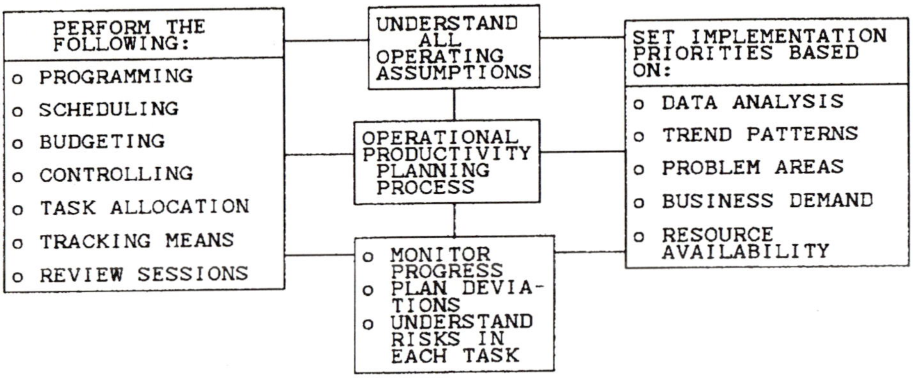

Figure 7. Elements in the OPPP

STEP 6: MEASUREMENT TECHNIQUES IMPLEMENTATION

In order to assess how well the plans are implemented, measurement techniques such as total, total factor, and partial productivities should be implemented. Edosomwan (1985) developed and tested a productivity measurement model that is unique in this type of situation. These measures should be used in conjuction with other types of financial measures such as: balance sheet rate of return on investment, risk assessment based on cash flow, and others. Record and take action on all potential problem areas.

STEP 7: IMPROVEMENT ACTIONS IMPLEMENTATION

In this step, improvement actions that address all potential problems identified in Steps 6 are implemented. Improvement action can vary from technical actions to people-oriented or customer-oriented actions. It could be task based or technology based. The key is to track the impact of all actions implemented and make modifications as necessary.

STEP 8: PROJECT MAINTENANCE

CPP is not a one-time action. It involves an on-going assessment of the organization requirements associated with the components of CPPC. Periodic reviews, training, competitive analysis, economic monitoring, and other forms of analyses have to be done continously. Data collection and information systems should be implemented to ease the burden of tracking historical issues.

COMMON PROBLEMS ENCOUNTERED DURING IMPLEMENTATION AND WAYS TO COUNTER THEM

When implementating and maintaining a CPPP, several basic problems are likely to occur. These problems vary depending on a number of factors such as the organizational setting, industry type, product mix, type of personnel, government and operating policies. Common problems that are usually encountered, and strategies for handling them are presented schematically in Figure 8. One key role of both the productivity planning council and the working teams is to develop a form that resolve these problem at the basic task levels.

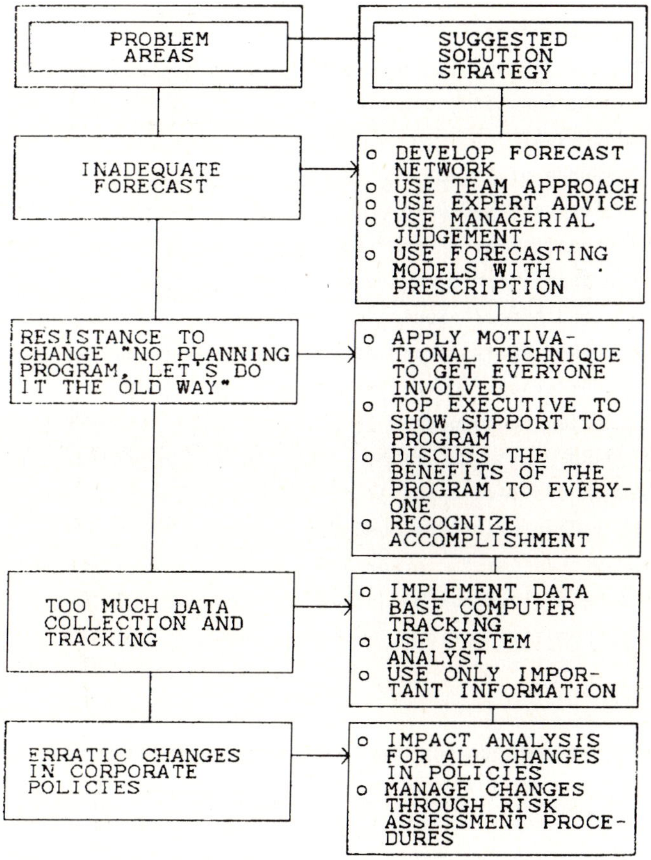

Figure 8. Basic Problems and Solutions in CPPP

COMPREHENSIVE PRODUCTIVITY PLANNING CHECKLIST

- Does the organization have a formal CPPP?
- What are the Components of CPPC within the organization?
- What are the strengths and weaknesses of the organization?
- What are the opportunities and threats facing the organization?
- What are the plan's central assumptions?
- What is the planning horizon (2 yrs, 5 yrs or 10 yrs)?
- What key risks do the organization face?
- What is the probability of occurrence for each risk?
- What is the impact of each risk on resources and net economy?
- What are the strategies to capitalize on the strengths and opportunities?
- What are the key objectives for the future?
- What planning awareness programs are in place?
- What are the mechanisms for resource allocation?
- What are the mechanisms for plan evaluation?
- What are the mechanisms for plan maintenance?
- What data base will the plan issues be tracked and maintained?
- What department, function, group of people or individual is responsible for which planning task?
- What is the source of forecasted input variables?
- Who does the forecast and with what method?
- What is the organization's understanding of the economic trend?
- What is the organization's understanding of the legal/government influence? Geographical conditions? Social conditions?
- What is the market analysis and competition? Distribution channels?
- Does the communication flow pattern support the plans?
- Does the technology, production processes, support the plan?
- What other input variables support the plan (energy, capital, materials, labor, research and development)?
- What are the mechanisms for scheduling activities?
- What are the mechanisms for programming activities?
- What are the mechanisms for budgeting and controlling?
- What are implementation priorities based upon?

CONCLUSIONS

The framework and methodology for CPP presented in this paper has wide application in both manufacturing and service organizations. Envisioning the future based on today's operating conditions and environment is often not an easy task. However, appropriate productivity planning is achievable, and manageable if supported by the total organization. The findings and issues addressed through the productivity planning can provide a substantial competitive advantage for the organization.

REFERENCES

Ackoff, Russell L., Creating the Corporate Future, New York, John Wiley and Sons, 1981.

Andrews, K. R., The Concept of Corporate Strategy, Homewood, Illinois: Dow Jones/Irwin, 1971.

Anthony, Robert. N., Planning and Control Systems: A Framework for Analysis, Boston, Harvard Business School, 1965.

Cotton, Frank, "In Productivity, Planning Is Everything," Industrial Engineering. November 1976.

Edosomwan, J. A., "A Conceptual Framework for Productivity Planning," Industrial Engineering, January, 1986.

Edosomwan, J. A., Productivity and Quality Management, Marcel Dekker Publishing Company, New York, 1987 in press.

Edosomwan, J. A., "A Task-Oriented Total Productivity Measurement Model For Electronic Printed Circuit Board Assembly," Proceedings for the first International Conference on Electronics Assembly, Santa Clara, California. October 7-9, 1985.

Jones, Harry, Preparing Company Plans: A Workbook for Effective Corporate Planning, New York, John Wiley and Sons, 1974.

Ohmae, Kenichi, The Mind of the Strategist: The Art of Japanese Business, McGraw-Hill, 1982

Quinn, James B., Strategies for change: Logical Incrementalism. Irwin 1980.

Sink, Scott. "Strategic Planning: A Crucial Step Toward a Successful Productivity Management Program" Industrial Engineering, January 1985.

Steiner, George A., Strategic Planning: What Every Manager Must Know, New York, Free Press, 1979.

Productivity Management Frontiers — I, edited by D.J. Sumanth
Elsevier Science Publishers B.V., Amsterdam, 1987 — Printed in The Netherlands

STRATEGIC PRODUCTIVITY, A MARKETING ADVANTAGE

MARSHALL N. GARTENLAUB
Corporate Consultant, Tenneco Inc., P.O. Box 2511
Houston, Texas 77001 (U.S.A.)

ABSTRACT

The secret of improving productivity is not how to solve reoccurring problems, but one of planning to avoid problems in the first place. Tenneco, a leading U.S. conglomerate, has devised innovating responses to maintain their competitive advantage as the low cost, value producer in the markets they serve.

To maintain advantage over the competition requires a dynamic system, strategically focused to continually improve productivity, comprised of four essential elements:

(1) Systems maintenance
(2) Error detection
(3) Goal checking
(4) Value building

At Tenneco we see the process of continually driving improvements as the fundamental essence to obtaining a marketing advantage.

INTRODUCTION

Tenneco Inc., a major conglomerate with 15 billion dollars in annual revenue, and many divisions in a wide diversification from Oil and Energy to Ship Building, Paper and Packaging, Agriculture, Tractors, Mining and Retailing, decided to pursue quality and productivity improvement as a company program in the mid 70's. We started Quality Circles as many other corporations did at that time. Several divisions, automotive, paper, and food, started first by establishing 40, 80, 200, and up over 300 Quality Circles. We organized, trained, facilitated and measured the growth of Quality Circles with help from our Corporate Consulting Group. The majority of Circle activity was very successful in obtaining people involvement and participative management. We also saved significant costs due to reduced waste, errors, and increased efficiencies.

After several years of activities, the Quality Circle effort was showing marginal impact. Our operating groups began to realize that significant improvement would never come by concentrating their efforts at the lowest levels of the organization. An effort was made to involve all levels of management. From Quality Circles, many of our divisions moved to a Total Quality emphasis. For several years, Total Quality became the focus for our divisional improvements. Tenneco divisions used the Crosby philosophy, the Juran process and the Deming methods, integrating each to fit their own needs.

Each of these quality gurus has a similar methodology of involving people to participate in problem solving groups. Crosby stressed attitude, Juran pushed detailed diagnosis, and Deming exuded statistics. No matter what words were used, it came out sounding alike. ". . . involve the people close to the job to solve the difficulties of that job and reduce error and waste by so doing."(a) In all, quality through problem solving was leading to better products and services being provided by Tenneco and concurrently increasing productivity.

Hundreds of successful projects brought in millions of dollars in cost savings. Quality in many divisions increased significantly, yet on a broad scale, Tenneco did not seem any better off then previously. How could that be? If we increased productivity and quality, we should have been getting better. At least that is what conventional wisdom tells us; but solving existing problems could yield productivity improvements only, if the basis of operations were fixed. When errors and mistakes keep destroying our operational base, increases in productivity, at best, just keep the operations even.(b)

How frustrating it became when we began to realize that most of the improvements these groups (Quality Circles and Improvement Teams) implemented were no more then corrections to the previous errors of those who came before! Let me explain, one very successful group had been improving a screw machine department for most of a year. They had found that the tooling was not to specification, the gears used were inconsistent, the cams were not efficient, and the thread rollers were not the best, etc. When each of these revelations was introduced, a major problem was solved. The results were praised and the group given recognition. They deserved it. They worked hard to uncover and solve the problems, but, when we analyzed the problems, we began to see that someone (or some group) let the screw machine department deteriorate some time ago.

Over and over again we found that many systems once put in place were neglected, thus quality and productivity were lowered, usually quite slowly, only to reach a point where we found the need for a team to solve the problems that had arisen due to this neglect. This was just as true in our factories as it was in our administrative and service areas. When I survey people about an existing problem, they would tell me, "well, in the past we did it so and so . . . and it was not causing any problems back then."

For example, in one of our consumer goods companies they have a life-time product guarantee. Any product, if returned to the dealer with a defect, will be replaced free. There are a number of ways to cheat the company by unscrupulous dealers and distributors. In the past two full-time Warranty Clerks were employed to monitor the integrity of the system. Then in a overhead reduction program the two clerks were removed. Warranty claims over the three ensuing years doubled, now they exceed nine million dollars annually.

A new project team was formed to solve this warranty problem. Guess what? They recommended installing a system very similar to what was in place three years ago.

The thought that the new Total Quality emphasis is a "problem solving" emphasis has been repeated often. Recent literature on productivity and Just-In-Time (JIT) refer to the Total Quality Concept (TQC) or company-wide quality system as the way to get problems solved.(c)(d) This may be true, but it represents only a partial solution to what is really needed. Problem solving inherently requires that a problem exist and that it be discovered. The difficulty with this approach is that for it to be successful a never ending supply of problems must be found to work on. Instead of directing our efforts to uncovering existing problems, after the fact, look how much more effective it would be to stop problems before they came into existence. Had we instituted proper precautionary systems in our business organizations it would have been the existence of problems we would be attacking not just their solutions.

The secret of improving productivity then becomes not how to solve reoccurring problems, but one of planning to avoid problems in the future. Thus, productivity improvements can be seen as a double edged sword.

(1) Resolve existing problems

(2) Prevent the creation of new problems

On a national basis much of our managerial energies in recent years have been spent on devising systems to resolve problems, with little effort spent on ways to prevent problems from coming into existence.(e) In their book, Restoring Our Competitive Edge, Hayes and Wheelwright state that: In order to gain a competitive edge in the market place, American businesses must focus on becoming the most productive suppliers of goods and services available.(f) How will this come about? Only when management systems are put in place capable of handling the dynamic nature of problem avoidance and resolution. As a leading U.S. corporation, Tenneco has devised innovative responses to maintain their competitive edge through becoming the low cost, value producer in the markets they serve.

In 1981, to assist the operating divisions of Tenneco world wide in improving productivity and quality, a corporate consulting group was formed. Since that time our department has been instrumental in devising responses to an ever changing business climate. What we have learned over the past five years is: maintaining an advantage over the competition in the markets we serve requires a comprehensive management system whose focus is to prevent problems from occurring rather than solving problems after they exist. Our system for problem avoidance requires a dynamic process, strategically focused on continual improvement in productivity, comprised of four essential elements.

PROBLEM AVOIDANCE

I. System Maintenance
II. Error Detection
III. Goal Checking
IV. Value Building

I. System Maintenance

The first element that became apparent in our corporate efforts to improve was that a continuing state of fitness must be imposed on all our systems. To do this, we had to maintain not only the hardware processes, but also the procedures and methods that we used to run our business.

Packaging Corporation of America, a Tenneco division with over 800 million dollars in annual sales, has been active in quality and productivity for several years. During this time, they have established 360 Corrective Action Teams (CAT) to address various levels of problems and opportunities. They estimate that between 70% and 80% of all problems under investigation were caused not by inherent design or misuse of the systems, but by decay or lack of maintaining the systems up to their original specifications. This exemplifies both the administrative and production areas of their business.

Throughout our corporation we hear the same argument many of you do, "if only we had newer or better equipment, we could have increased productivity." But, the conclusion is false. It does not take the latest and greatest innovation to obtain productivity, it takes consistency and dependability. Remember, every new piece of equipment, every new automated machine, was produced across used equipment! System Maintenance is the underlying strength to all improvement. If we do not keep our business operations capable, then we direct too much of management's energies repairing the errors and fighting fires.(g)

Maintaining all our processes and procedures up to standard does not guarantee that some errors will not creep into our systems. What our consulting group saw even with significant increases in training and preventative maintenance was that some errors did occur. We realized that what we needed was a way to detect mistakes and errors at the source, before they became real problems. Our efforts were then turned to integrating elements into our improvement process which could monitor and detect errors. Error Detection is the sum of all actions required to keep a constant vigilance on the process. This could take the form of Statistical Process Control (SPC), audits, and measurements. Our goal was to counteract the idea that today no one seems to be minding the store. Detection is what Dr. Deming refers to as the Control Element. It is the constant sampling of output in order to adjust the system in real time.(h)

II. Error Detection

At Tenneco, we have built huge libraries of Standard Operating Procedures (SOP). We maintained them faithfully, adding updates and reviews, yet meanwhile very few managers were following what those procedures said! We found machines that cost millions to build and install, well maintained, yet not being used for their intended purposes. Either being under utilized in capacity or over stressing their capabilities.

Detection can take many forms, in manufacturing it tends to follow SPC. In many of our divisions we added statistical control charts to hundreds of processes. We trained whole divisions from laborers to presidents in the use of statistics and process controls. Several of our major factories in the auto parts market earned Q-1 ratings from Ford Motor Company. Newport News Ship Building has been granted numerous citations for quality and productivity.

Error Detection also takes the form of audits. An audit is simply a periodic inspection, made to assure that a process is conforming to its requirements. It is an inspection of the process - not an inspection of output. At Tenneco we have a very elaborate system of audits. Our divisions have financial audits, safety audits, human resource audits, planning audits, budget audits, security audits, and audits on audit systems. We use both internal and external staffs to verify, and check the paperwork and computer systems we use throughout Tenneco.

Error Detection also means knowing when there is significant shifts in the outputs of a process or function. It requires us to measure ourselves as never before. We have always had detailed measurements in all production areas. It became simple to use these measurements in an error detection mode, but our office services and administrative areas had little to no measures of their output. Our Oil Exploration and Production division was one of the first in developing and measuring work elements for their accounting and administrative staffs. They first established department goals and objectives that everyone could align with. Then they measured success or failure of the department to meet those objectives in concrete terms.

One group in our corporate headquarters was approached several times. Each time they refused to establish clear objectives and stated, "what we do cannot be measured." After numerous frustrating attempts to get the group to clearly state their goals and objectives, the senior manager stepped in and said, "If the department cannot be measured then we cannot know if they are accomplishing their mission." The senior manager then disbanded the department and placed eight professionals into other positions, and never once did he miss their contribution.

We now had participative problem solving, systems maintenance and error detection functions in our corporation but we still occasionally found some divisions or departments that were not achieving expected productivity and quality improvements. We had procedures that gave inadequate results and existing product lines that were not acceptable in the market place. We found whole departments that seemed to fill no useful function, and capital projects for which we could not justify their existence.

An analysis of the misplaced problems showed that all these difficulties had a common thread. It became evident that what was once useful and viable could quickly change to irrelevance and uselessness due to the vicissitudes of the market place. What we needed was a method to verify if the original assumptions for the project or process was still valid over a long period of time. We needed to see if the current goals were in line with the outputs of the process.

III. Goal Checking

Goal checking became the third leg in our problem avoidance program. It required that any existing system be reviewed on a periodic basis to determine if the original objectives had changed to the point of making the operating system obsolete or error prone. Even if a process is in control and in a good state of repair does not assure that the output will conform to the requirements of the customer if the requirements have changed.(i)

We live in a dynamic world; too often engineers design systems with excellent reliability only to have the process outlive the function for which it is intended. Often we continue using equipment which does not conform to customer needs, simply because it is economically difficult to justify scraping them. Our focus must change from short term profit maximization to long term customer satisfaction. It is the constant appraisal of the customers need that should dictate the measures by which we judge our true productivity.

Tenneco and others had built and put in operation a multi-million dollar project for coal gasification. Its total cost was nearly a billion dollars. The project's purpose was to economically produce commercial pipeline quality gas directly from coal shale. The systems worked well, it was very capable of producing gas from coal, the facility was well maintained and the personnel well trained and quality oriented. Yet last year we closed the operation! Why? The goals originally establishing the need for the existence of a coal gasification plant were no longer relevant. Gas prices were dropping and the government was no longer seeking alternative energy sources.

The real difficulty in problem avoidance comes in this step because so many of our goals today are short term. Business decisions today are made to maximize cash flow and return on investment (ROI). Top management is focused on the quarterly dividend and stock performance rather than building sound business practices. When management tends to these fleeting goals then the processes developed to meet those goals may soon be irrelevant to changing business conditions. It becomes obvious that we can never avoid problems if management is not tending to long term development. Therefore, in order to have a system for problem avoidance, we must establish and pursue long term strategic goals.

Are long term strategic goals enough? Steps one, two and three are those needed to protect the status quo. By maintaining our existing systems, detecting any slippage from the course we are headed in, and reviewing our position against objectives, we can insure that we have a process which will be as good as it was originally intended to be. But is that enough? Shouldn't a quality system not only prevent us from getting in trouble but also keep us out of trouble in the future? As we debated that issue our consulting group came to the eventual conclusion that there is yet another step to achieve excellence through problem avoidance. That final element in our process came into being when we recognized that no matter how good a company we are, there was still significant room for improvement. In his book American Spirit Larry Miller states that, "Excellence is not an accomplishment. It is a spirit that dominates the life and soul of a person or corporation. It is the never-ending process of learning that provides its own satisfaction."(k) We felt we had to provide that spirit throughout Tenneco in such a way as to increase the value of our outputs to all customers.

IV. Value Building

In pursuing this goal Tenneco developed a national program whose theme is "Building on Quality." The idea being that Tenneco is already a quality organization in every sense of the word and we could build on that existing structure to achieve greater value in the markets we serve. Top corporate management provided the impetus for all managers and employees to seek the answer to the strategic question, "How much better an organization can we be?" As compared to the older strategic emphasis which asked divisions, "How good are you compared to your competition?"(l)

This fourth element we labeled "Value Building" because it represented a thrust that would allow us to increase our value in the market place. Not through hype or advertisements but through the tried and true procedure of increasing quality and productivity.

The challenge then became to look at our well run organizations and make sure that all their processes and procedures were obtaining maximum potential. This improvement started at the individual contributor level and worked its way through the organization. Personal achievement and goals become critical in obtaining even more improved productivity. By stretching goals beyond individual comfort levels, managers rendered vast increases in quality and productivity to their organization. A Lawrence Foster writing in the National Productivity Review states that "managers who impose realistic but high goals on their employees will be rewarded with high levels of productivity."(m) Motivation research has found irony in that those individuals who are most concerned with accomplishments, those who have the highest need to achieve, are not the individuals who set the highest goals for themselves. Put another way, participative goal setting as opposed to top-down goal setting, results in overall lower performance; other factors being equal.

The old adage that "you get what you expect" is especially true in regards to what a manager expects from his subordinate. If individual and company expectations are below what is achievable then that is all they will get. Yet, that is not to say that management can legislate excellence. The key being that effective goal setting must be realistic. Lee Iacocca in his autobiography relates that though companies are striving for excellence they must realize that top-down mandates are never enough to achieve true excellence.(n) To have an excellent company requires not platitudes but that the components of the organization be excellent. An excellent company has to have excellent people, excellent management systems, excellent accountants, excellent product development, excellent financial procedures, excellent marketing, etc.

The Tenneco Corporate Consulting Group believes strongly that the future prosperity of Tenneco and American business in general will be measured by their abilities to reach new levels of productivity and quality. American business is faced with unmerciful pressure to improve methods of operations and efficiencies in order to compete in world markets. This competition demands that we build excellent organizations capable of fulfilling customer needs at reasonable cost with perfect quality and do this in a dynamic market place that is always changing.(o)

CONCLUSION

To achieve these allusive goals of productivity, quality and customer satisfaction, companies must put forth the energies and resources necessary to drive improvements, improvements not only of today's operations but tommorrow's as well.

We must move our strategic focus from the next quarter to the long range plans of tomorrow. Solving today's problems will only gain us short term solutions. We must, as business units and as a nation, take a longer view on strategic issues. Our concentration should be on achieving maximum potential from our existing resources, human, technical and financial. This can be achieved by:

(1) Maintaining our existing system of operation and assuring that they are capable at all times.

(2) Detecting errors before they become a major destructive force.

(3) Verifying our goals to insure that all outputs are meeting our customers expectations.

(4) Building values in our markets by consistently striving to improve; to never being satisfied with the status quo.

These strategic directions will allow American business and Tenneco to achieve productivity and quality levels capable of gaining for us a competitive advantage in all the markets we serve.

REFERENCES

(a) Teresko, John. "Quality Whose Job Is It?" Industry Week, October 18, 1982.

(b) Young, John A. "Teamwork is More Easily Praised than Practiced", Quality Progress, August 1985.

(c) Schonberger, Richard J. Just In Time. Norcross, Georgia: Industry Engineering and Management Press, 1984.

(d) Juran, J. M. Management Breakthrough. New York: McGraw Hill, 1964.

(e) Crosby, Philip B. Quality is Free. New York: New American Library, 1979, p. 54-69.

(f) Hayes, Robert H., Wheelwright, Steve C. Restoring Our Competitive Edge. New York: John Wiley & Sons, 1984.

(g) Hayes, Robert H. "Why Japanese Factories Work." Harvard Business Review, July-Aug. 1981.

(h) Deming, Edwards W. Quality, Productivity and Competitive Position. Cambridge, Mass: MIT Center for Advanced Engineering Study, 1982.

(i) Crosby, Philip B. Quality without Tears. New York: McGraw Hill Book Company, 1984.

(j) Hall, Robert W. Zero Inventories. Homewood, Illinois: Dow Jones - Irwin, 1983.

(k) Miller, Lawrence W. American Spirit. New York: William Morrow and Company, 1984, p. 59.

(l) Many divisions did this by establishing corporate mission statements and building those into their long term strategies. See Exhibit I for an example by Monroe Auto Equipment Company.

(m) Foster, Lawrence W. "Better Management for Better Productivity. National Productivity Review, Vol. 2, Spring, 1983.

(n) Iacocca, Lee. IACOCCA New York: Bantam Book, 1985.

(o) Porter, Michael E. Competitive Advantage. New York: The Free Press, 1985.

EXHIBIT I
Page I-1 1987-1991 Long Range Plan

Monroe Corporate Mission Statement:

The following represents the Corporate Mission for the Monroe Auto Equipment Company:

> "Manufacture, distribute, and market the world's highest quality ride improvement products and systems, while generating a significant value perception to our customers and the ultimate consumer. Consistently achieving higher productivity levels in all areas of our business is an important ingredient to our market success. Obtaining these objectives will result in Monroe Auto Equipment Company's ability to achieve consistent long term growth in sales, profits, and profitability."

Note: Monroe Auto Equipment Company has an advantage far surpassing most "excellent" companies.

(a) They presently hold 57% of the aftermarket for shock absorbers.

(b) They have an 84.7% brand recognition in consumer testing.

(c) Their financial rate of return on assets employed exceeds 50%.

F. PRODUCTIVITY IMPROVEMENT APPROACHES

Productivity Management Frontiers — I, edited by D.J. Sumanth
Elsevier Science Publishers B.V., Amsterdam, 1987 — Printed in The Netherlands

AN APPROACH TO THE ANALYSIS OF PRODUCTIVITY CHANGES IN SPINNING DEPARTMENTS

M.F. HASSAN, M.A. SHALABY*, and M.K. ABDEL-HAMID**
*Mech Design & Production Department, Cairo University, ** Production Engineering Department, Helwan University, Egypt.

ABSTRACT
Hassan, M.F., Shalaby, M.A., and M.K. Abdel-Hamid, 1987. An Approach To The Analysis Of Productivity Changes In Spinning Departments.

An approach for analysing productivity changes in Textile companies is presented. A revision of Bela Gold approach is adopted and applied to spinning departments. Productivity ratios are derived and evaluated for two textile companies in Egypt. Capacity utilization rate, productivity of fixed investment, and fixed investment per manhour are among these ratios. Analysis shows that changes in labor productivity for each company examined are largely attributed to changes in the derived ratios. A microcomputer package is designed to automatically evaluate productivity ratios on weekly and yearly basis. The approach can be extended to other textile companies as well.

1 INTRODUCTION

The welfare of nations, enterprises, and even individuals is largely dependent on their relative productivity levels. Hence, it is commonly accepted that one of the major targets of any nation, enterprise, or individual is to improve productivity. The achievement of such target requires planned and intensive efforts in the areas: Productivity measurement, analysis, and improvement. Productivity analysis can be looked at as a bridge between productivity measurement and improvement. Productivity changes can not be interpreted unless reliable productivity analysis techniques are used. Thus the availability of these techniques is a pre-condition for carrying out any productivity improvement program.

This paper presents a model for analysing productivity in spinning departments The model was primarily developed to accomodate the available data, and the needs of the Egyptian Public Sector textile companies. A microcomputer package was also developed for conducting necessary calculations automatically. The Package has features to monitor productivity on weekly and yearly basis

2 PRODUCTIVITY ANALYSIS TECHNIQUES

During the last three decades, a number of productivity analysis techniques were developed. These techniques include:

a) Regression analysis approach [2],[7],and [8]

b) Production function approach [1]

c) Structural changes approach [7] and [9]

d) Productivity monitoring approach [7] and [10]

e) Managerial and technological ratios approach [3],[5], and [6]

The first three techniques are statistical in nature, while the last two are based on the estimation of partial and total productivity indices, and some managerial and technological ratios. By comparing these indices and ratios over time, or among similar firms or departments, causes of productivity changes can be revealed. Clearly, the application of the first group of techniques is relatively difficult. Also, it requires personnel with statistical background, so its application is limited, particularly in developing countries. The last two techniques are more favourable since they are easy to apply and to interpret.

In view of the above discussion, the proposed model is based on the managerial and technological ratios approach.

3 DEVELOPMENT OF MANAGERIAL RATIOS IN SPINNING

The approach used here is based on the Bela Gold model [5], where labor productivity is expressed as a product of three managerial ratios:

$$\text{Labor Productivity} = \begin{pmatrix}\text{Capacity}\\ \text{Utilization}\\ \text{Rate}\end{pmatrix} \begin{pmatrix}\text{Productivity}\\ \text{of fixed}\\ \text{Investment}\end{pmatrix} \begin{pmatrix}\text{Fixed}\\ \text{Investment}\\ \text{per manhour}\end{pmatrix}$$

These ratios have to be redefined to accomodate the available data and the working conditions in the intended spinning departments. The majority of public sector textile industries in Egypt operate under the direct supervision of the "General Public Sector Authority of Spinning and Weaving". Thus, the textile companies experience the same conditions as regard to working periods, accounting and costing systems, basic wage and incentive rates,...etc. This fact makes it possible to develop a set of ratios for spinning departments that is equally applicable to most companies of the textile sector. The recommended ratios are given as:

i- Capacity Utilization Rate

This ratio is decomposed into two factors; working days factor and utilization factor. Since the spinning departments normally operate for 350 days/year, and 24 hours/day,

Capacity Utilization Rate=Working Days Factor x Utilization Factor

$$= \left(\frac{\text{Number of Working Days/year}}{350}\right)\left(\frac{\text{Net Working Time /year}}{\text{Estimated Working Time/year}}\right)$$

$$= \left(\frac{\text{Estimated Working Time/year}}{\text{Number of Spindles x 24 x 350}}\right)\left(\frac{\text{Net Working Time /year}}{\text{Estimated Working Time/year}}\right)$$

ii- Productivity of Fixed Investment

Productivity of spindle/year is taken here as a measure of productivity of fixed investment in spinning departments.

Productivity of fixed investment= Productivity of spindle/year

$$= \frac{\text{Actual production} \times 24 \times 350}{\text{Net working Time/year} \times 1000}$$

iii- Fixed investment per manhour

This ratio will be replaced by the following one:

$$\left(\frac{\text{Number of spindles}}{\text{Number of production workers in final spinning}} \right)$$

To express the above ratios into mathematical forms, let

S= Number of spindles

W= Number of production workers in final spinning

P= Actual yearly production in ton-count

E= Estimated working time per year in 1000 spindle hour

N= Net working time per year in 1000 spindle hours, then:

$$\text{working days factor} = \frac{E \times 1000}{S \times 24 \times 350} = 0.119\ (E/S)$$

$$\text{Utilization factor} = N/E$$

$$\text{Productivity of Spindle/year} = \frac{P \times 24 \times 350}{N \times 1000} = 8.4\ (P/N)$$

Number of Spindles per production worker = (S/W)

Based on the above discussion, labor productivity in spinning departments (PR) can be estimated as follows:

$$PR = 0.119\left(\frac{E}{S}\right) \times \left(\frac{N}{E}\right) \times 8.4\left(\frac{P}{N}\right) \times \left(\frac{S}{W}\right)$$

4 APPLICATION OF THE PROPOSED APPROACH

The proposed model has been applied to analyse productivity in two selected companies with code names "SHARCO" and"EGYPCO". Data required were obtained from the company's annual reports of performance evaluation. The results of study are presented next.

4.1 SHARCO company

Table (1) shows the number of spindles, number of production workers in final spinning, actual production, estimated working time/year,and net working time/year in SHARCO company during the period 1974-1980/81. Table (2) summarizes the results of analysis, and Fig(1) exhibits their graphical form. As shown in fig(1) labor productivity in spinning department has a remarkable increase during 1976-1977, followed by a sharp decline during 78 and 79, and rose again in 80/81. Changes in labor productivity can be attributed to changes in the number of spindles per production worker, and productivity of spindle. Accordingly, management should be directed to investigate reasons behind the drop in number of spindles/production worker, and productivity of spindle.

4.2 EGYPCO company

Table (3) shows data collected from the spinning department in EGYPCO.

TABLE 1

Collected data for SHARCO company

Year	Number of spindles (S)	Number of production workers in final spinning (W)	Actual production per year (ton.count) (P)	Estimated working time/year (in 1000 spindle hr) (E)	Net working time/year (in 1000 spindle hr) (N)
1974	28273	564	60044	228111	209666
1975	28555	624	63537	231081	215991
1976	34795	644	76613	276164	257747
1977	34685	631	80049	280532	262534
1978	43727	780	84719	335212	299001
1979	43727	890	76992	352024	295789
80/81	39848	772	76749	349293	290268

TABLE 2

Results of Analysis for SHARCO company

Year	Working days factor		Utilization factor		Productivity of spindle per year*		Number of spindles per production worker*		Labor productivity per year (PR)	
	factor	Index no.	factor	Index no.	factor	Index no.	factor	Index no.	factor	Index no.
1974	0.960	100.0	0.919	100.0	2.41	100.0	50.13	100.0	106.59	100.0
1975	0.963	100.3	0.935	101.7	2.47	102.8	45.76	91.0	101.78	95.5
1976	0.944	98.3	0.933	101.5	2.50	103.8	54.03	107.4	118.92	111.5
1977	0.962	100.3	0.936	101.7	2.56	106.5	54.97	109.7	126.71	118.9
1978	0.912	95.0	0.892	97.0	2.38	99.0	56.06	111.5	108.60	101.8
1979	0.960	99.8	0.840	91.5	2.19	91.0	49.13	97.7	86.47	81.1
80/81	1.040	108.7	0.830	90.3	2.22	92.4	51.62	102.6	99.38	93.2

*Units are in ton.count

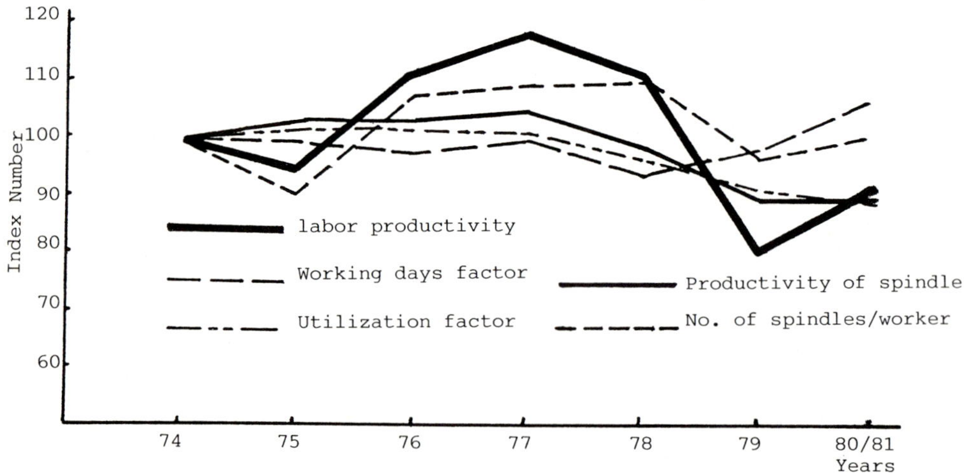

Fig.(1) Changes in labor productivity and factors affecting it - SHARCO company

Table (4) exhibits the results of analysis. Fig(2) presents changes in the index numbers of working days factor, utilization factor, productivity of spindle and number of spindles per production worker. A sharp drop in labor productivity during 75,76,and thereafter fluctuated around the 78% level,is observed in the figure. The drop in labor productivity level during the first period can be attributed to the decline in the number of spindles/production worker. The stability of productivity at about 78% level during the second period is due to the reduction of utilization and productivity of spindle.

4.3 Diagnoses of Probable Causes

Changes in the above productivity factors are found to be attributed to several reasons. Changes in utilization are mainly due to excessive down time, spare parts and raw material availability, labor attendence, and power shortage. Productivity of spindle is affected by spindle speed, labor skill, and quality of incoming yarns. The number of spindles/worker factor is affected on the short range by the number of production workers. Further investigation is needed to reveal which of the above causes are applicable for each year. The data required for such an investigation is normally available in Textile factories. A computerized data base will facilitate fast data handling and analysis.

5 PRODUCTIVITY ANALYSIS PACKAGE

The ability of the developed managerial ratios to reflect and interpret productivity changes encouraged management of the concerned companies to request a computerized analysis package. A major requirement is to analyse productivity on weekly as well as yearly basis. Management believed that weekly analysis will provide basis for taking corrective actions in time. Furthermore, management formulated other package requirements as :

1- The company's own data processing team should be able to run the package without any outside consultancy support. Top management should be able to comprehend and analyse all output results.

2- The package should utilize raw data in its current available format to minimize the need to train the company's staff to use new formats. Input/Output processes should eventually be in arabic.

3- Weekly and yearly reports should be classified, and designed differently for different management levels.

4- The package should run on a microcomputer of the IBM or compatibles variety.

5- The package should be expandable to include other departments as well.

6- The package should run in an interactive easy-to-use environment, with no prior computer knowledge assumed.

The General Public Sector Authority for Spinning & Weaving Industries requested that the package should include a data file for each spinning company

TABLE 3

Collected data for EGYPCO company

Year	Number of spindles (S)	Number of production workers in final spinning (W)	Actual production per year (ton.count) (P)	Estimated working time/year (in 1000 spindle hr) (E)	Net working time/year (in 1000 spindle hr) (N)
1974	118808	1627	292781	936120	880838
1975	118808	1838	305670	938702	891836
1976	118808	2101	304225	946662	899888
1977	119364	2104	296282	948228	875414
1978	120080	1890	255090	960400	802014
1979	119468	1579	221981	955745	707275
80/81	118808	1748	239407	869002	809668

TABLE 4

Results of Analysis for EGYPCO company

Year	Working days factor		Utilization factor		Productivity of spindle per year*		Number of spindles per production worker*		Labor productivity per year (PR)	
	factor	index no.	factor	index no.	factor	index no.	factor	index no.	factor	index no.
1974	0.938	100.0	0.94	100.0	2.79	100.0	73.02	100.0	179.95	100.0
1975	0.941	100.3	0.95	101.1	2.88	103.2	64.64	88.5	166.42	92.5
1976	0.949	102.2	0.95	101.1	2.84	101.8	56.66	77.4	144.8	80.5
1977	0.945	100.7	0.92	97.9	2.84	101.8	56.47	77.3	140.82	78.3
1978	0.952	101.5	0.84	88.8	2.67	95.7	63.53	87.0	134.86	74.9
1979	0.951	101.4	0.74	78.7	2.64	94.6	75.77	103.8	140.77	78.2
80/81	0.893	95.2	0.93	98.9	2.48	88.9	67.97	93.1	140.00	77.8

*units are in ton.count

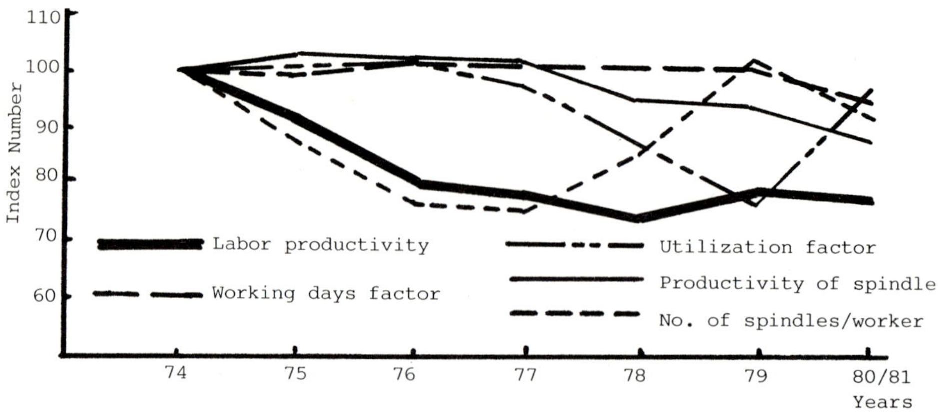

Fig.(2) Changes in labor productivity and factors affecting it - EGYPCO company

in the public sector. The objective here is to be able to monitor and evaluate the performance of several spinning companies on periodic basis. For each company, five managerial ratios are available: working days factor,utilization factor, productivity of spindlc, number of spindles per production worker, and labor productivity. Yearly ratios can be obtained by simple transformation from the weekly ratios.

Data files for this package are created and edited using a text editor (or any special program for storing random files). The fields in this file include: company name, department name,and input parameters S,W,P,E,and N (as in sec. 3). The record number in this file is related to the year and the week of the year by the relationship: record #=(year-base year) x 52 +week number. A general flow chart is given in Fig(3). There is one data file for each company, termed FILE1.DAT for SHARCO and FILE2.DAT for EGYPCO . The data entry section utilizes any text editor of the user's choice. All data are accessed by their record number. Direct accessing makes it easy to enquire about a certain week in a year or even a set of weeks. The package is written in BASIC and needs less than 64K of memory configuration.

The package screens and reports are designed to be user friendly. A one day of training is estimated to be sufficient for a middle manager to run the package. The main menu shows six options; one for labor productivity , 4 for the other factors, and a comparison option. Within each option, other menues can be called. The comparison option allows a user to compare yearly ratios for a given company over a period of years. A printed report can be obtained for all screen displays.

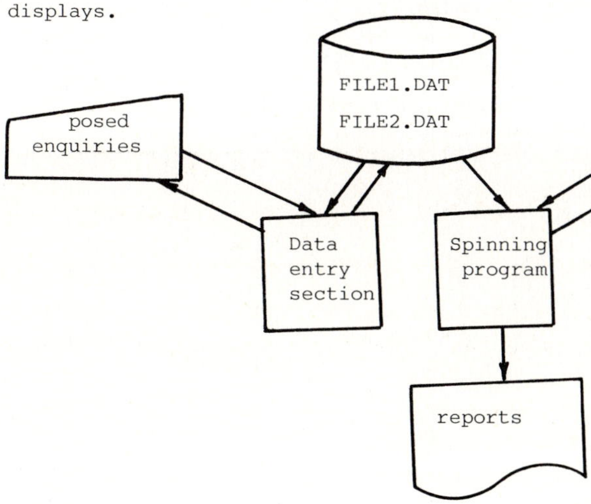

Fig.(3) Flow Chart for labor productivity analysis package.

6 CONCLUSIONS

The presented approach relies on deriving appropriate managerial ratios and using a computerized productivity analysis package. The approach has the advantages of being simple and not requiring statistical background. Also, the approach as applied is proved to be useful in monitoring specific factors that hinder productivity level in spinning departments. This can facilitate taking necessary measures for improving productivity. Successful application of the approach should consider the nature of the industrial activity and the type of available data while developing the appropriate productivity ratios. Since the approach requires extensive data handling, a computerized data base system will enhance the implementation.

7 REFERENCES

1 Boucher,T.,1981. Capital Investment and Productivity in US Metal Working Industries. In:A. Dogramatci et al(Editor), Aggregate and Industry Level Productivity. Martinus Nyhof , Boston.

2 Dogramaci, A. (Editor), 1981. Productivity Analysis - A Range of Perspectives. Martinus Nyhof, Boston.

3 Eilon, S., Gold, B.,and J. Soesan, 1976. Applied Productivity Analysis For Industry. Pergamon Press, Oxford.

4 Eilon, S.1985. A Framework for Profitability and Productivity Measurements. Interfaces, 15(3). pp31-40.

5 Gold, B.,1979. Productivity, Technology, and Capital. Lexington Books, Lexington.

6 Gold, B.,1985. Foundations for Strategic Planning for Productivity Improvement. Interfaces, 15(3). pp 15-30

7 Hassan,M.F., Shalaby, M.A.,and P.M. Ghare,1985. A Study for Measuring and Improving Productivity in a Basic Industrial Sector in Egypt. Tech Report FRCU 81019, Cairo University. Cairo,Egypt.

8 Hayes, R.H., and K.B. Clark, 1985. Explaining Observed Productivity Differentials Between Plants: Implications for Operations Research. Interfaces, 15(6). pp 3-14

9 Horn, K.H., 1967. Structural Changes and Productivity. Tech Report 773, The Institute of National Planning. Cairo, Egypt.

10 Sumanth, D.J.,1983. Productivity Engineering and Management. McGraw Hill, NJ.

Productivity Management Frontiers — I, edited by D.J. Sumanth
Elsevier Science Publishers B.V., Amsterdam, 1987 — Printed in The Netherlands

PROMIG: A MODEL TO BOOST TOTAL PRODUCTIVITY THROUGH EMPLOYEES INVOLVEMENT AND SYSTEMATIC PROGRAMS

By:
J.I. LOPEZ ARRIORTUA
General Motors España, Industrial Organization Department Director, P.O. Box 375 ZARAGOZA (Spain)

ABSTRACT

López Arriortua J.I., General Motors España, 1987. Promig: A model to boost total productivity through employees involvement and systematic programs.

PROMIG is a model and organizative system, specially created for today's industry. Its objective is the optimization of business management and as a consequence the improvement of company's profitability.

It is based on the sinergy of the stregths generated by:

- The systematic application of specially designed techniques, directed towards improvement of quality, total productivity and product value.

and at the same time by:

- Usage of designed programs for the improvements of participation, integration and motivation of the company's personnel.
  Techniques and prograsm which jointly applied, generate creativity, improvements and in overall spectacular results.

1. INTRODUCTION

In today's business world, managerial quality becomes more critical everyday. The market opportunities are available for everyone. The technology is also accessible to any company. But the main factor for success or failure depends, to a high degree, on a the quality of business management.

In all the studies performed in the main industrialized countries, there is a common conclusion: the companies with success, are those which have a model and an overall management system, better than those of its competition.

Due to the above mentioned we proposed the consideration of a new concept named: Quality Management Assurance, which guarantees the perfection of overall management in all levels and company fields, and therefore the improvement of total productivity. In this line, we have developed a working tool in order to systematize this new concept: Promig model which has been applied with success in different companies in Spain.

2. DEFINITION OF PROMIG

(Program for the overall improvement of management)

Promig is a model and organizative system, specially created by us for today's industry. Its objective is the optimization of business management of each company and as a consequence the improvement of company's profitability.

It is based on the sinergy of the strengths generated by:

- The systematic application of specially designed techniques, directed towards improvements of quality, total productivity and product value.

and at the same time by:

- Usage of designed programs for the improvement of participation, integration and motivation of the company's personnel.

Techniques and programs which jointly applied, generate creativity, improvements and in overall spectacular results.

In figure 1 representing in the axle OY the degree of emphasis placed on improving the total productivity and quality, and in the axle OX the emphasis placed on reaching the employees participation, integration and commitment, and dividing the graphic in four areas, it is in the right upper quadrant, where most of the companies with success are placed. This is the area where the optimization of quality, productivity and people involvement drives the companies to the excellence.

We have developed a group of applicable techniques for the improvement of total productivity and quality that applied systematically move any company to the upper positions in the OY axle.

Techniques for:

Optimization of purchase prices with programs taylored for Productivity Improvement at Suppliers, Make or Buy studies, Value Analysis, Purchasing Price Evaluation, etc.

Optimization of quality, through: Quality Assurance, Statistical Process Control, Reliability, Auto Control, Zero Defects, Improvement on Suppliers Quality, etc.

Rationalization of the productive process by making MTM studies included in Mergomet system, Ergonomics studies, Robotics, Layout, etc.

Optimization of the usage of material and stocks by applying Value Analysis techniques, Material studies, Juts-in-Time approach, etc.

Rationalization of the administrative work using in all these areas the Overhead Value Analysis techniques, etc.

Application of systematic programs for cost reduction.

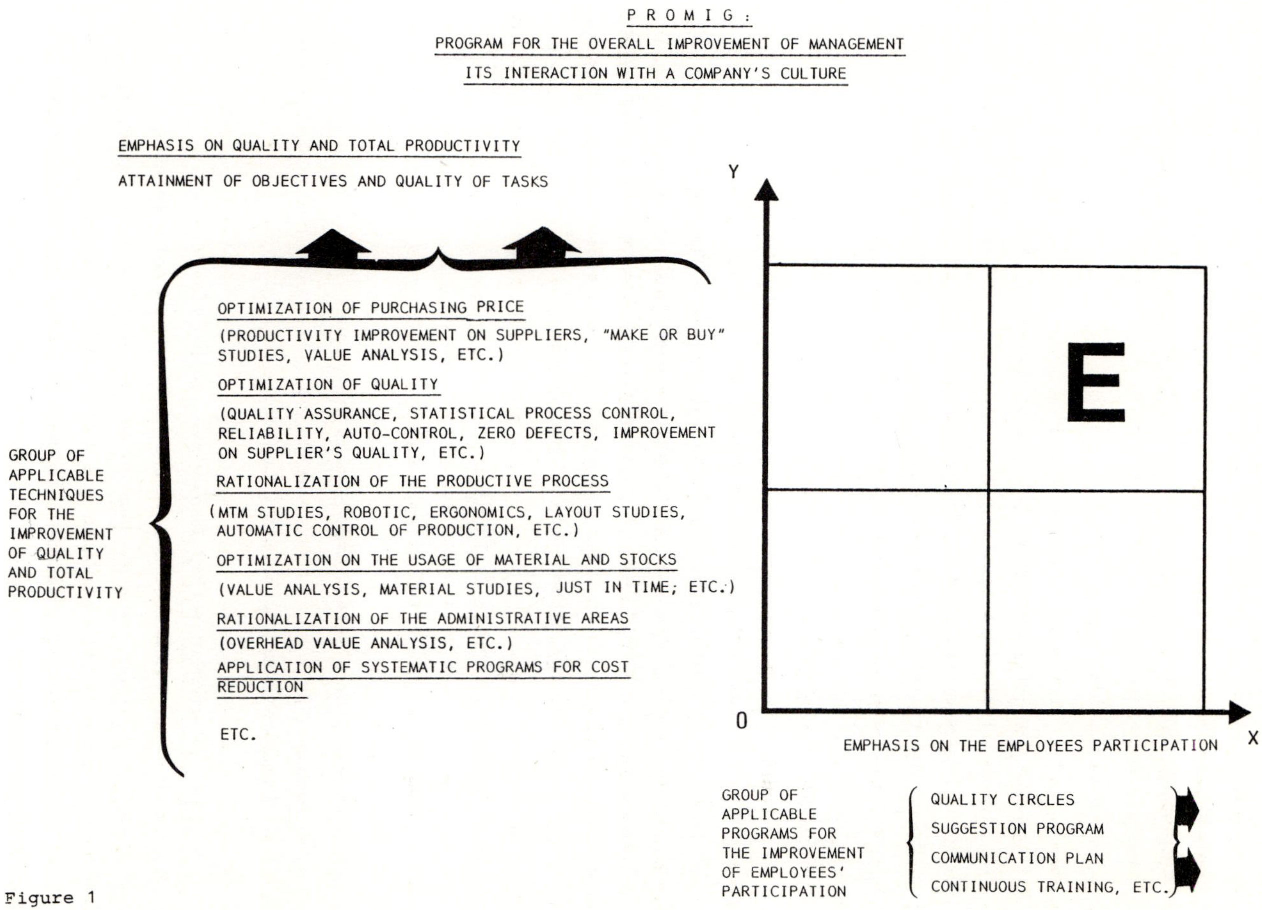
P R O M I G :
PROGRAM FOR THE OVERALL IMPROVEMENT OF MANAGEMENT
ITS INTERACTION WITH A COMPANY'S CULTURE
EMPHASIS ON QUALITY AND TOTAL PRODUCTIVITY
ATTAINMENT OF OBJECTIVES AND QUALITY OF TASKS
Y
GROUP OF APPLICABLE TECHNIQUES FOR THE IMPROVEMENT OF QUALITY AND TOTAL PRODUCTIVITY
OPTIMIZATION OF PURCHASING PRICE
(PRODUCTIVITY IMPROVEMENT ON SUPPLIERS, "MAKE OR BUY" STUDIES, VALUE ANALYSIS, ETC.)
OPTIMIZATION OF QUALITY
(QUALITY ASSURANCE, STATISTICAL PROCESS CONTROL, RELIABILITY, AUTO-CONTROL, ZERO DEFECTS, IMPROVEMENT ON SUPPLIER'S QUALITY, ETC.)
RATIONALIZATION OF THE PRODUCTIVE PROCESS
(MTM STUDIES, ROBOTIC, ERGONOMICS, LAYOUT STUDIES, AUTOMATIC CONTROL OF PRODUCTION, ETC.)
OPTIMIZATION ON THE USAGE OF MATERIAL AND STOCKS
(VALUE ANALYSIS, MATERIAL STUDIES, JUST IN TIME; ETC.)
RATIONALIZATION OF THE ADMINISTRATIVE AREAS
(OVERHEAD VALUE ANALYSIS, ETC.)
APPLICATION OF SYSTEMATIC PROGRAMS FOR COST REDUCTION
ETC.
E
0
X
EMPHASIS ON THE EMPLOYEES PARTICIPATION
GROUP OF APPLICABLE PROGRAMS FOR THE IMPROVEMENT OF EMPLOYEES' PARTICIPATION
QUALITY CIRCLES
SUGGESTION PROGRAM
COMMUNICATION PLAN
CONTINUOUS TRAINING, ETC.

Figure 1

But the singularity of our concept is the application, at the same time, of a group of programs for the improvement of employees participation. programs such as:

- Quality Circles or Employees Participation Groups
- Suggestion Program
- Communication Plans
- Continuous training
- Etc.

that tend to move the company to the right side of OX axle.

We have realized, that applying this model and concept, it is generated a sinergy of the stregths comming from the simultaneous application of these techniques and these programs, that reinforce, as a catalyzer, the power of each technique and allows a continuous improvement in the company fields.

3. METHODOLOGY OF PROMIG (See figure 2)

In any company, we have objectives on one side and on the other side we are looking for the best global results.

In order to reach this, we have designed programs and techniques (compiled in Promig) specially taylored for each objective and we are applying them in a closed loop. We are measuring the results, taking new decisions and acting through our programs again and again without stopping. This is in summary the methodology of our Promig.

4. BRIEF SUMMARY OF TWO MAIN PROGRAMS

Now we are presenting a brief summary of two main programs, selected from each of the before presented axles X and Y.

The first is:

- Mergomet (Ergonomic Method Improvement Technique)

and the second:

- E.P.G. (Employees Participation Groups)

4.1 Mergomet (Ergonomic Method Improvement Technique)

Mergomet is an integrated analysis system for the work methods improvement, based on the breakdown of every operation into micromovements and on the study of body position and the muscles involved.

It is focused towards methods improvements, ergonomic improvements of work places, to the work content measurement, planning and design of equipment and tools, (see figure 3).

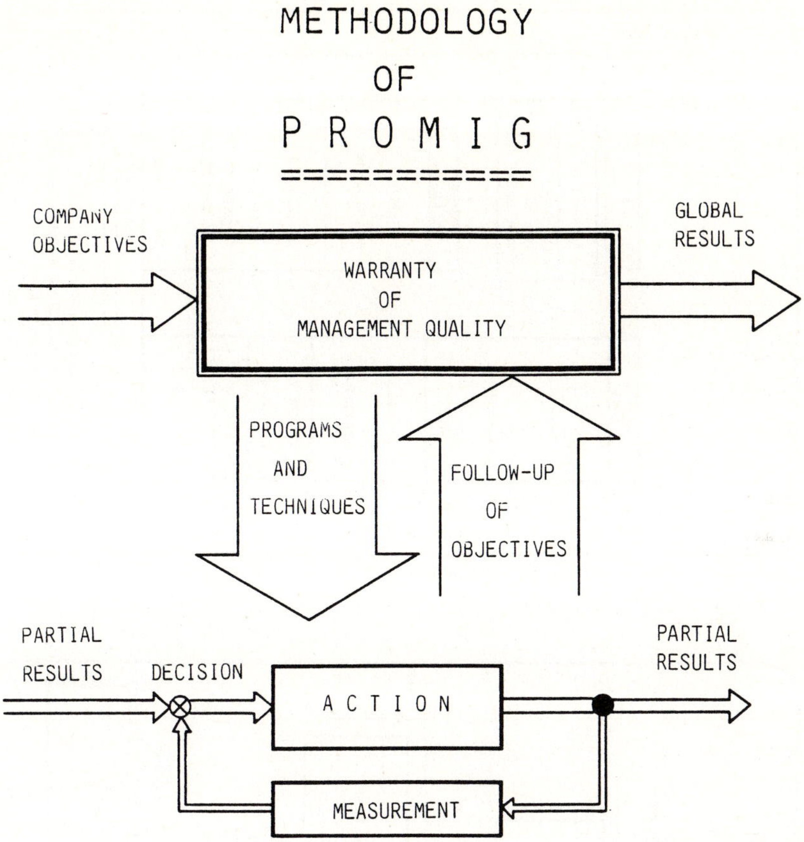

<u>OBJECTIVES</u>	<u>PROGRAMS</u>
- PRODUCTIVITY	- ERGONOMIC IMPROVEMENT OF METHODS BY MICROMOVEMENTS STUDY
- QUALITY	- STATISTICAL PROCESS CONTROL/AUTOCONTROL
- MOTIVATION	- E. P. G.
- WORK CONDITIONS	- ERGONOMICS
- PRODUCTION	- JUST IN TIME APPROACH
- ADMINISTRATIVE EFFICIENCY	- OVERHEAD VALUE ANALYSIS
- COSTS	- VALUE ANALYSIS / MAKE OR BUY

Figure 2

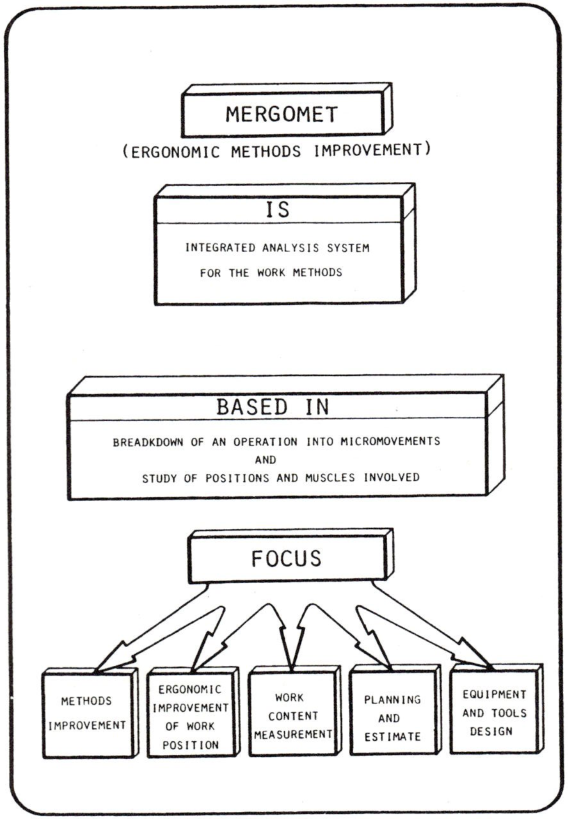

Figure 3

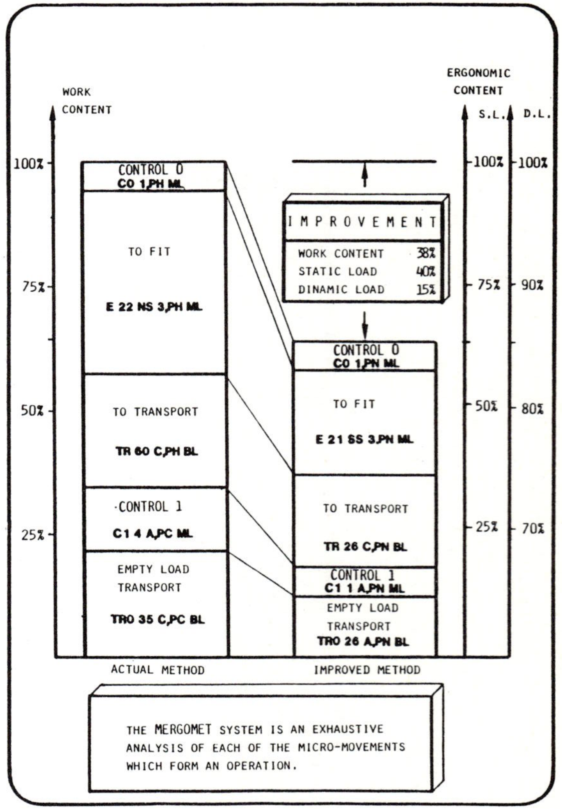

Figure 4

Figure 4 shows one example of the comparison between the work method before and after its analysis by the Mergomet system, where it can be appreciated 38% reduction in the work content and at the same time, a substantial improvement in the ergonomic conditions, by reducing 40% the static load and 15% the dynamic load.

This result is obtained by the analysis and optimization of the variables of each micromovement of the method, and building a new method with improved micromovements.

A computer program is available to assist and support the engineers with the identification of the points to be improved in the different areas under study.

Its content will be briefly explained: (see figure 5)

Once performed the Mergomet analysis of the present work process, the analyst will introduce the information in the computer.

Through an established algorhythm and using the work content data bank and the ergonomic data bank, the computer will offer two different outputs:

- The present work content and recommendations for the improvement of this work content.
- The present ergonomic content and recommendations for the improvement of this ergonomic content.

Using both outputs as a base line, the engineer will modify the various involved factors of the analyzed operative method, to achieve an improved method.

These improvements feed-back the system, in order to obtain the best operating method.

This is a powerful tool that allows us to improve systematically our working procedures from the point of view of ergonomic conditions and the work content.

4.1.1 Mergomet System Advantages

Mergomet system offers a lot of advantages when compared with other traditional analysis systems.

As an example we will point out some of them:

- Focus the attention of the engineers on the method improvement, considering time and ergonomics.
- Drives to the performance of the operating methods analysis, achieving optimum results.
- Allows the usage of time formula and the computer data processing.
- Facilitates the design optimization for the tools and auxiliary production elements.

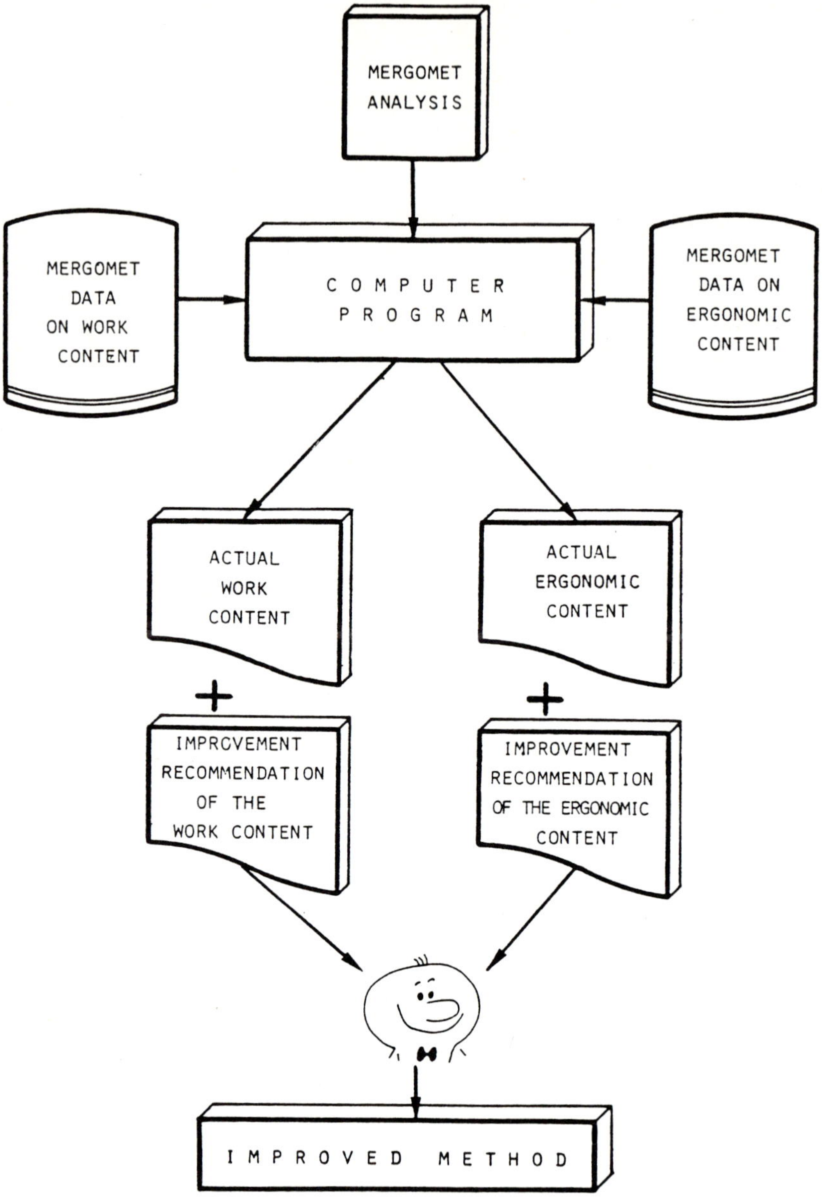
MERGOMET ANALYSIS
MERGOMET DATA ON WORK CONTENT
COMPUTER PROGRAM
MERGOMET DATA ON ERGONOMIC CONTENT
ACTUAL WORK CONTENT
ACTUAL ERGONOMIC CONTENT
+
+
IMPROVEMENT RECOMMENDATION OF THE WORK CONTENT
IMPROVEMENT RECOMMENDATION OF THE ERGONOMIC CONTENT
IMPROVED METHOD

Figure 5

- Facilitates the introduction of new methods, since they are perfectly defined.
- Eliminates the subjectivity, component of other systems. The results are more objective.
- Allows the study of brief operations, which are overlooked by other systems.
- Offers the possibility of obtaining a reliable time during the planning phase.

4.2 E.P.G. (Quality Circles or Employees Participation Groups)

The program is related to participation, motivation and integration of the people in order to improve the creativity which is an important aspect of Value Management. This program is Quality Circles or Employees Partipation Groups (E.P.G.), as we call it in our company. The example we are showing is related to a technical department.

The personnel participation concept can be contemplated under two points of view: (see figure 6)

a) As an objective:
Which means the organization commitment for making compatible the company objectives (such as productivity, quality costs, image, etc.), and individual needs and objectives of the personnel (such as professional self-development, work satisfaction, etc.)

b) As a process:
Which means the planned and systematic effort, to achieve the active personnel participation.

E.P.G. promote creativity and ideas, that - turned into quality and productivity improvements - are a source of motivation for the group, increasing their conciousness to generate and implement new ideas, participating in the company objectives (see figure 7).

The E.P.G. are teams of 6 to 12 people, joined in order to select, analyze and solve together problems about:
- Quality
- Productivity
- Cost reduction
- Procedures
- Etc.

The prerequisites for a successful introduction of E.P.G. in any company are:
- Adaptation of the company philosophy to the E.P.G. participative concept.
- Commitment and support program from all hierarchy levels.

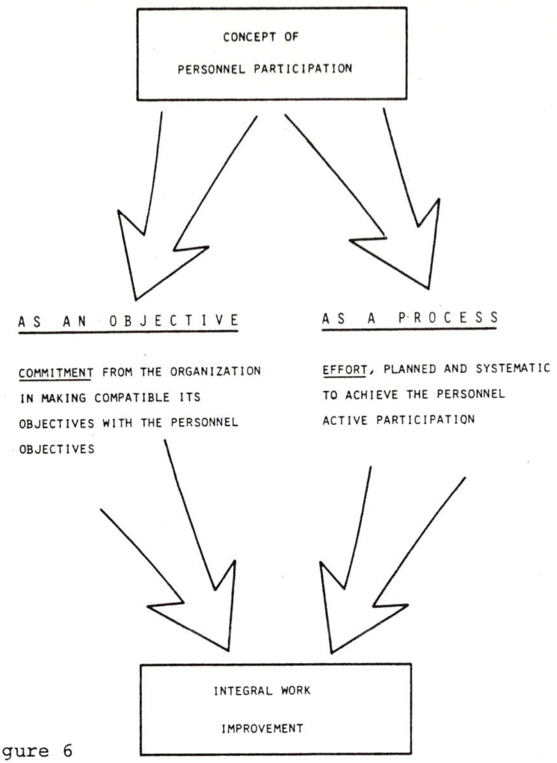

Figure 6

INTERATION BETWEEN THE E.P.G. AND THE TOTAL PRODUCTIVITY QUALITY IMPROVEMENT

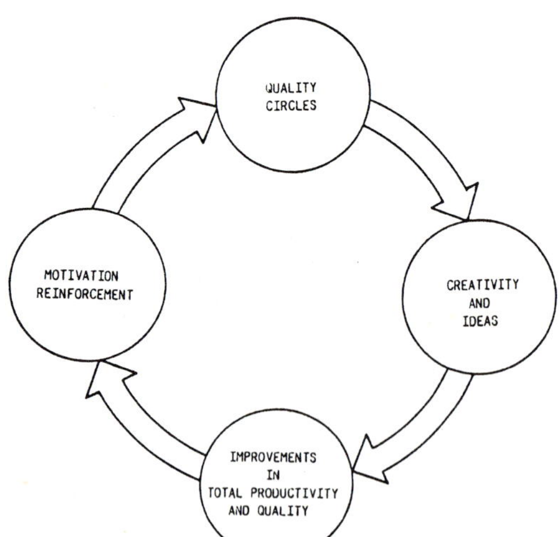

Figure 7

- Training to each E.P.G. member.
- Systematic process and patience.

A twenty hours training course on E.P.G. principles and problem solving techniques was given to all department members-divided into four groups.

As a result, seven E.P.G. groups were created with the voluntary participation of all the members of our technical department. The structure is based on the matrix organization chart of the department, but with the following special characteristics (see figure 8):

Each member can participate in two E.P.G.'s: the one belonging to his natural working group and another one of his choice, taking into account his professional interests as well as the educational enlargement needs and havings the opportunity of rotating annually among the different E.P.G. groups.

For example, a member X from area 4 belongs to the E.P.G. from his functional area and at the same time can participate in one from any of the others operative or functional areas, at his free choice.

This is how all members involved can become acquainted with all areas and functions within the department, improving their professional knowledge, the communication between groups and the motivation of the individuals.

In figure 9 we show the summary of the solution of a typical case solved by an E.P.G.

More important than the savings obtained, are the benefits for the teams by working together, improving their communications and enhancing the interest of the people for the quality of their daily work.

5. NEED OF PROMIG APPLICATION TO THE ACTUAL COMPANIES

Figure 10 summarizes the cyclical nature concept of a company's activities.

Currently we could place the majority of our industry within a transitional or critical period or even in the decline area.

The only way to come out successfully from this situation is getting into a creative phase. Promig has just been designed for the reinforcement of creativity within a systematic and practical way.

If there is not such a reaction, the decline will continue until the company's disappearance.

6. IMPROVEMENT OF COMPANIES COMPETITIVE SITUATION (See figure 11)

Promig improves the competitiveness of any company by:

- Reinforcing the obstacles for the entrance of competitors obtaining high productivity, low cost, high quality ect.

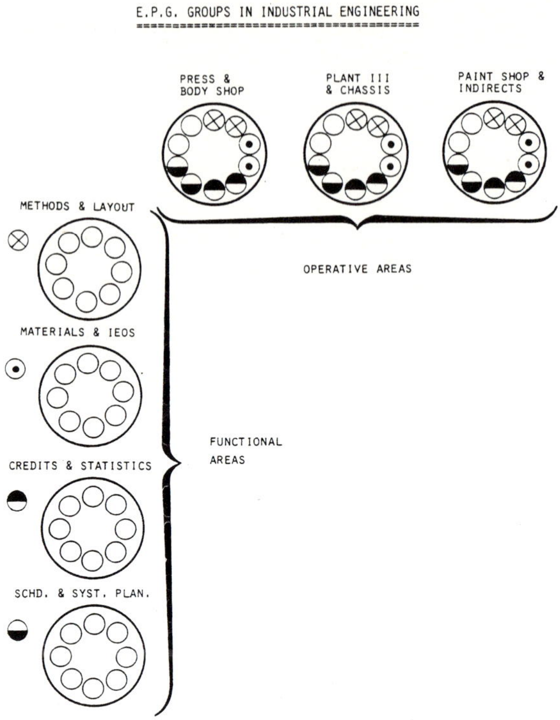

Figure 8

MEETING NR.	DATE	NUMBER OF PARTICIPANTS	MEETING DURATION	ACTIVITY	RESULTS
1		10	1.5 HR.	BRAINSTORMING	22 NEW IDEAS
2		9	1.0 HR.	ANALYSIS, CRITIQUE AND SELECTION OF IDEAS UPON CRITERION BASED ON: - APPLICATION FACILITY - REALISTIC POSSIBILITY - ORIGINALITY	SELECTION OF 9 IDEAS
3		12	1.5 HR.	CLASSIFICATION AND VALUATION OF IDEAS BY: - SERIOUSNESS, IMPORTANCE - URGENCY - TENDENCY	SELECTION OF THE IDEA TO BE DEVELOPED
4		10	2.0 HRS.	METHOD DEFINITION	ACTIVITIES DIAGRAM
5		12	1.0 HR.	TO COLLECT INFORMATION	ATTAINMENT OF THE NECESSARY DATA.
6		12	2.0 HRS.	PRESENTATION OF THE COMPILED INFORMATION	ATTAINMENT OF TECHNICAL DATA AND MTM STUDIES.
7		10	1.5 HRS.	PROPOSALS, ANALYSIS, SELECTION AND ELABORATION OF IMPLANTATION PROGRAM.	START-UP PROGRAM
8		12	1.0 HR.	COMMUNICATION OF IMPLANTATION AND OBTAINED RESULTS.	SAVINS OF 1.550.319 PTS./YEAR.

Figure 9

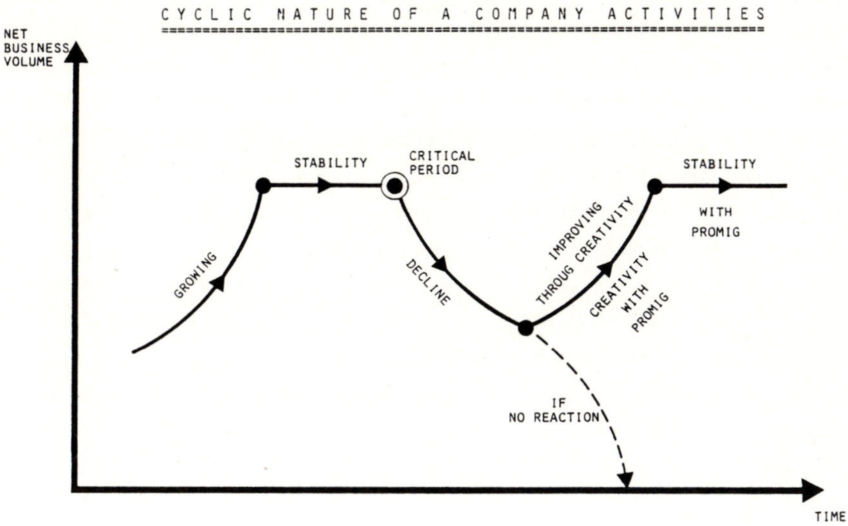

Figure 10

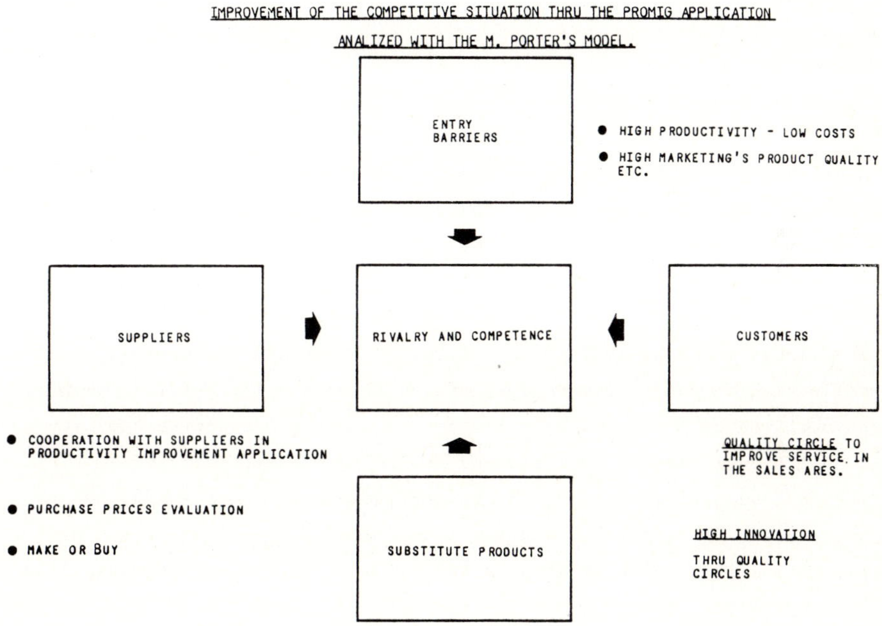

Figure 11

- Optimizing the suppliers situation (application of a program for the optimization of purchasing price)
- Improving the perspectives and customer's acceptance (application of quality circles with customers)
- Introducing substitute products (improving the innovation thru Value Analysis, Quality Circles, etc.)

Definitively, Promig improves all and each one of the variables which take part in the competitive companies position.

7. THE COMPANY'S INTEGRATION ON ITS SOCIO-ECOLOGICAL ENVIRONMENT

Over the past years social evolution shows a greater sensitivity of public opinion with regards to company interaction with its socio-ecological environment. The importance of this concept will be increased within the coming years.

Promig has developed an integration model, as indicated in figure 12, where is recommended the application of our philosophy for quality circles in the interphases of the company with different organisms, such as:

- University
- Local authorities
- Consumers
- Government's organisms
- etc.

Working together with these groups, the image of the company is improving and at the same time there is an enhancement of mixed economies that at the end is for the benefit of all the participants.

8. THE GLOBAL STRATEGY IMPLIED IN PROMIG (See figure 13)

The global strategy implied in Promig, is to obtain the sinergy of all significative company's dimensions, basing it on the integration process, which utilizes up to the maximum all of the company's creative capacity, as well as its environment, applying the following strategies

- Reinforce and optimize the internal performance of the company, by improvement of productivity, quality and product value; improvement of participation and personnel integration, as well as the correct adaptaion of technology to the product and process.
- Improve the company's competitive position by reinforcing the barriers to competition, improving the suppliers, as well as the customer's service and product's distribution.
- Obtain the company's integration in the environment, improving its image and collaborating with the different existing organisms in the area.

PROMIG AND COMPANY'S INTEGRATION ON ITS SOCIO-ECONOMICAL ENVIRONMENT

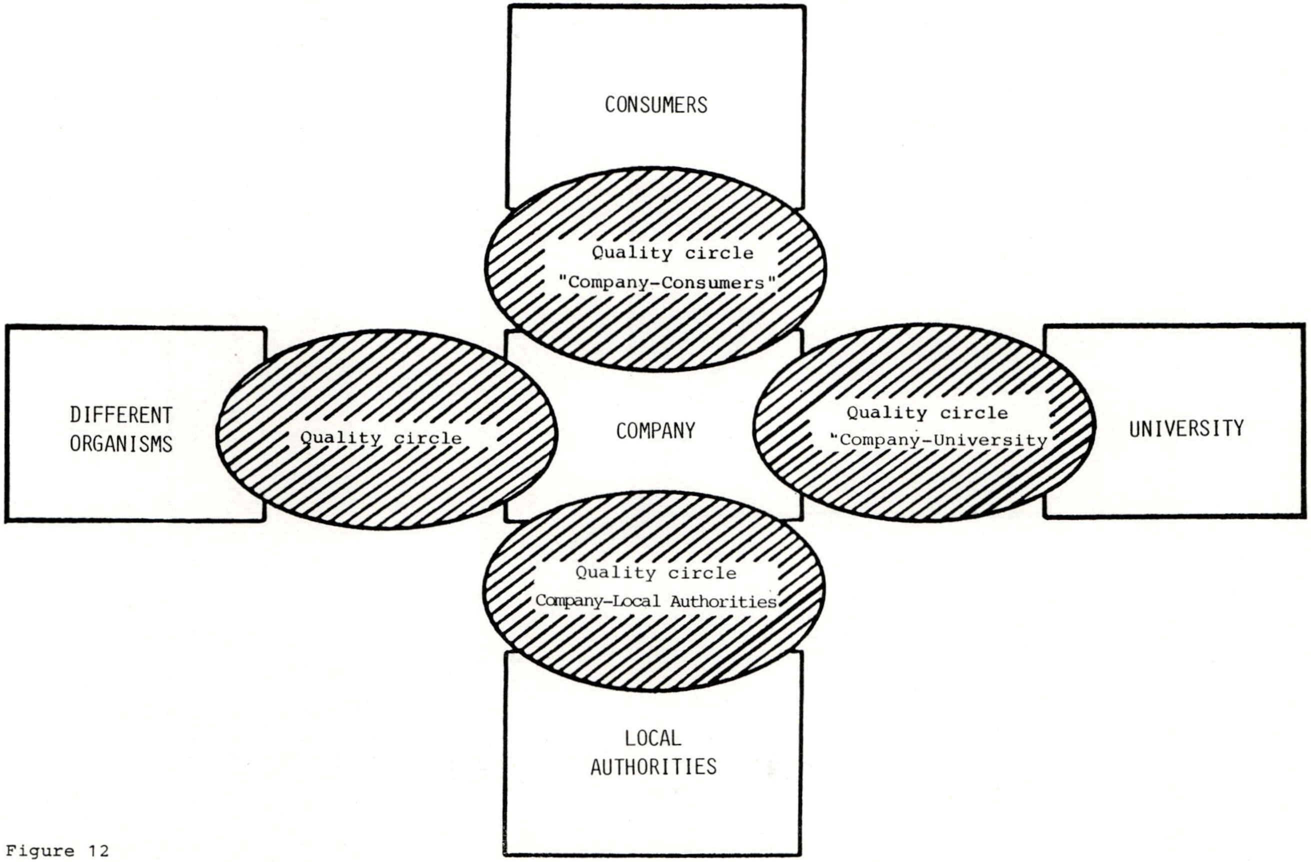

Figure 12

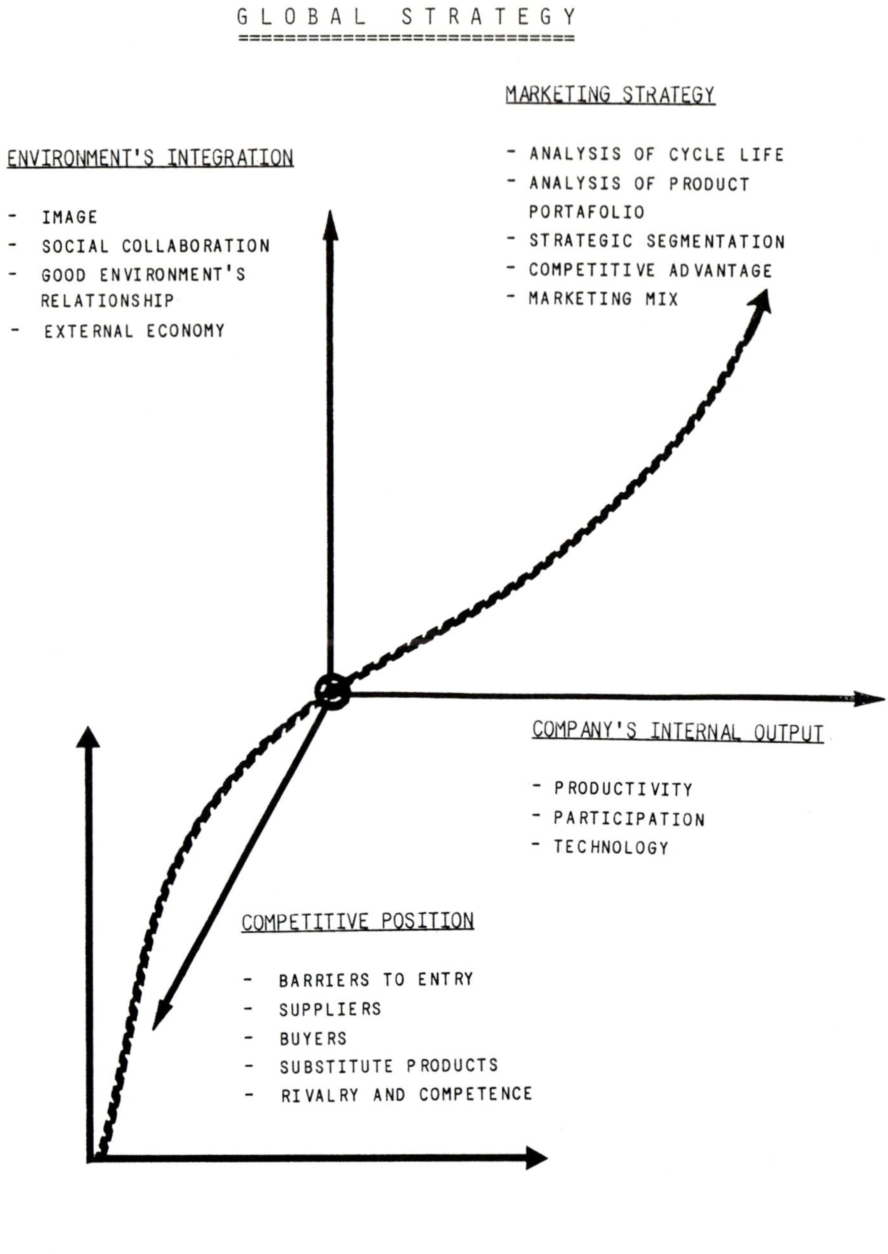
GLOBAL STRATEGY
MARKETING STRATEGY
- ANALYSIS OF CYCLE LIFE
- ANALYSIS OF PRODUCT PORTAFOLIO
- STRATEGIC SEGMENTATION
- COMPETITIVE ADVANTAGE
- MARKETING MIX
ENVIRONMENT'S INTEGRATION
- IMAGE
- SOCIAL COLLABORATION
- GOOD ENVIRONMENT'S RELATIONSHIP
- EXTERNAL ECONOMY
COMPANY'S INTERNAL OUTPUT
- PRODUCTIVITY
- PARTICIPATION
- TECHNOLOGY
COMPETITIVE POSITION
- BARRIERS TO ENTRY
- SUPPLIERS
- BUYERS
- SUBSTITUTE PRODUCTS
- RIVALRY AND COMPETENCE

Figure 13

And always guided by the marketing strategy, that based on the analysis of cycle life and product portafolio through a strategic segmentation of the market and the application of an optimum marketing mix shoots for a competitive advantage that assures the company success.

Promig has the necessary techniques and programs to reach all these objectives through the creativity in each of these fields.

9. THE MULTIPLIER EFFECT (See figure 14)

Promig's final objective is the generation of a profitable multiplier effect for the whole industry of a country.

If Promig is applied to the leading industries in a community, we obtain a first optimization. However, this benefitial effect can also be applied through the multiplier effect on the leading industries as these transmit the new programs and techniques of management to their suppliers and auxiliary companies, obtaining with this an improvement which will be shared by both parts.

10. SUMMARY (See figure 15)

And now, as a summary I am showing you again the managing philosophy included in our Promig, that consists in placing the company where there is the maximum emphasis in improving productivity and quality through the attainment of objectives and the quality of tasks performed by each employee, and at the same time also maximum emphasis in increasing the level of employees participation, commitment and integration. In this way it is possible to make coincide the personal interest of people with the overall interest of the company.

In order to do this we recommend to develop programs and techniques and apply them systematically wihtout stopping, knowing that the perfection is never achieved, and there is always room for improvement.

We are happy to say that in our own experience we could realize the meaning of the sinergy resulting from the simultaneous application of these different programs and techniques and demonstrate that two plus two is more than four with this model. The rule is to taylor the necessary tools to the specific situation environment and apply them with faith and perseverance.

MULTIPLIER EFFECT OF THE OVERALL IMPROVEMENT MANAGEMENT'S PROGRAM

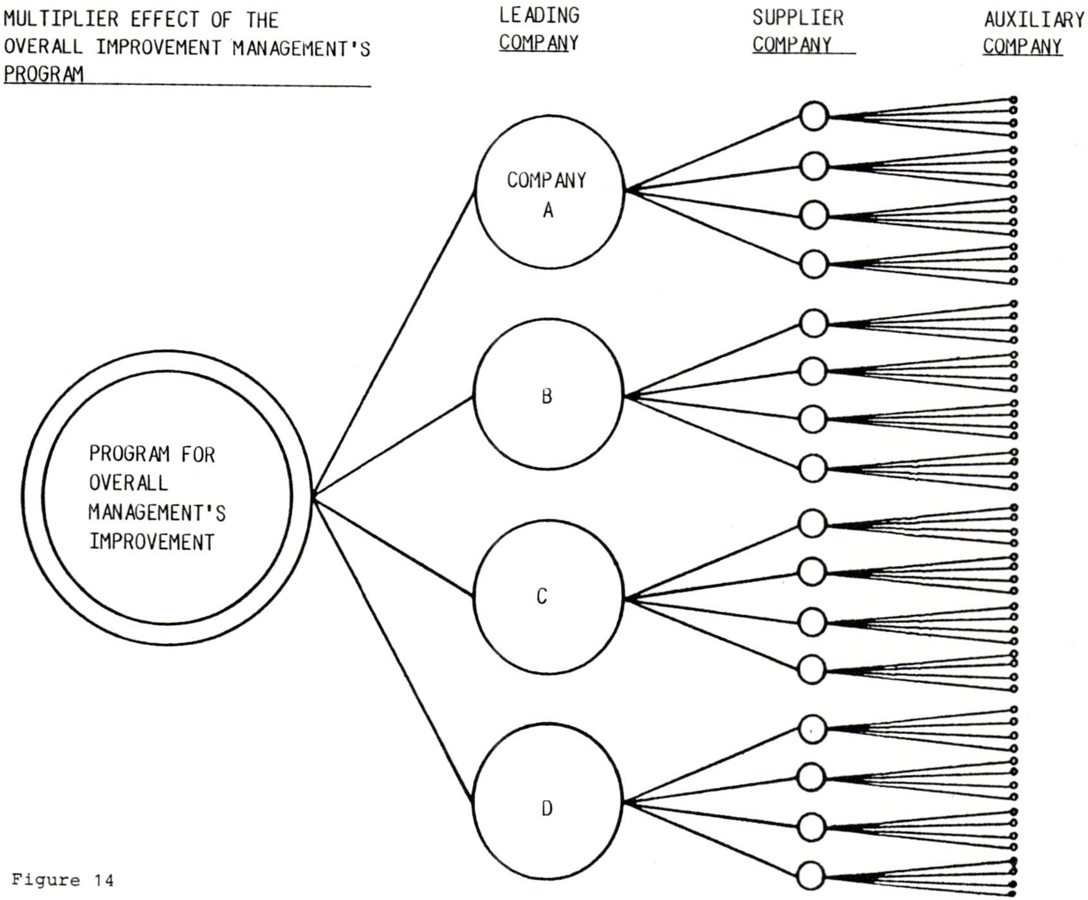

Figure 14

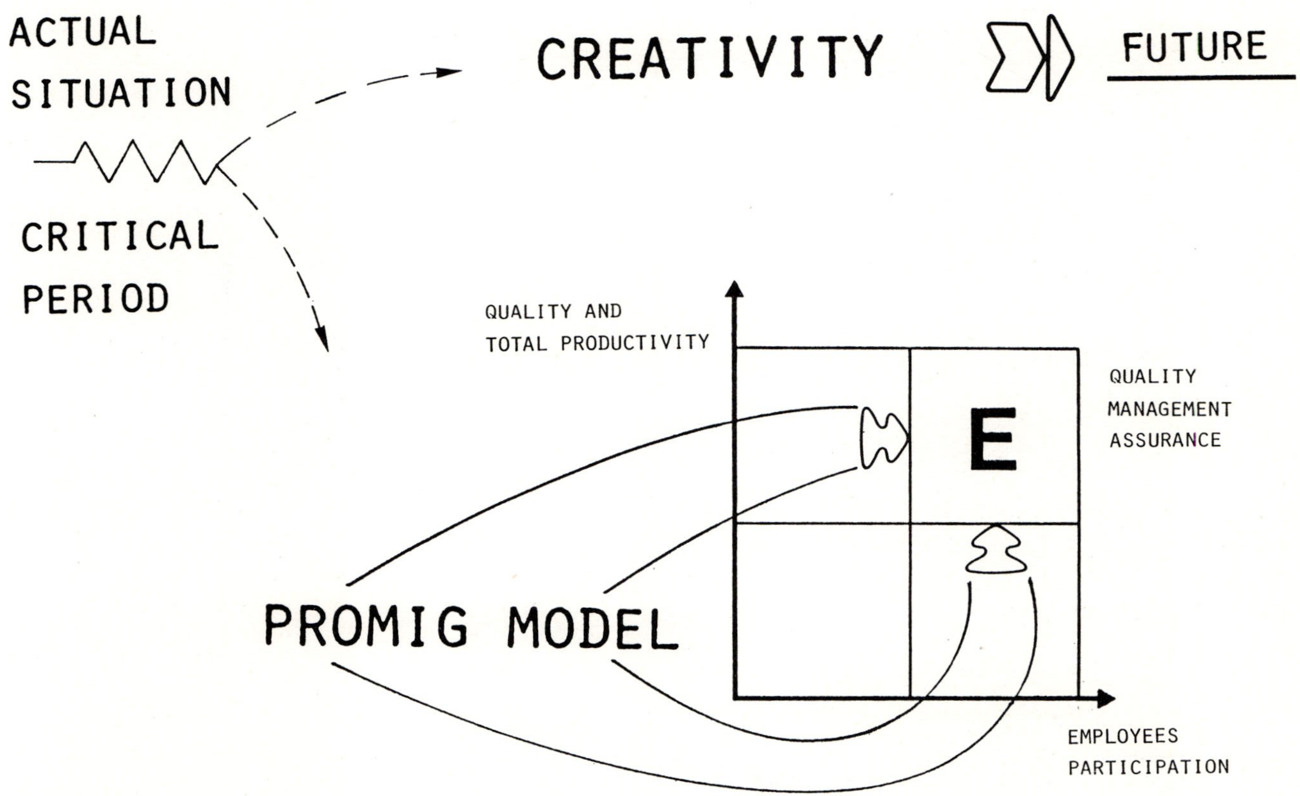
ACTUAL
SITUATION
CRITICAL
PERIOD
CREATIVITY
FUTURE
QUALITY AND
TOTAL PRODUCTIVITY
E
QUALITY
MANAGEMENT
ASSURANCE
PROMIG MODEL
EMPLOYEES
PARTICIPATION

Figure 15

G. PRODUCTIVITY IMPROVEMENT THROUGH PARTICIPATION AND GAINSHARING

Productivity Management Frontiers — I, edited by D.J. Sumanth
Elsevier Science Publishers B.V., Amsterdam, 1987 — Printed in The Netherlands

GOAL SETTING, NEED FOR ACHIEVEMENT AND THE HARD-IMPOSSIBLE THRESHOLD: CONTRADICTION OR PRODUCTIVITY OPPORTUNITY?

Paul F. Du Mont
Dept. of Administrative Sciences
Kent State University
Kent, Ohio 44242

Rosemary Ruhig Du Mont
School of Library Science
Kent State University
Kent, Ohio 44242

ABSTRACT

Need achievement theory and goal setting techniques have separately achieved wide use in organizations as motivational devices. However, the two motivation approaches make contradictory, and heretofore irreconcilible, predictions concerning how job, task or goal difficulty impacts performance. This paper explores these two models and uses the fundamental concepts of goal acceptance and goal commitment to develop the Hard-Impossible Threshold as a reconciling construct. Reconcilement will permit managers to achieve greater productivity benefits through revised applications which come closer to realizing the maximum potential of both approaches.

INTRODUCTION

Most contemporary research into improving employee productivity includes a consideration of human behavior and particularly the role motivation plays in improving worker performance. Indeed, the extent to which motivation theory has become a concern for managers interested in the behavioral aspects of productivity, is reflected in the amount of attention paid to motivation theories by management textbooks. For example, Tosi, Rizzo, and Carrol (1986) devote several chapters to theories of motivation.

While motivation theories have proliferated, considerable controversy has developed in the literature about the underlying nature of motivational forces. Over the past two decades or so, researchers have attempted to clarify the nature of these forces by establishing statistical curves representing the relationship to performance as motivational forces vary. In so doing, they have often reached contradictory conclusions about basic questions, such as whether performance increases continually or peaks and declines as work tasks become more difficult. For example, goal setting studies have repeatedly found that as goals get harder performance continues to increase, if the goals are accepted. In contrast, the theory of need for achievement shows the curve relating difficulty to output to be an inverted U, increasing until some peak is achieved and then declining (Locke, 1968).

Clearly, there is need for clarification. This paper examines some of the issues that underlie investigations of work motivation. The discussion examines two basic models of motivation: Locke's Goal Model and the Theory of Need for Achievement. Locke's Goal Model (Locke, 1968) focuses upon the motivating properties of goal setting. The model has gained popularity because of its appealing advantage of simplicity and improvement in productivity for many organizations (Latham & Locke, 1979; Locke, Shaw, Saari & Latham, 1981). The Goal Model (GM) predicts that individual performance improves more or less linearly as specific goals become more difficult, SO LONG AS THE GOALS ARE ACCEPTED (acceptance caveat). The foundation of the model's predictions is "that an individual's conscious ideas regulate his action" (Locke, 1968, p. 157). The interactions among goal difficulty, intention and effort are: 1) hard goals are classified by the individual as requiring more effort in order to be achieved than easier goals; and 2) assuming the individual ACCEPTS the goal and is COMMITED to trying to reach it, the individual's intentions should then be to expend the higher level of effort deemed necessary (Locke, 1968). In short, hard goals (when accepted) result in increased effort and presumably improved performance. The practical implication of the theory is that one way to improve worker productivity is to set specific and increasingly difficult, but acceptable goals (Latham & Yukl, 1975; Locke, 1968; Terborg, 1978). Perhaps the model's main limitation is the extent to which individual differences (e.g. need for achievement, etc.) are not accommodated (Ivancevich, 1978).

In contrast, the Theory of Need for Achievement (n Ach), makes other predictions. According to n Ach theory, performance is related curvilinearly to increases in goal difficulty. Persons with a high level of n Ach typically experience the performance and goal difficulty relationship as an inverted U. Theoretically, the peak of the curve occurs when the probability of success is 50 percent (Atkinson, 1958). For persons low in need for achievement the performance curve is U-shaped (Kukla, 1975). According to Korman (1977), the minimum point of the U, for these persons, occurs at the same (0.50) probability of success.

These GM and n Ach contradictions have been explicitly identified in the literature (Locke, 1968, p. 168), and are highlighted in Figure 1. The pivotal consideration in this paper is whether or not the goal acceptance caveat is associated with the maximums and minimums in the performance vs goal difficulty curves for n Ach. If n Ach can be shown to be theoretically compatible with the GM, then empirical tests of these theoretical issues may resolve the contradictions between the GM and n Ach. This may provide a foundation for a more general theory of motivation, incorporating heretofore disparate models.

MODEL DIFFERENCES

Several issues may underlie the contradictory predictions of the GM and n Ach theory. First, important conceptual differences between GM and n Ach research concerning specific variables may account for disparity in research results. Table 1 identifies five factors and their respective consideration and treatment within each model. The remainder of this paper discusses the variables listed on Table 1. One factor, goal difficulty can be eliminated since both models utilize similar ranges to produce contradictory results.

The models could be appropriate for different conditions of goal acceptance and goal commitment. An analysis by Campbell and Pritchard (1976) draws a theoretical distinction between acceptance and commitment. Goal ACCEPTANCE influences choice of the DIRECTION of BEHAVIOR...e.g. acceptance of a task suggesting a decision to attempt that task. The extent of goal COMMITMENT determines the LEVEL of EFFORT expended. Goal acceptance is viewed as a categorical variable rather than as a true variable, i.e. acceptance is dicotomous with an all or nothing connotation (acceptance or rejection). In contrast, commitment is used as a variable with potential be be measured in gradations. A weak level of goal commitment may evoke a minimal effort response and the goal may be easily abandoned, while strong goal commitment implies continued persistence and sustained effort to accomplish the goal. Moderate levels of commitment might result in fluctuating effort.

Within this framework, goal commitment decreases across the goal difficulty spectrum leading to a point in the area of extreme goal difficulty where the goal is rejected because it is seen to be impossible (Du Mont, Worrell, Watson, and Grimes, 1986). Impossibility may cause subjects to "abandon their goals if they become too difficult" (Locke, Shaw, Saari, and Latham, 1981, p. 132.). Stedry and Kay (1966) report that goals perceived to be impossible will result in worse performance than easy goals, a view consistent with GM theory only if the goals are not accepted.

The impact on performance curves of goals perceived to be impossible is documented in several studies (Erez & Zidon, 1984; Stedry, 1960; Stedry & Kay, 1966). These studies indicate that some threshold boundary point distinctively separates the very hard from the impossible goals. Goals which are objectively almost impossible, but which are accepted with high commitment will lead to extremely good performance. When the same goals are seen as impossible and are rejected, (no commitment due to no acceptance) performance drops to levels below those attained through normal work assignment goals.

Need for achievement research also addresses the phenomena of impossibly difficult tasks, but explores the effects of impossibility upon task preference. Kukla (1975) reports that both high and low n Ach subjects prefer tasks with

Table 1
Conceptual Differences on Selected Variables
Goal Model vs. Need for Achievement

Variable	Goal Model	Need for Achievement
Goal Difficulty	•Only easy to moderately hard goals used and therefore Goal Model research deals only with the upward slope, or left hand side, of n Ach curves (Campbell & Pritchard, 1976; Locke, 1975; Steers & Porter, 1974) •Full range of goals used, from easy to extremely hard, including objectively impossible goals (Locke, 1968; Locke, 1982).	•Full range of goals used from easy to very hard (Klinger & McNelly, 1969)
Goal Acceptance	•Goal caveat invoked, e.g. hard goals must be accepted to generate better performance than easier goals (Locke, 1968)	•Not considered as a factor in the theory (Becker, 1978)
Goal Commitment	•Identified as control variable over effort (Campbell & Pritchard, 1976) but not yet recognized in Goal Model (Locke, Shaw, Saari & Latham, 1981)	•High n Ach subjects prefer challenging goals in achievement situations and strive hardest (persistance or commitment) under these conditions (Klinger & McNelly, 1969) •Low n Ach subjects prefer easy or unrealistically hard challenges and strive hardest under these conditions (Korman, 1977)
Individual Differences	•Rarely considered •Need for achievement, locus of control, need for affiliation act as intervening variables. Data not separated on these dimensions produce linear GM predicted relations between goal difficulty and performance. The same data when separated on these motives, produced inverted "U" relationship (Von Bergen, 1974)	•High n Ach subjects peak on tasks at approximately .3 objective probability of success (OPs). (Atkinson, 1957) •Low n Ach subjects peak on tasks at .1 OPs or lower (Klinger & McNelly, 1969)
Impossible Goals	•Objectively impossible goals (no one actually reaches the goal) produce results similar to very hard goals (performance increases) when subjects are NOT aware that the goals are actually impossible (Locke, 1982) •Subjectively impossible goals result in performance decreases to levels below those attained under easy and "no goal" conditions (Stedry & Kay, 1966; Erez and Zidon, 1984; Garland, 1983)	•Rarely considered. •Hard tasks preferred over easy or impossible tasks by both high and low n Ach subjects (Kukla, 1975)

high perceived difficulty when paired with objectively impossible or trivially easy tasks.

One conclusion to be drawn from these studies is that there is a critical relationship between goal difficulty and performance. As long as the perception is that goals can be achieved, more difficult goals will cause more effort to be exerted and higher performance will result, at least on simple tasks. A performance peak occurs within a critical region of perceived difficulty. For levels of difficulty exceeding that of the critical area, effort and performance are reduced (Erez and Zidon, 1984; Garland, 1983; Stedry and Kay, 1966).

THE HARD-IMPOSSIBLE THRESHOLD: THE COMMON LINK

The general concept of a critical region (hereafter labeled the Hard-Impossible Threshold, or HIT) is a plausible link between the GM and n Ach theory. The issue to be addressed concerns the hypothetical shape of the HIT region...e.g. whether the region contains a continuous function (as shown in Figure 2) or another relationship. The distinction between impossible goals and low acceptance goals is a matter of degree, but goals viewed as impossible to achieve are rejected. The person with an impossible goal consciously assesses the probability of the success versus effort relationship. A perceptually impossible goal is thought to be so extremely difficult that not even the most extraordinary effort has any chance of achieving goal success. In such a situation, the goal is rejected with the result being substantially reduced effort (Kukla, 1972; Erez and Zidon, 1984). The point at which rejection occurs may vary by individual and be influenced by personality factors (such as n Ach).

If the degree of difficulty is not so extreme as to result in perceived impossibility (e.g. only slightly easier), a different goal difficulty versus acceptance and commitment relationship may operate. Individuals may consciously accept the goal, but for reasons which they may not even be aware of, they may not be totally committed to the goal. As goals become more difficult, and if the anticipated satisfaction derived from potential goal accomplishment does not increase accordingly, then effort may slowly decline. Figure 2 illustrates this concept. At point W of curve G, acceptance is complete. A plotting of similar W points (or other points on the individual's curves to the LEFT of the HIT region) for persons A through H results in Locke's linear relationship. However, as the goal for person G becomes slightly more difficult, the degree of goal commitment may further decrease, resulting in points X and Y. At some point Z (the HIT point itself), the degree of goal commitment approaches zero and goal acceptance changes to rejection.

FIGURE 1

PERFORMANCE CONTRADICTIONS:
GOAL MODEL VS NEED FOR ACHIEVEMENT

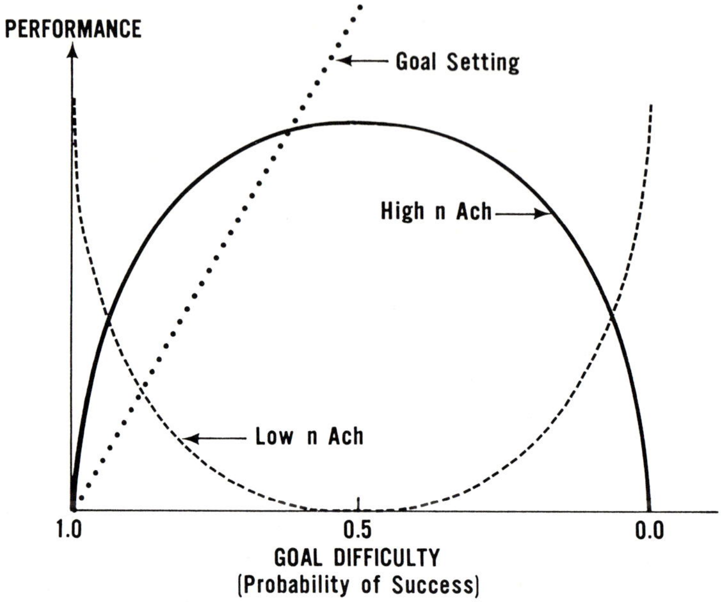

FIGURE 2

HYPOTHETICAL EFFECTS OF THE HIT POINT
ON THE GOAL DIFFICULTY-PERFORMANCE CURVES

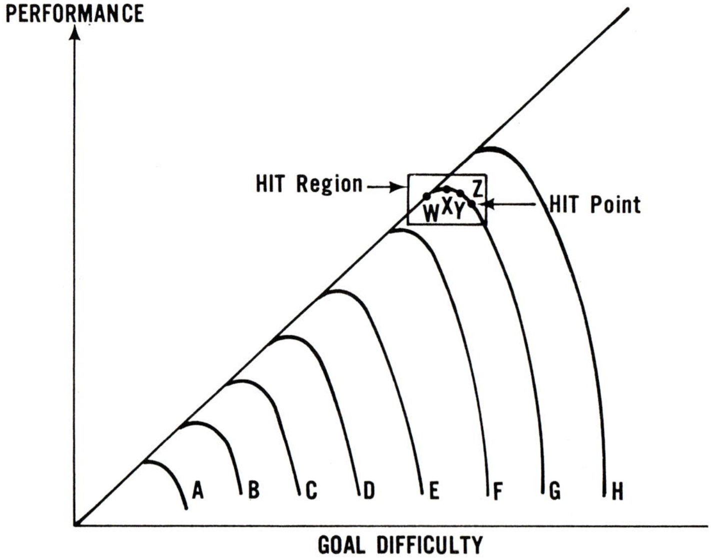

Research supports these possibilities. The summation of data from one group of subjects has generated both the n Ach curve when the data was segmented on n Ach and the GM line when the data was combined (Von Bergen, 1974). Although the region of declining commitment preceding rejection is shown in Figure 2 as being very narrow, this is only conjecture. The actual range will have to be established empirically.

THE GOAL MODEL, NEED FOR ACHIEVEMENT AND THE HIT

It is apparent, although research results are not extensive, individual differences (e.g. n Ach) do affect the goal setting process. However, the exact character of the interaction processes has yet to be determined (Mitchell, 1979). Atkinson (1975) suggests, for example that the n Ach performance curve could be skewed as shown in Figure 2. Skewed curves could accommodate both GM and n Ach findings.

One reason skewed n Ach curves have not yet been recognized can be attributed to design limitations. Many n Ach studies (and GM studies) rely upon three or four points to span the range of goal difficulty (Klinger and McNelly, 1969; Locke, 1982), providing grounds for an argument as to whether so few data points are sufficient to reliably predict the goal difficulty vs. performance relationships. It is suggested that the normally reported inverted U needs modification. Figure 3(a) depicts the normal GM linear relationship and Figure 3(b) shows the relationship advocated by n Ach. Figure 3(c) demonstrates how the data points from both models can be included using skewed curves. Current research (Locke, 1982; Erez and Zidon, 1984) suggests that the skewed relationships shown in Figure 3(c) can be used to reconcile the GM and n Ach research findings.

Very precise relationships must hold for the n Ach curves to accommodate both the GM relationship when added together and n Ach theory when separated. Figure 4 shows these relationships. Low n Ach subjects must perform in a modified relationship which has point A lower than point C and point B only slightly lower than point A. This sloping "J" relationship is shown in Figure 4(a). Similarly, the performance curves of high n Ach subjects must follow an inverted and reversed "J" curve. This is shown in Figure 4(b) where point D is substantially below point E but point F is only slightly below E. Such "J" and inverted and reversed "J" curves could be combined to yield an increasing linear relationship of Figure 4(c). This model would be compatible with the Von Bergen (1974) findings. (See Du Mont, 1981, for an extended discussion of these concepts.) Some research (Atkinson, 1957; Atkinson and Litwin, 1960; Erez and Zidon, 1984) supports these revisions that persons high in n Ach experience a shift of the performance curve to the right with a peak occurring approximately at 0.3 OBJECTIVE probability of success,

FIGURE 3

GM VS N ACH:
PERFORMANCE CURVE EXTRAPOLATION

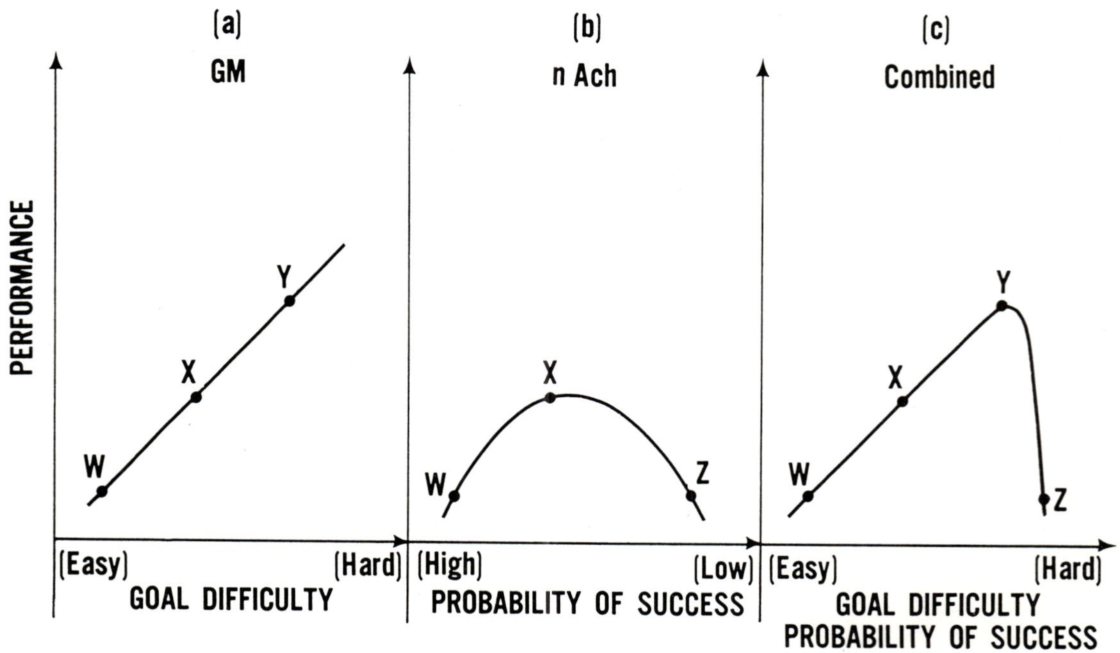

because of high self confidence, as shown in Figure 4(b). Klinger and McNelly (1964) suggest that a similar LEFTWARD shift of the performance minimum to approximately .1 objective probability of success, as shown in Figure 4(a), occurs for persons low in n Ach.

SUMMARY AND CONCLUSION

Of practical concern is the effect of task and personality variables upon individual productivity. In this paper we argue that goal difficulty can be used, up to some region, labeled the Hard-Impossible Threshold (HIT), to promote productivity. It is also argued that personality factors (such as n Ach) exert influence on the point where the HIT will occur. The discussion draws upon concepts from two heretofore contradictory motivational models and has developed some common ground for reconciling these models.

A reconcilement enhances understanding of fundamental personality dimensions underlying goal setting processes and can lead to a more unified theory of motivation. The establishment of the HIT as a viable concept and its relationship to n Ach parameters provides practical insights for unraveling the important question of "How hard is hard?" The coupling of n Ach and the HIT provides practicing managers with a general means to convert individual employee potential into performance by modifying goal setting techniques to the pecularities of specific individuals. Individual task assignment is an activity with which managers have long struggled without the appropriate tools or knowledge base.

It is clear that locating the HIT point can be of considerable benefit to managers in the quest to improve productivity. The HIT concept can be empirically supported by locating where on the goal difficulty scale the HIT occurs for various categories of individuals. "Who will perform how high? and how high will they perform? Only by having BOTH kinds of information will accurate predictions of the level of individual performance be possible" (Schneider, 1978, p. 302). These questions can be answered by determining the HIT location, as influenced by different personality dimensions. There is, however, one important qualification which must be considered. The HIT is a PERCEPTUAL variable and as such can be influenced by differences in individual characteristics and situational elements (Ivancevich, 1978; Kukla, 1972).

Empirical support for these proposals will provide a firmer base upon which prescriptions for work assignments can rest. In the longer term, research can extend the usefulness of the HIT construct by exploring the effects of other personality profiles with task characteristics to yield HIT location norms. The challenge will then be to use such norms to set individual employee goals

FIGURE 4

MODIFIED N ACH PERFORMANCE
CURVE PREDICTIONS

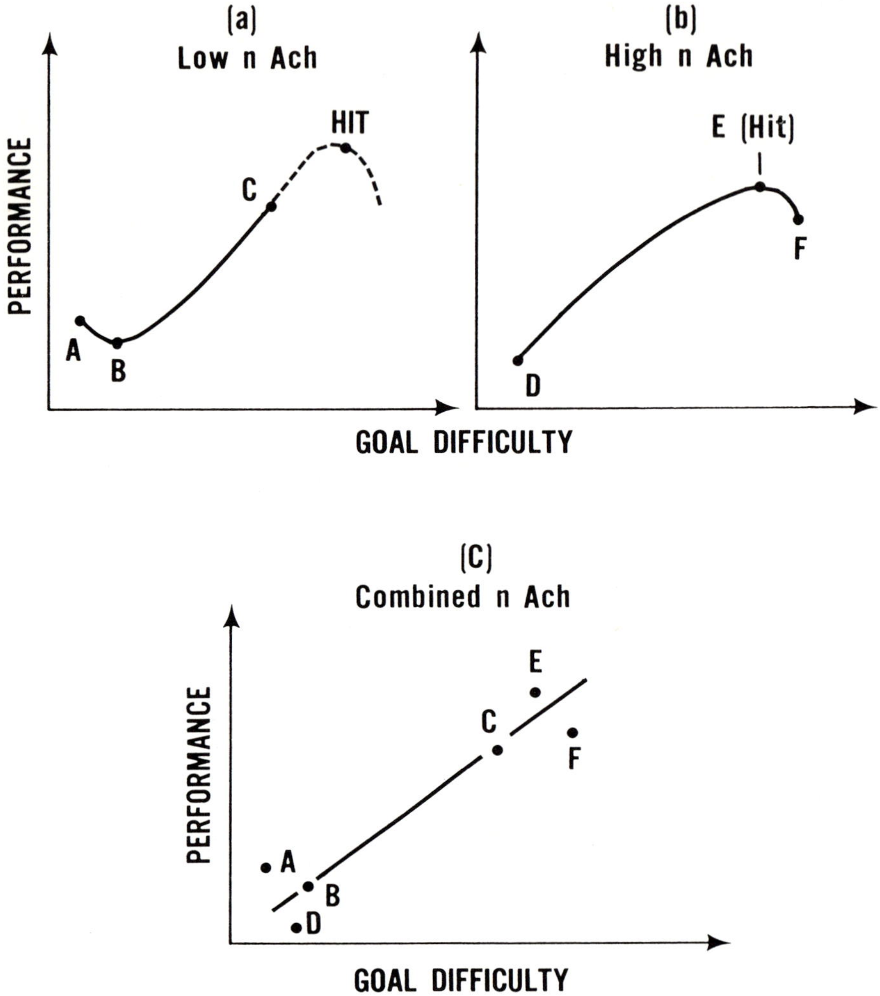

close to, but below the HIT area. When this can be done, a potent tool will be available for unlocking individual employee potential, thereby enhancing overall productivity.

We want to emphasize that we do NOT advocate that practicing managers classify individuals as being high, low, etc. in n Ach (or other personality variables that demonstrate significant impact on goal setting). This can be easily done using standardized instruments. The classification can even be conducted as part of the pre-employment, or orientation activities since many personality factors exhibit relative stability. Managers can then be given this employee data, along with performance norms for each classification. The "individualized" goal setting then requires the manager to adopt only the empirically developed general performance norms to the tasks at hand and to set goals near, but below, the norm HIT points. We suggest setting goals safely below the norm HIT point since the concern for managers is productivity improvement, not the identification of each employee's theoretical absolute potential.

REFERENCES

Atkinson, J.W., 1957. Motivational determinants of risk-taking behavior. PSYCHOLOGICAL REVIEW, 64, 339-372.

Atkinson, J.W., 1958. Toward experimental analysis of human motivation in terms of motives, expectancies and incentives. In J.S. Atkinson (ed.), MOTIVES IN FANTASY, ACTION AND SOCIETY. New York: Van Nostrand.

Atkinson, J.W. and Litwin, G.H., 1960. Achievement motive and test anxiety conceived as motive to approach success and motive to avoid failure. JOURNAL OF ABNORMAL AND SOCIAL PSYCHOLOGY, 60, 52-63.

Becker, L.J., 1978. Joint effect of feedback and goal setting on performance: field study of residential energy consumption. JOURNAL OF APPLIED PSYCHOLOGY, 63, 428-433.

Campbell, J.P. and Pritchard, R.D., 1976. Motivation theory in industrial and organizational psychology. In M.D. Dunnette (ed.), HANDBOOK OF INDUSTRIAL AND ORGANIZATIONAL PSYCHOLOGY. Chicago: Rand McNally.

Du Mont, P.F., 1981. THE HARD IMPOSSIBLE THRESHOLD: AN EXPERIMENTAL STUDY OF PERFORMANCE PREDICTIONS OF THE TASK GOAL MODEL AND THE THEORY OF NEED FOR ACHIEVEMENT. Unpublished D.B.A. dissertation, University of Kentucky.

Du Mont, P.F.; Worrell, D.; Watson, W.E., and Grimes, A.J., 1986. The Hard-Impossible Threshold: A pragmatic limit for goal setting. NORTHEAST AIDS 1986 PROCEEDINGS. Fifteenth Annual Regional Conference, American Institute for Decision Sciences, Williamsburg, 266-268.

Erez, M. and Zidon, I., 1984. Effect of goal acceptance on the relationship of goal difficulty to performance. JOURNAL OF APPLIED PSYCHOLOGY, 69, 69-78.

Garland, H., 1983. Influence of ability, assigned goals, and normative information on personal goals and performance: a challenge to the goal attainability assumption. JOURNAL OF APPLIED PSYCHOLOGY, 69, 20-30.

Ivancevich, J.M., 1978. Individual differences and organizational goal settings: issues, research and new directions. Paper presented at the 38th annual meeting of the Academy of Management, San Francisco.

Klinger, E. and McNelly, F.W., 1969. Fantasy, need achievement and performance: a role analysis. PSYCHOLOGY TODAY, 76, 574-591.

Korman, A., 1977. ORGANIZATIONAL BEHAVIOR. Englewood Cliffs, N.J.: Prentice-Hall.

Kukla, A., 1972. Foundations of an attributional theory of performance. PSYCHOLOGY REVIEW, 79, 454-470.

Kukla, A., 1975. Preferences among impossibly difficult and trivially easy tasks: a revision of Atkinson's theory of choice. JOURNAL OF PERSONALITY AND SOCIAL PSYCHOLOGY, 32, 338-385.

Latham, G.P. and Locke, E.A., 1979. Goal setting - a motivation technique that works. ORGANIZATIONAL DYNAMICS, 8, 68-80.

Latham, G.P., and Yukl, B.A., 1975. A review of research on the application of goal setting in organizations. ACADEMY OF MANAGEMENT JOURNAL, 18, 824-845.

Locke, E.A., 1968. Toward a theory of task motivation and incentive. ORGANIZATIONAL BEHAVIOR AND HUMAN PERFORMANCE, 3, 157-189.

Locke, E.A., 1975. Personnel attitudes and motivation. ANNUAL REVIEW OF PSYCHOLOGY, 26, 457-480.

Locke, E.A., 1982. Relation of goal levels to performance with a short work period and multiple goal levels. JOURNAL OF APPLIED PSYCHOLOGY, 67, 512-514.

Locke, E.A.; Shaw, K.N.; Saari, L.M.; and Latham, G.P., 1981. Goal setting and task performance: 1969-1980. PSYCHOLOGICAL BULLETIN, 90, 125-152.

Mitchell, T.R., 1979. Organizational behavior. In M.R. Rosenzweig and L.W. Porter (eds.), ANNUAL REVIEW OF PSYCHOLOGY, 30, 243-281.

Schneider, R., 1978. Implications on the conference: a personal view. PERSONNEL PSYCHOLOGY, 31, 299-304.

Stedry, A.C., 1960. BUDGET CONTROL AND COST OF BEHAVIOR. Englewood Cliffs, N.J.: Prentice-Hall.

Stedry, A.C. and Kay E., 1966. The effects of goal difficulty on performance: a field experiment. BEHAVIORAL SCIENCE, 11, 459-470.

Steers, R.M. and Porter, L.W., 1974. The role of task goal attributes in employee performance. PSYCHOLOGICAL BULLETIN, 81, 434-452.

Terborg, J.R., 1978. Motivation and the goal setting process: an attempt at clarification. Paper presented at the 38th annual meeting of the Academy of Management, San Francisco.

Tosi, H.L.; Rizzo, J.R. and Carroll, S.J., 1986. MANAGING ORGANIZATIONAL BEHAVIOR. Marshfield, Ma: Pitman.

Von Bergen, C.W., 1974. INDIVIDUAL DIFFERENCE VARIABLES AND PERFORMANCE GOALS AS DETERMINANTS OF LEVEL OF PERFORMANCE. Unpublished Ph.D. dissertation, Purdue University, 1974.

Productivity Management Frontiers — I, edited by D.J. Sumanth
Elsevier Science Publishers B.V., Amsterdam, 1987 — Printed in The Netherlands

KNOWLÉDGE WORKER GAINSHARING

CARL G. THOR
American Productivity Center, 123 North Post Oak Lane, Houston, Texas 77024
United States of America

INTRODUCTION

The evolution of two popular topics in productivity research has now reached the point where they must be considered together. Current work on both group incentive plans (gainsharing) and white collar productivity measurement suffers from an overly narrow experience and literature base.

Extensions of gainsharing from very successful factory floor application to the office or laboratory stops abruptly when the "output" of the off-line process is no longer clear and when higher labor productivity becomes only one of several desired outcomes. Also it is often discovered that office and laboratory workers do not have the tradition of accepting and working toward "incentives"; anything that sounds like factory piecework or sales "fluff" is beneath them and destructive of required team work.

Measurement of the productivity of white collar workers is common in "back rooms" of service institutions for example, but as the work becomes less repetitive and more creative in its requirements, measuring the apparent work flow becomes less and less meaningful. "Knowledge workers" need special treatment. A lawyer handling utility right-of-way issues might be measured on rights-of-way obtained per month, but on a multi-year bet-the-company liability case, what does "productive legal work" mean?

The problem with both traditional gainsharing systems and white collar productivity measurement is that two front-end issues have been oversimplified:

-- What do we really want the outcome of the group's work to be?

-- How do we stimulate the group members to work toward that outcome?

The new answers are respectively:

-- We recognize the inevitable need for multiple measured outcomes, using a balanced "family" of measures of performance

-- We involve the group members themselves in the creation of the measurement basis to activate their interest and "buy-in" to the underlying improvement process and we consider giving tangible reward for improved measured results.

I. THE EVOLUTION OF GAINSHARING

A. Traditional plans

The pioneering gainsharing programs generally carry two key components. One is a simple pay-for-performance formula with which covered employees receive pre-determined reward if improvement in that formula occurs. The other is some sort of idea-handling mechanism that becomes the engine for improvement; this can be simply special attention for employee suggestions within the formal organizational structure, but more often it involves some sort of parallel structure such as quality circles or special task forces.

The oldest and simplest formulas are piecework; if the individual worker makes more than a target level of product, a bonus is paid for each extra item. The oldest large group plans, Scanlon and Rucker, work off partial financial ratios, comparing sales or value added with labor expense. Improshare compares earned hours of output with actual hours of input. Profit sharing uses operating profits at the plant, division or company level.

B. Problems

Each of these plans attempts to represent the desired outcome through only one formula, and each partly fails. The formulas that involve labor productivity only, do not reward for improvement in use of other production factors such as raw materials yield, scrap, energy, inventory, maintenance materials, etc. Dollar-based formulas typically do not exclude the disruptive effects of inflation. Some physical formulas do not recognize the equally disruptive effects of product mix shift. Even deflated multi-factor formulas do not explicitly recognize other key outcomes such as product quality improvement, on-time-delivery performance, or employee development.

Perhaps it is unfair to expect a single formula to cover all these performance elements, but if it doesn't there is a clear risk that well-meaning managers and employees alike will seek to satisfy the formula at the expense of other performance variables.

A single complex formula that tries to meet the above problems will, in the process, tend to destroy the major basis of the whole gainsharing idea, namely that people will work to improve their immediate performance with the stimulation of extra reward. If there is not a reasonable link of controllability between employee effort and improvement of formula, why put out that effort? The formula outcome is only a complex "accident", apparently more affected by accountant and engineer manipulation than by work habits or improvement ideas. Thus a carefully crafted profit sharing plan, automatically covering all the input factors, and perhaps adjusted for inflation effects, may nevertheless be too remote from the average worker to be a stimulant unless the profit center is quite small.

The nature of the local idea-handling system also influences the choice of formula. The traditional emphasis on suggestion systems fits well with the single, simple overall formula. Previously, once the employee's suggestion was processed, management was largely responsible for making improvement happen through technology investment or better official methods of work.

As the idea handling shifts to a more grass roots basis (quality circle or task force) where the workers do most of the improvement work themselves, there is not as much willingness to sit back and wait for the official overall improvement number to be issued from headquarters. Their grass roots efforts will be largely unlocatable in the official statistics. It would be better to have local specific indicators where improvement progress can be seen by the improvers and directly stimulate further improvement.

Another key trend is that plans are now usually being co-designed by management and the work force/union. The traditional approaches had many of the issues pre-decided. The newer plans will leave more of the detail for at least joint discussion if not joint decision. Organized work forces tend to prefer formula elements they can understand and perhaps even "see" on a meter or production report rather than in arbitrary aggregates.

C. New Solution

What is recommended, especially for large operating situations emphasizing grass roots involvement, is a family of measures, balanced to reflect several of the key factors of production but also considering quality, customer delivery and other factors beyond the conventional definition of productivity or efficiency.

For example, Motorola uses six factors for its base gainsharing plan:

- Production cost
- Quality
- Delivery vs. schedule
- Inventory
- Housekeeping and safety
- Methods improvement

A large chemical company uses:

- Continuous product yield
- Continuous product error reduction
- Batch product cycle time
- Batch product recycle percent
- Energy use

-- Maintenance salvage
-- Tools and supplies use
-- Waste water cost

A large metal processor uses:

-- Direct labor productivity
-- Scrap rate
-- Special ingredients use
-- Packaging use
-- Truck fleet turnaround time
-- Workman's compensation cost

The advantages of this sort of "family" approach are both technical and behavioral:

-- Technically, the desired outcomes of the process being measured (and rewarded) can be balanced and given their proper relative weighting. Nothing will be gained by speeding up at the end of the month to fatten the production count if it is offset by decline in the quality or safety statistics. The family members can each be simple, physical measures so that technical complexity is avoided, but the existence of several partially offsetting measures keeps employee response balanced and seeks to reduce unproductive game playing.

-- Behaviorally, the measures are all of local origin, readily understood by everyone and designed and calculated by local people. There are no corporate overrides, special inventory debits or inflation adjustments. Employees can at least partly control the improvement of the measures by improving their work, and they can see the results directly.

D. The Knowledge Worker as a Gain Sharer

What does this have to do with office and laboratory staff, the knowledge workers? Traditionally, not much. If knowledge workers were included in a gainsharing plan at all, it was as beneficiaries of what the operating people they "support" were able to accomplish. If the number of tons of output per direct labor hour or the local operations profit went up, not only did the direct labor get rewarded, but in some cases so did everyone else at the location.

This makes some sense for those who are truly support people such as maintenance persons, production planners, purchasers, and plant engineers, but it becomes indefensible for white collar staff such as research people, systems programmers or public affairs officers who are essentially independent in the short run of operations fluctuations. It isn't that these people don't have outputs, but their outputs are not the mainstream output, and these outputs are ignored or lost in the footnotes. It isn't that these people cannot be incentivated; they are among the hardest workers in the company during crisis situations. How can that effort be made normal?

II. THE EVOLUTION OF WHITE COLLAR MEASUREMENT

Measurement of white collar groups by industrial engineering methods has a history of success where the output is clear and tangible and where effort is made to ensure employee acceptance of the process (and even participation in the standard setting). But as the work becomes non-routine, intermediate and intangible, direct measurement efforts have largely ended and measurement has become hopelessly indirect. Staff departments are "controlled" by actual vs. budget performance with no attention to the changing levels of output of their staff groups. Worse yet, budgets or staffing are ratioed with the company's operating output. Research budgets are cut because the $Sales/$R&D expense ratio has declined, even though sales$ and R&D$ are dealing in completely different time frames and product strategies.

To break out of this problem it is necessary to recognize that all knowledge worker groups <u>do</u> have outputs that can be measured, at least with a little subjective touch, and also that the productivity ratios these outputs help generate need to be supplemented with other performance measures including quality, timeliness, utilization and customer satisfaction. Once again, the need is for a balanced family of grass roots, department specific examples of performance. And once again, these have little value for stimulation unless they are locally understood and accepted.

Since many of the hard-to-measure knowledge workers are located in staff functions at plant, division or corporate level, another difference develops between them and their indirect factory support colleagues. In staff areas the user or "customer" for the product or (more often) service that is their output is internal to the company and thus rarely recognized explicitly. Factory direct and support workers have a clearly identified customer and set of customer requirements.

In most cases, the most important indicator of a high performance staff group is that their service is effective from the user's point of view. Productivity/efficiency and on-time-delivery may also be important, but if the service is not what the user needs, it is worthless. Thus knowledge worker measurement needs to consider both internal and external performance, performance against their own internal procedures and experience and performance against the user's expectations and needs. This automatically generates the need for a family of measures; no simple measure could reflect both the internal and external point of view and still give analytic value.

Selection of the appropriate family of measures for a given department might be guided by the matrix illustrated in Figure 1.

FIGURE 1

A FAMILY OF MEASURES

	Internal Performance	User Effectiveness
Productivity	X	X
Quality	X	X
Timeliness	X	X

The internal measures are often already available, e.g., productivity or efficiency of any key processes, quality in the sense of errors or rework, and timeliness versus the internal schedule agreed upon. The external measures are more complex: productivity in the sense that the user is satisfied with the cost of the service (either versus estimate or the services market place outside the firm); quality in the sense of user satisfaction with the nature of the delivered service and whether it meets specifications; and timeliness against the user's own needs.

Other potential family members in some circumstances are equipment utilization, innovation, attendance, turnover, and behavioral change. Some of these are impossible to measure directly but interesting surrogates have been developed.

Change is occurring not only in broadening the scope of what is measured but also in the way measures are developed. There is solid experience in using participative methods such as the Nominal Group Technique to allow the measurees to nominate potential measurement family members. Done carefully, this tends to provide a substantially higher level of acceptance and interest among the measured employees to use the results to drive improvement initiatives.

III. INTEGRATION OF GAINSHARING AND WHITE COLLAR MEASUREMENT

The main things holding back development of gainsharing plans in knowledge worker groups have been (a) the inability to measure "output" and (b) the recognition that an output/input productivity ratio is not the only, or even the most important, performance indicator. These issues are removed by application of the family of measures methodology outlined above. If there is a measurable routine process within a knowledge worker department, a traditional output/input measure can be included as a member of the family and given whatever relative weight is appropriate. Any difficulty that an analyst may find in identifying such processes, can be significantly reduced by including the measurees in the processes; after all, they know their own workplace better than anyone. But in a group where other-than-productivity measures are of major importance, they can be admitted to the family and weighted appropriately.

With a weighted (and measuree-approved) family of measures in place, it becomes relatively easy to fashion a gainsharing mechanism that pays off according to balanced improvement in the family of measures. Figure 2 illustrates two different approaches with a hypothetical family of measures; a weighted Part A connecting overall family performance directly with a single reward formula and Part B rewarding separately for each element and thus weighting indirectly. A uses the Objectives Matrix concept that is becoming the standard scaling approach. Many other variations of details are possible. Thus pay for a specific group's performance can become a routine feature of cooperation for the knowledge worker also.

With the strong history of success shown in factory gainsharing, it is likely that even partial success in the knowledge worker area represents very substantial improvement potential in the competitiveness of U.S. industry.

REFERENCES

Felix, Glenn H., 1983. Productivity Measurement With the Objectives Matrix, Corvallis, OR, O/P/C Press (Oregon Productivity Center).

Graham-Moore, Brian E. and Ross, Timothy L., 1983. Productivity Gainsharing: How Employee Incentive Programs Can Improve Business Performance, Englewood Cliffs, NJ, Prentice-Hall Inc.

O'Dell, Carla S., 1981. Gainsharing: Incentives, Involvement, and Productivity, New York, AMACOM.

Sink, D. Scott, Spring 1983. "Using the Nominal Group Technique Effectively", National Productivity Review, pp. 173-184.

Thor, Carl G., April 1986. "Using Nominal Group Technique to Establish a White Collar Productivity Measurement System", Productivity Brief, American Productivity Center.

FIGURE 2.A

Productivity Criteria	Weighted Tests / Man Hour	Zaptest Machine Use / Actual vs Estimate	Materials Substitutions Accepted (Value Weighted)	On-Time Delivery	Customer Satisfaction Survey
Performance	3.43	92	5.8	92	98
Scores: 10	5.00	100	12.0	100	121
9	4.70	99	10.5	99	118
8	4.50	98	9.0	97	115
7	4.30	96	7.5	95	112
6	4.15	94	6.0	93	109
5	4.00	92	5.0	91	106
4	3.85	90	4.0	89	103
3	3.71	88	3.1	87	100
2	3.55	86	2.0	85	97
1	3.40	84	1.0	83	93
0	3.25	82	0	81	90
Score	1.2	5.0	5.8	5.5	2.3
Weight	40	20	15	15	10
Value	48	100	87	82	23

INDEX

340

Pool accumulates at $1,000 per point.

300 BASE

340 CURRENT

40 Points x $1,000 = $40,000 Pool

FIGURE 2.B

MEASURE	BASE	GAIN FORUMLA	CURRENT DATA	$000 BONUS POOL GAIN/LOSS
WEIGHTED TESTS PER MAN HOUR	3.71	$4,000/.01	3.43	(72)
ZAPTEST MACHINE USE: ACTUAL VS. ESTIMATE	88	$10,000/1	92	40
MATERIALS SUBSTITUTIONS ACCEPTED (VALUE) WEIGHTED	3.1	$1,500/0.1	5.8	41
ON-TIME DELIVERY	87	$7,500/1	92	38
CUSTOMER SATISFACTION SURVEY	100	$3,500/1	98	(7)
				40

Productivity Management Frontiers — I, edited by D.J. Sumanth
Elsevier Science Publishers B.V., Amsterdam, 1987 — Printed in The Netherlands

QUALITY CIRCLES AND PRODUCTIVITY IMPROVEMENT IN A NEW FACTORY - HOLDEN CANADA LTD.

B. PORTIS and D. FULLERTON
School of Business Administration, The University of Western Ontario, London, Ontario, Canada, N6A 3K7

ABSTRACT

For five years Holden Canada Ltd., Scientific Instruments Division (company and industry disguised), has used quality circles to achieve productivity improvements and quality control. The quality circle program was begun shortly after the plant opened and the first manufacturing line started in 1980. The quality assurance supervisor served as the first facilitator for the program and met with employees weekly to teach problem solving techniques and their application to the manufacturing process. As further lines were added so were more circles.

Much success was noted in terms of waste reduction and productivity. In 1982 the first circle formed was noted as the outstanding circle in Holden Canada, a large Canadian company. As a result of the early success and increased work load for the part-time facilitator a decision was made to put quality circles entirely on their own. A levelling off was noted on productivity and quality measures so a new approach was needed.

The plant manager asked the production supervisor and a process engineer to serve as facilitators to revive the five quality circles. These groups now work on both problems identified by the groups and suggestions from the facilitators or the plant manager. Their projects include process-related matters (such as changeover times, machine-oriented features, methods improvements) and organization of work (such as groups setting their own production targets and conducting their own inspection of quality).

Productivity and quality have again risen sharply under these new arrangements for the quality circles. The plant manager would like to broaden the work of circles to have them work with suppliers to the plant and also its customers. This would further reduce lead times and inventory while improving product quality.

HOLDEN CANADA LTD - SCIENTIFIC INSTRUMENTS DIVISION

Kent Davies, plant manager of the Scientific Instruments Division, initiated a quality circle program in 1980 upon opening of the new plant. Mr. Davies utilized quality circles as part of the training of employees for the new division. He later worked with supervisors and quality circle groups to achieve work improvement through objective setting, downtime reduction, increased productivity, and individual quality inspection. The incorporation of other measures such as Total Quality Control and Just-in-Time management theory were further considerations. In 1985, Mr. Davies was reviewing quality circle performance and considering what was required for future direction and support of the program.

THE FIRST QUALITY CIRCLE

The Scientific Instruments Division in St. Catharines, Ontario began construction in 1978. A chart is provided to show the division's position within Holden Canada Ltd. (Exhibit 1). The division was later to consist of five manufacturing areas: Oscilloscope, Radio Spectrometer, Photon Counter, Audio Oscillator, and Microwave Receiver Manufacturing. Oscilloscope Manufacturing commenced operations in 1978. In 1980, Radio Spectrometer Manufacturing was established. Subsequent growth in production areas between the years 1982-84, necessitated the formation of Photon Counter, Audio Oscillator, and Microwave Receiver Manufacturing, in that order. A recent organization chart is provided in Exhibit 2.

Radio Spectrometer Manufacturing commenced operations in the summer of 1980 with five people on each of two shifts. At the same time, a quality circle was instituted at Radio Spectrometer on an experimental basis with the hope of improving communication, providing team work in the generation of new ideas, and improving quality. Mr. Davies utilized the first quality circle as a test for further expansion of the quality circle program.

IMPLEMENTATION OF THE QUALITY CIRCLE PROGRAM

- Plant Manager's Objectives -

Mr. Davies possessed certain objectives for the new plant and the use of quality circles. The plant manager introduced a management-by- objectives system with flexible targets to motivate and reward employee performance. The importance of recognizing small achievements was immediately understood. He believed the presence of complicated technical difficulties at the outset would provide momentum for the first quality circle group, and relieve initial management operating problems. Quality circles seemed to fit the participative management style desired by the plant manager. Mr. Davies viewed the program as a forum for employees to voice opinions and frustrations about their new surroundings.

The initiation of quality circles was supported by a push throughout Holden to commence training of a quality circle program. The General Manager of the Scientific Instruments Division at head office granted approval for the program. Mr. Davies faithfully supported the new concept. The plant manager believed successful implementation would require initial assistance from Holden Canada Ltd., continued management support, and effective coordination of the quality circle program.

- Selection of a Facilitator -

The plant manager was responsible for selecting a facilitator, or coordinator, for the quality circle program. Nancy Wilson possessed the necessary qualifications: availability, knowledge of quality circle programs from prior research, strong interpersonal skills, and the title of Quality Assurance Supervisor. She reported to the General Manager of the Scientific Instruments Division at the head office in Toronto, but was located in St. Catharines. Nancy attended a three day company training course in Toronto to learn how to teach the quality circle concepts to the new staff.

Mr. Davies announced an afternoon meeting for employees to explain the purpose of quality circles and employee involvement in groups. Nancy was introduced as the program facilitator and aided in the presentation. Mr. Davies received unanimous consent from Radio Spectrometer employees for the voluntary quality circle meetings to be held on company time.

The plant manager delegated responsibility to the facilitator for quality circle group training. The facilitator utilized a quality control circle handbook consisting of ten sessions as part of the training process (Exhibit 3). Group members showed a liking for brain-storming and cause-and-effect diagrams. However, they felt certain concepts were complicated (pareto diagrams, histograms, graphs) and not practical as problem solving tools. The facilitator conducted meetings at shift change since only two shifts were in operation at that time. She allotted twenty hours of training over the ten week course, to be split equally between learning of concepts and application of principles to the job. In retrospect, the facilitator believed that too much introductory training was provided.

The facilitator viewed the necessary components of a successful quality circle program to include a formal regimented introduction process, strong facilitator support for the program, and a great amount of enthusiasm on the part of both group leaders and management. She felt management presence at circle meetings along with the preparation of accurate minutes contributed to employee involvement. Nancy Wilson was able to achieve initial widespread enthusiasm and employee involvement for the quality circle program.

- Group Development -

In the first two years of the program steady progress was made regarding improvements in performance and employee attitudes. Reaction to the program was positive by all participants. After initial success with the Radio Spectrometer circle, Mr. Davies decided to start a quality circle within the Oscilloscope area. At this time two supervisors, John Banks and Steve Jackson, were responsible for two large quality circles. In 1982, the Radio Spectrometer circle achieved the status of best circle within Holden Canada as a result of employee effort and strong management support for the program.

In the same year, Mr. Davies attempted to allow the program to run on its own. At this time, Nancy withdrew as facilitator because of increased workload with the program and her regular job. The size of the circle groups were particularly large: an additional shift had been adopted, and rapid growth was occurring in different production areas. Formal quality circle meetings abruptly ended because of a lack of interest from the participants. Implementation of suggestions and the rate of quality improvement slowed at this stage of the program.

REORGANIZATION OF THE QUALITY CIRCLE PROGRAM

In early 1983, Mr. Davies decided to revive the quality circle program. He noticed a dip in performance and recognized the inability of the quality circle groups to function independently. The plant manager approached Steve Jackson and asked him to assist with the quality circle program. Steve was to preside as chairman at four of the five group meetings, prepare minutes, motivate supervisors to plan meetings, aid underproductive groups, and form a problem solving task force. Mr. Jackson and Phil Stanley (Process Engineer), both previous maintenance employees, were asked by Mr. Davies to provide information to the groups for problem solving. The plant manager required quality circle minutes and copies of management presentation to keep informed of quality circle activities. Mr. Davies suggested some problems to quality circle teams, such as shortening changeover time for the spectrometer fabrication line.

Subsequent growth and expansion in production areas necessitated the formation of a quality circle team for each product line by late 1984. At this time, Mr. Davies gave responsibility to five supervisors for leadership of the various quality circle. John Banks and Steve Jackson were responsible for the Audio Oscillator and Oscilliscope quality circles. Phil Stanley took charge of the Microwave Receiver circle. His strong support for quality circles rekindled enthusiasm for the program. Bill Stevenson and Jill Larson were responsible for the Radio Spectrometer and Photon Counter quality circles. These two supervisors lacked experience at the plant and knowledge concerning the production process. They received assistance from Steve Jackson in the quality circle meetings.

On May 17, 1985 five quality circle teams were still in operation. Groups ranged in numbers from five to nine and generally contained a mix of production, maintenance, and quality assurance staff. The groups were meeting sporadically, with Steve Jackson and Phil Stanley assuming the role of the facilitators. Many problems undertaken were process-related (i.e. changeover times, machine- oriented features, methods improvements). The quality circle teams acquired assistance for implementation of new ideas intra-company from maintenance, process, and engineering departments. Mr. Davies recognized quality circle achievements by including pictures and achievement notices in the Holden newspaper.

MONITORING

- Plant Measures -

The plant manager established a Program for Profit system focusing on productivity improvement and waste reduction. Several productivity and waste measures were utilized by Mr. Davies to record plant activity. Radio Spectrometer Manufacturing comprised fifty percent of total business activity. Monitoring measures for Models 27 and 36 spectrometers included spectrometers per shift, waste measured per 1,000 spectrometers, and percentage run time. Close monitoring of this segment of the division was essential for plant success. Photon Counter Manufacturing had experienced gains in both productivity (Photon Counters/shift) and percentage waste reduction of the Model 1200 Photon Counter. The 1984 dip in productivity could have been caused by injury of the regular operator and adoption of a third shift. Audio Oscillator Manufacturing had experienced substantial productivity gains and waste reduction in the past two years. Many product design modifications enabling easier assembly had contributed to these gains. Oscilloscope Manufacturing produced eight different types of oscilloscopes. Mr. Davies measured the total number of oscilloscopes per total labour hours. He felt this was not a very efficient measure, since production mix in this department fluctuated regularly. Mr. Davies possessed very little information concerning monitoring measures for the Microwave Receiver Product. He estimated similar productivity and waste reduction measures for both the Photon Counter and Microwave Receiver products. Graphs identifying trends in productivity and waste measures are provided in Exhibit 4.

Mr. Davies compared the departmental output average with respect to targets to evaluate plant performance. He required an explanation if targets did not reach expectations. The plant manager believed a large number of factors contributed to the continued evolution of productivity improvement: design changes, learning curves, quality circles, and various others. The quality circles were not solely responsible for the rapid advance in productivity and waste reduction.

- Quality Circle Activity -

Mr. Davies desired participating employees to take an active part in the quality circle program. Since 1983, Mr. Davies had filed all quality circle minutes and management presentations to monitor group progress and development (Exhibit 5). Lately, he was anxious to determine employee reaction to the quality circle program at various stages of development. The plant manager felt that employee attitudes and suggestions would provide insight for planning the future of the program. Mr. Davies chose long-standing employees who had experience in various production areas, and who were active in the quality circle program.

Jennifer Young had worked in almost all areas of production, and was one of the first employees hired at the plant. Jennifer was initially skeptical of the idea, but indicated a positive reaction to increased responsibility resulting from the quality circle program. She felt management provided definite support for the program by clearly stating the purposes of the program, providing necessary training devices, and committing time and effort towards the initial quality circle groups. The meetings increased self-respect, strengthened leadership qualities, and improved communications among all participants. She felt the program lost credibility in 1982. Many group members had lost enthusiasm and viewed quality circles as a waste of time. Several employees thought that management was occasionally burdening them with work as a result of the quality circle program. A large number of problems were not pursued to the implementation stage, and management-directed problems had formed a substantial portion of the program. Jennifer credited Steve Jackson and Phil Stanley with only partial revival of the program in 1983. She summarized her thoughts concerning the quality circle program at Holden: "In the beginning the program was a success. Quality circles had to be productive and free of finger pointing to succeed. They had lost their usefulness as an effective tool. The life of a quality circle is approximately two years."

Elizabeth MacDonald had not been exposed to the early stages of the quality circle program. She was a very competent employee who had spent two years with the company as a Radio Spectrometer operator. Elizabeth indicated an enthusiastic reaction to increased responsibility as a result of the quality circle program. She felt the program encountered little initial skepticism, as management was very supportive in the early stages. Membership in the program enabled her to get involved and input new ideas. She believed quality circles increased the worker's understanding of the difficulties faced by supervisors, increased management's respect for the employees, and involved group members more actively in their jobs. Recently, meetings had become less regular, and generally occurred once a month. Elizabeth viewed management-directed problems as useful contributions to the program. The only improvements deemed necessary were occasional switching of supervisors and group leaders between departments, and swifter management action regarding follow-up of ideas. Elizabeth believed the program was a success. It generated useful ideas, kept people informed, and increased communication.

Sheila Gibson started at Holden as a Radio Spectrometer operator several months after commencement of the training program. She soon was promoted to Quality Control Inspector, and later to the position of Oscilloscope Operator. Sheila felt management was definitely supportive of the program. At first, she was quiet and intimidated by quality circles. However, she experienced a strong sense of achievement when management recognized the group's accomplishments with $20 cash or drinks. In 1982 she lost interest and motivation for attending quality circle meetings. Mr. Davies pushed to revitalize the quality circle program in 1983. Meetings were resumed on a weekly basis, but soon became bi-weekly sessions because of the reduced input of ideas. In the past eight months only three meetings had been scheduled for the Oscilloscope circle. Management was currently not following through with the backlog of ideas. Sheila expressed her views concerning the program: "In the beginning the program was successful, due to the implementation of numerous ideas. Lately meetings had dwindled and became less productive. It was beneficial to bring quality circles to the new plant. However, they needed constant commitment from both management and employees."

Mr. Davies believed these views to be representative of general employee attitudes throughout quality circle development. He recognized that the Oscilloscope circle had stalled, but the other circles were meeting regularly.

FUTURE CONSIDERATIONS

Mr. Davies encouraged senior employees to provide input regarding the future of the quality circle program. Steve Jackson wanted employees to take more responsibility in implementation of ideas, greater use of weekend quality circle sessions, and additional cross-training of employees. He thought the development of existing corrective action teams would provide management direction to the program. Nancy Wilson felt management had not been active enough in the early years of the program. She believed the program could have been more successful through extensive monitoring measures, management commitment, and greater facilitator support. She stated that productivity may not have been a good measure of quality circle success.

A new factor for consideration was union involvement in the quality circle program. The employees had recently voted to be supported by the same union that represented production employees in Toronto. The union did not desire to be involved in quality circles in Toronto, and Mr. Davies doubted if they wanted to be active in the program at the St. Catharines' plant. The quality circles would not be a threat to jobs because the plant was experiencing continued growth, and company policy discouraged layoffs.

Mr. Davies expected the program to evolve by using techniques such as Crosbie's philosophy of Total Quality and Just-in-Time management theory. Crosbie emphasized greater employee involvement in quality inspection and improvement. Just-in-Time was a Japanese philosophy which involved curtailing lead times, shortening changeover periods, and reducing inventory. The plant manager aimed to move "closer to the customer" by including marketing employees in circles to review group interviews with customers. Mr. Davies believed the future entailed continued automation of plant operations, and the formation of additional quality circle teams to review work process and design product modifications. He was pleased with the results of the quality circle program. Productivity and achievement were far ahead of his expectations.

EXHIBIT 1
HOLDEN CANADA LTD. - SCIENTIFIC INSTRUMENTS DIVISION

POSITION WITH RESPECT TO HOLDEN CANADA

POSITION WITH RESPECT TO HOLDEN CANADA

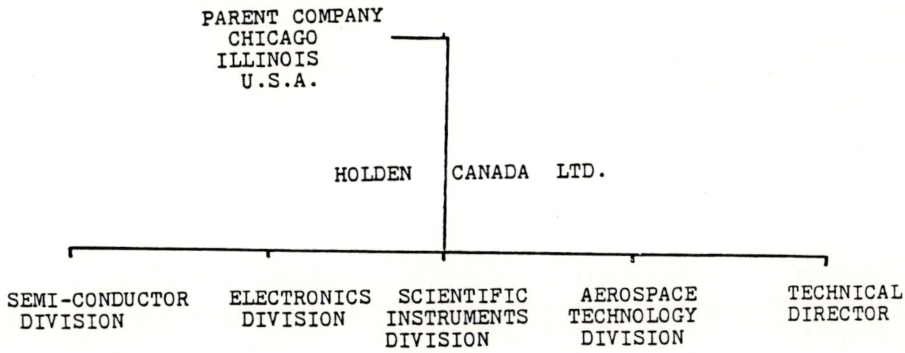

EXHIBIT 2
HOLDEN CANADA LTD. - SCIENTIFIC INSTRUMENTS DIVISION

ORGANIZATION CHART

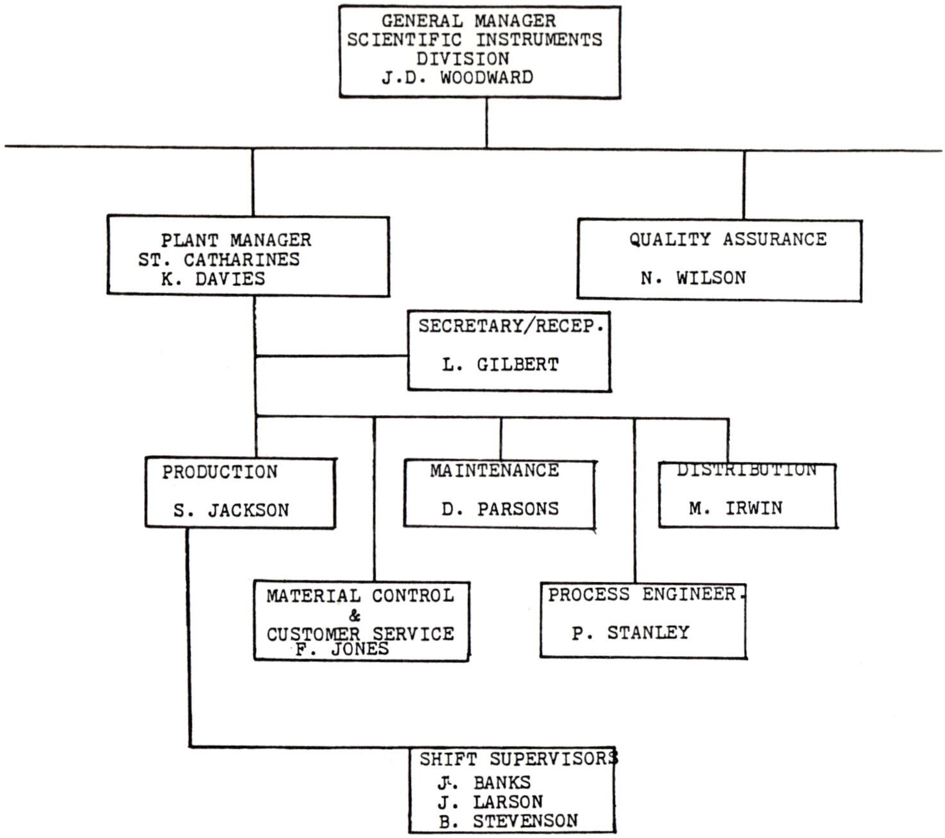

EXHIBIT 3
HOLDEN CANADA LTD. - SCIENTIFIC INSTRUMENTS DIVISION

EXCERPT FROM THE QUALITY CONTROL CIRCLE HANDBOOK

BASIC TECHNIQUES

The nine Basic techniques are presented to the Leaders in their training class. The techniques are then studies by the members during regular QC Circle meetings.

INTRODUCTION TO QC CIRCLES
BRAINSTORMING
CAUSE AND EFFECT DIAGRAMS - INTRO.
CAUSE AND EFFECT DIAGRAMS - II
PARETO DIAGRAMS
HISTOGRAMS
CHECK SHEETS
CASE STUDY I
GRAPHS
MANAGEMENT PRESENTATIONS

ADVANCED TECHNIQUES

SAMPLING
DATA COLLECTION
DATA ARRANGEMENT
CONTROL CHARTS #1; $\bar{X}$-R
CONTROL CHARTS #2: Pn
STRATIFICATION
SCATTER DIAGRAMS
CASE STUDY - II

The Advanced techniques are presented in the Leader Advanced Training. The techniques are studied by the Circle members when required, as determined by the Leader and the Facilitator.

Audio-visual packages are available to assist the presentation of all Basic and Advanced Techniques

EXHIBIT 4
HOLDEN CANADA LTD. - SCIENTIFIC INSTRUMENTS DIVISION

TRENDS IN PRODUCTIVITY AND WASTE MEASURES

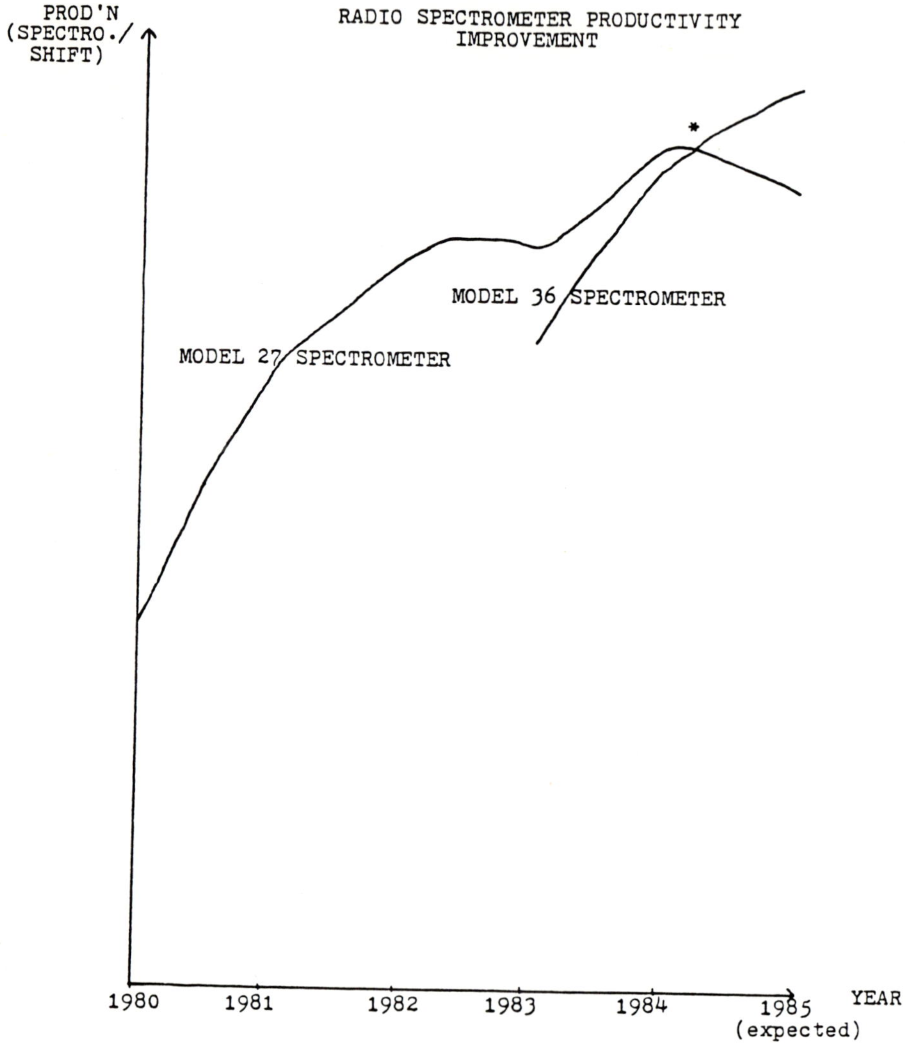

* changed mix proportions (Model 36 comprised 60% of spectrometers produced)

EXHIBIT 4 (continued)

TRENDS IN PRODUCTIVITY AND WASTE MEASURES

PLANT WASTE IMPROVEMENTS

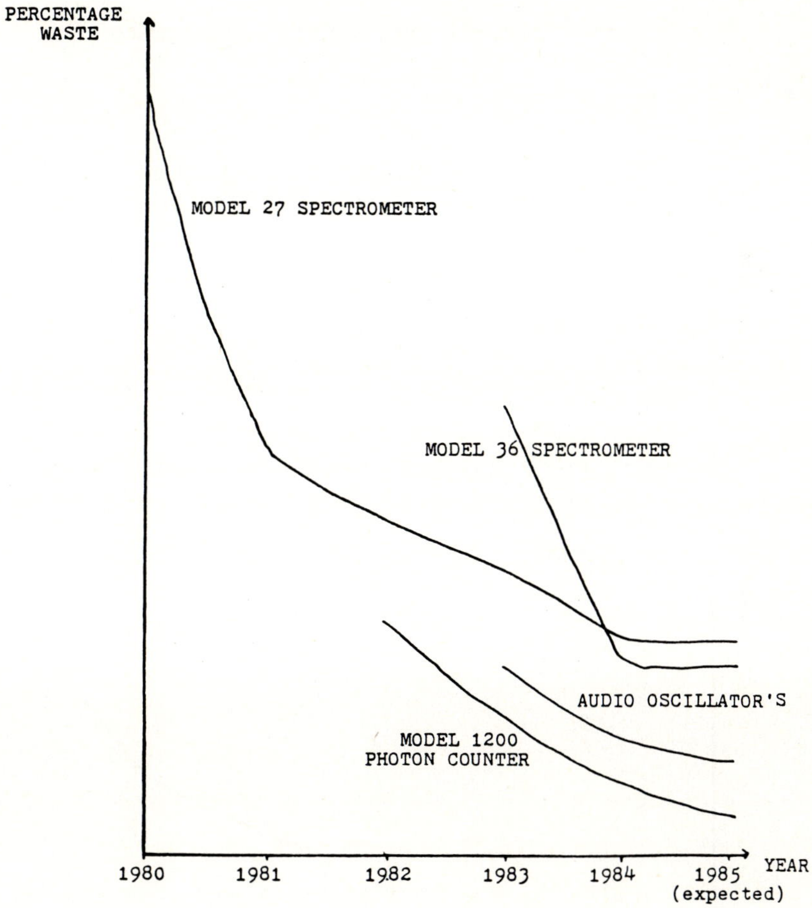

EXHIBIT 4 (continued)

TRENDS IN PRODUCTIVITY AND WASTE MEASURES

PHOTON COUNTER PRODUCTIVITY IMPROVEMENT

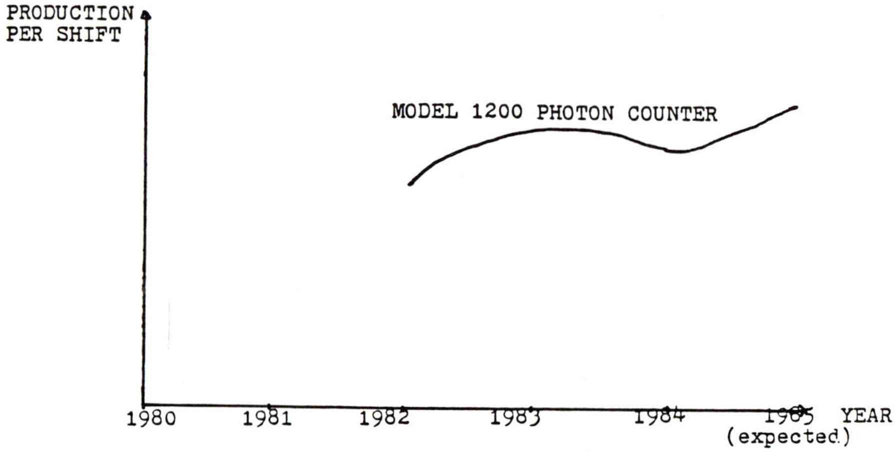

AUDIO OSCILLATOR PRODUCTIVITY IMPROVEMENT

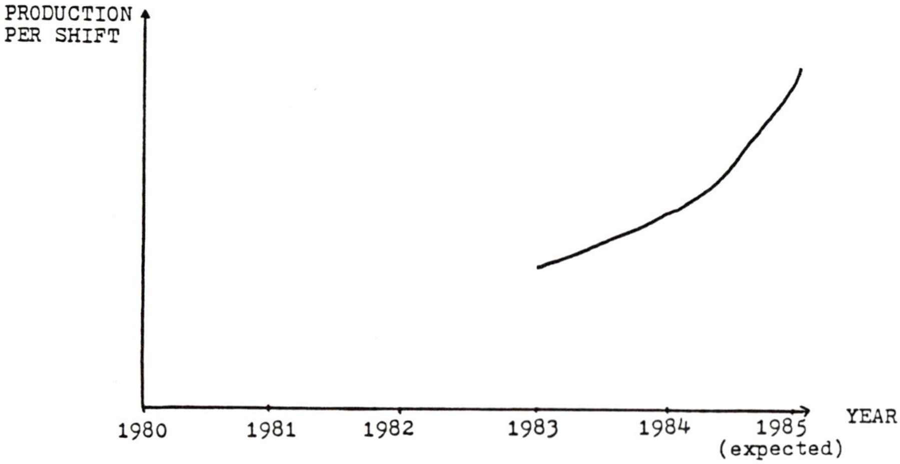

EXHIBIT 5
HOLDEN CANADA LTD. - SCIENTIFIC INSTRUMENTS DIVISION

MINUTES FROM A RECENT OSCILLISCOPE QUALITY CIRCLE MEETING

APRIL 18th, 1985

Discussion centered around having smaller runs, more often, by increasing output, faster set-ups and reducing line downtime through equipment improvements.

ACTION AND SUGGESTIONS:

(1) More vigorous preventive maintenance program needed to prevent unexpected breakdowns. Henry will treat this as a priority but realizes it's hard to find the time.

(2) Improve Set-Ups:
Sheila and Francine will work with maintenance to help design "standard set-up blocks" to be used for speeding up routine set-up functions.

(3) Missing Labels:
Everyone in agreement that some sort of sensing device to alert the packager that an oscilliscope passed by without a label is required. Phil, Dave, and Steve are looking into the best way to do this.

(4) Downtime Analysis:
Sheila and Kathy will begin using their run analysis sheets to document major downtime during runs. Maintenance will use this information to adjust priorities.

Task force is being set up to eliminate material discrepencies. A redesign of production reports as well as manner in which material is charged out will be looked at.

Productivity Management Frontiers — I, edited by D.J. Sumanth
Elsevier Science Publishers B.V., Amsterdam, 1987 — Printed in The Netherlands

THE IMPACT OF A PRODUCTION-ORIENTED QUALITY CIRCLE ON PRODUCTIVITY

JOHNSON AIMIE EDOSOMWAN
Dept. 146, Bldg. 012, International Business Machines Corporation, P.O. Box 950, Poughkeepsie, New York 12602, USA

ABSTRACT
Edosomwan, J. A., "The Impact of a Production-Oriented Quality Circle on Productivity."

This paper presents a step-by step methodology for implementing quality error removal (QER) technique in a manufacturing environment. The QER quality circle technique provides guidelines for work groups in resolving quality problems at the source of production. A case study that assesses the impact of the introduction and use of the QER technique on total productivity is presented. Two groups of five subjects each were studied. The groups initially performed the manual method of assembling printed circuit boards without any formal training on the QER technique. After a period of five weeks, the subjects were trained on how to use the QER technique and then performed the same task. A task-oriented total productivity measurement model was used to compute the partial and total productivities periodically, before and after the use of QER technique. The results of the study suggest that the implementation of the QER technique in a production task has positive impact on labor and total productivity. The problems encountered in setting up and implementing the QER technique are discussed as well as ways to correct them.

INTRODUCTION

In the early 1960s, the Japanese started the Quality Circle concept, based on the motivational techniques developed by American researchers and practitioners. According to Barra (1983), in April of 1962, the Japanese Union of Scientists and Engineers (JUSE) launched a new magazine QC for the foreman. In it's first issue it suggested the formation of workshop level groups headed by foreman and participated in by his subordinate workers. Naisbitt (1982) reported that the first QC was registered by JUSE in May of 1962. In the late 1970s, the Quality Circle concept began to receive unprecedented worldwide attention. Naisbitt (1982) pointed out that the number of participants has subsequently grown to over six million workers participating in over 600,000 active QC in Japan alone. In recent years, many companies and organizations throughout the world have adopted the QC technique. There are also several publications on the subject of QC, such as that of Dewar (1980), Robinson

(1982), Doyle et al. (1978), IBM (1983), JUSE (1980), Ishikawa (1968, 1972), Muramatsu et al. (1981) and Edosomwan (1985a). The quality error removal technique (QER) developed by Edosomwan (1983, 1987) is a unique type of QC that has the following benefits:

- QER technique improves quality and productivity at the source of production or service.
- QER technique provides the basis for dividing large work contents into small achievable tasks that are manageable by work groups. Organizational goals are clearly simplified into small tasks at the operational level. Work groups assume total responsibility for their quality and productivity, and feel an integral part of the goal setting process and the end results.
- QER technique improves employees morale and job satisfaction. It provides the basis for employees to assume total control at the source of production. Problems are solved through a team brainstorming approach at a much faster pace. Sources of help are readily available to work group team members.
- QER technique improves communication channels at all levels of the organization. Management is viewed as part of the team effort that removes the obstacles that affect the quality of production and service.
- The greater sense of control provided through the QER technique improves employee commitment, promotes a sense of dedication, less turnovers, lower absenteeism rate and provides good feelings about job security.
- The improved quality and productivity obtained through QER technique translates to more profit for the organization through product/service cost reduction, and lower prices which are paid by consumers for goods and services offered.

QUALITY ERROR REMOVAL (QER) TECHNIQUE DEFINED

Quality error removal technique provides a framework, principles and guidelines to a group of employees who voluntarily work together to select and solve key problems affecting an organization's work unit or task. Organizational goals are broken down into small tasks at the operational level and continuous effort is applied to improve productivity and quality within each work unit.

QER KEY PRINCIPLES AND GUIDELINES

The following are the key principles and guidelines used by the Quality

Circle team:

1. Teach Basic Job Skills and Cross-Train for Multiple Skills

   Both employee and management will be convinced that productivity and quality improvement is possible in an organization and within each task unit, once everyone involved understands how the job should be performed to deliver a good product. Everyone also understands the nature of the product or services being delivered. Two types of training are encouraged: A formal training through company educational program or an outside institution and training by peers.

2. Encourage Decision Making and Problem Solving at the Source of Production or Services

   At the operational level, the QER teams select their own focus problems and decide how to attack the bottleneck issues that prevent productivity and quality improvement. The team identifies problems which are controllable. All members of the team freely contribute ideas and reach a consensus on the best possible solution. The brainstorming technique is highly encouraged. The following steps are followed to understand the cause and effect of a key problem.

   a. Precisely define the problem.

   b. Draw a flow diagram and label all categories.

   c. Brainstorm to identify root cause of problem.

   d. Use pareto diagram to select a focus problem.

   e. Obtain group consensus on a solution.

   f. Implement the solution.

3. Encourage Productivity and Quality Improvement at the Source of Production or Service

   Each product or service inspection is done at the source. Inspection is on the line at the task level. Members of the QER team perform inspection of own work. Quality and productivity errors are resolved by joint effort among the team. Teams have checkpoints to review all outgoing finished products. Any member of the team can call for group help to reach a consensus on key problems quickly. Feedback mechanisms between each QC team member are immediate. Team members are viewed as teachers and error-correction masters.

4. Encourage the Following QC Team Characteristics

All the members of the team should clearly understand and agree on common goals. Create a forum for the sharing of ideas, encourage a high level of communication and participation among team members. Supervisors or managers support the team on problems when called upon. Reward and recognition are shared among team members.

5. Clear Definition of Roles and Responsibilities of the QC Team

The definition of the roles and responsibilities are as follows:

The Teacher: The person who teaches team members the skills needed to perform the task. At one stage or another, everyone on the QC team is a teacher.

Team Leader: The person who leads the team through the process of resolving a key problem. The team leader is chosen by voluntary means. Each member of the QC team picks up leadership on a rotational basis.

Team Members: A group of people who participate in the problem solving and resolution process. The team meets regularly at specified times and locations to resolve problems related to their tasks.

Facilitator: A team member that records all the team's progress, keeps records of the agenda and assists the team leader to accomplish stated objectives and goals of the QC team.

Team Correction Master: Any member of the team that identifies a quality error in the output produced by other team members. Such a member ensures that the error is corrected at the source of production or service. All team members participate in random and revolving inspection processes to help detect quality errors.

Team Presentations: Presentations to management on team accomplishments is rotated among team members. The presenter introduces team members at the start of the presentation. The agenda and the presentation package is jointly prepared by all members of the team.

IMPLEMENTATION METHODOLOGY

The following six steps, shown in Figure 1, are suggested for the implementation of the QER technique.

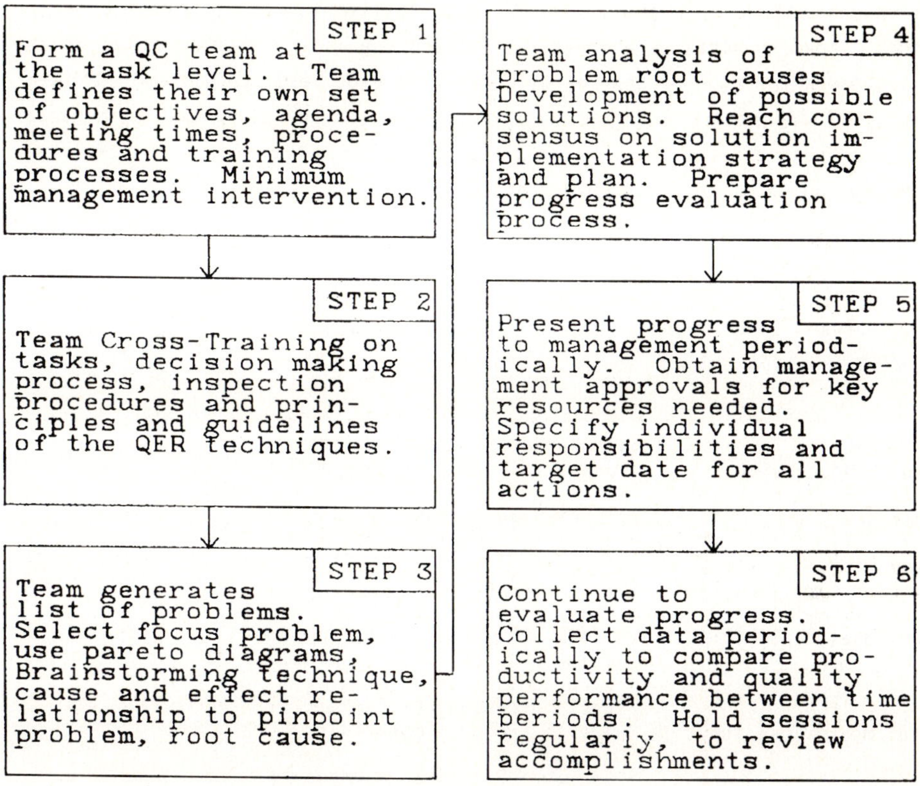

Figure 1. Implementation steps for the QER technique.
Source: Edosomwan, J. A., (1987).

CASE STUDY

The quality error removal technique (QER) was implemented at a manufacturing plant that produces printed circuit boards to customer orders. Two groups of five subjects each were studied. The groups initially performed the printed circuit board assembly task without any formal training on QER technique. After a period of five weeks, the subjects were trained on how to use the QER technique. The assembly process of printed circuit boards involved the normal insertion of components such as transistors, diodes, modules, resistors and capacitors into an empty printed circuit board (PCB).

The work area was in an approved electrostatic discharge area that consisted of a grounded work bench, chair, tools, and a light box. The manual assembly process started with the verification of paperwork and parts. The component placement list (CPL), picklist, routing, part numbers, engineering change level, serial numbers, templates and any special instructions were checked before assembly. The foil, which was a blueprint of component insertions, was placed on the lightbox. The foil contained the outline of two boards. The components were manually inserted one at a time, according to the foil. Upon completion of both cards, they were each verified with the CPL template. If any rework or scrap was to be done, it was performed in a manner similar to that of insertion. When the printed circuit boards were completely assembled and verified, they were manually placed into trays.

MEASUREMENT METHOD FOR PRODUCTIVITY

The task-oriented total productivity measurement model (TOTPM) developed by Edosomwan (1985b, 1985c) was used in the case study. A brief description of TOTPM is given below:

$$\text{Total productivity of task i, in site j, in period t} = \frac{\text{Total measurable output of task i, performed in site j, in period t}}{\text{Total measurable input of task i, performed in site j, in period t}} \quad \ldots(P1)$$

where:
(i = 1,2,...,n; j = 1.2....,m;; t = 1,2,...,k)

Total Measurable Output = Value of (finished units produced
+ Value of partial units produced
+ other output associated with units produced).

Total Measurable Input = Value of (Labor + materials +$ Capital
+ Energy + Robotics Expenses +$ Computer
Operating Expenses + other
Administrative Expenses + Data
Processing Expenses).

Total-Factor productivity is the ratio of total measurable output to the associated labor and capital input.

$$\text{Total-Factor Productivity of task i, in site j, in period t} = \frac{\text{Total measurable labor and capital output of task i, performed in site j, in period t}}{\text{Total measurable labor and capital input of task i, performed in site j, in period t}} \quad \ldots.(P2)$$

Partial Productivity is the ratio of total measurable output to one class of input. For example:

$$\text{Partial Productivity of task i, with respect to labor input in site j, in period t} = \frac{\text{Total measurable output of task i, performed in site j, in period t}}{\text{Measurable labor input of task i, performed in site j, in period t}} \quad \text{....(P3)}$$

The values of the output and input are expressed in constant monetary terms of a reference period. In the case study, complexity factor of printed circuit boards and the proportional contribution to total number of printed circuit board insertions was used as an allocation criteria for overhead expenses. Deflators were not required, because the study period of ten weeks was not long enough to be affected by significant price changes for input components.

RESULTS

As shown in Table 1, the implementation of the QER technique in a manufacturing task improved labor and total productivities. Perhaps the success can be attributed to the application of the QER principles and guidelines, the teamwork approach adopted by the QC team, willingness to train each other on skills required and strong ability to quickly resolve problems at the source of production. Equally important was the high motivational level experienced by all team members. The feeling that members of the team had total control over what they do, and were able to take total responsibility for decision making, also contributed to the gains in productivity.

Problems Encountered During Implementation

The problems associated with the implementation of the QER technique could vary from one organization to another, depending on the nature of task performed, QC team composition and levels of management support. Based on the implementation in a manufacturing task, it was observed that friction among team members arose in some situations.

Interestingly, the team resolved all frictions. It was also difficult for team members to accept and tolerate rejection of ideas from management. This problem was resolved through on-going communication between the QC team and management. They also faced situations in which the team had too many problems to resolve, or too little to keep them busy. However, the team agenda improvement process helped resolve the problem.

CONCLUSIONS

As demonstrated from the case study, the QER technique is a powerful tool for the improvement of productivity and quality at the source of production or service. The technique has great potential for application in both manufacturing and service work environment. It is very inexpensive to set-up and maintain the QER technique. QER basic strategy rests on providing decision latitude, control and ownership to those charged with task responsibility. Management support, training and other resources required, are necessary for the successful implementation of the QER technique. The QER is unique and works best when the QC team is limited to a small group.

TABLE 1: Labor and Total Productivity Values for Group 1 & Group 2

	BEFORE TRAINING ON QER TECHNIQUE				AFTER TRAINING ON QER TECHNIQUE			
	GROUP ONE		GROUP TWO		GROUP ONE		GROUP TWO	
PERIOD	Labor Productivity Values	Total Productivity Values	Labor Productivity Values	Total Productivity Values	Labor Productivity Values	Total Productivity Values	Labor Productivity Values	Total Productivity Values
WEEK 1	2.13	0.96	2.96	1.21	4.78	1.89	5.00	2.00
WEEK 2	1.78	0.72	2.13	0.99	5.22	2.10	5.16	2.23
WEEK 3	1.91	0.76	1.72	0.68	4.82	1.90	4.96	2.08
WEEK 4	2.00	0.90	1.99	0.79	4.94	2.00	4.99	2.60
WEEK 5	1.69	0.70	2.21	1.07	5.1	2.10	5.06	2.01
The difference between labor and total productivity values before and after training on QER technique is statistically significant at $\alpha = 0.05$ using ANOVA technique.								

REFERENCES

Barra, Ralph, (1983). Putting Quality Circle to Work, McGraw-Hill New York.
Dewar, Donald L., (1980). The Quality Circle Handbook, Quality Circle Red Bluff, California.
Doyle, M., Straus, D., (1978). How to Make Meetings Work, Playboy Paperbacks, New York.
Edosomwan, J. A., (1985a). Productivity and Quality Improvement at the Source of Production. IBM Technical Working Paper.
Edosomwan, J. A., Productivity and Quality Management, Marcel Dekker, 1987, in press.
Edosomwan, J. A., (1983). Quality Error Removal Technique. Unpublished manual, IBM Data Systems Division, New York.
Edosomwan, J. A., (1985b). "A methodology for assessing the impact of of computer technology on productivity, production quality, job satisfaction and psychological stress in a specific assembly task". Doctoral Dissertation, Department of Engineering Administration, the George Washington University, Washington D.C. 20052. Social Science Research Council (US Department of labor) Grant #SS-36-83-21 and IBM Corporation Grant #IBM 2J2/2K5-722271/83/85.
Edosomwan, J. A. "A task-oriented total produvtivity model for electronic printed circuit board assembly," Proceedings for the First International Conference on Electronic Assembly, October 7-9, 1985c, Santa Clara, California.
Ishikawa, Karou, (1968). QC Circle Activities, JUSE, Tokyo, Japan.
Ishikawa, Karou, (1972). "Quality Control and Ends with Education," Quality Progress, August, 1972.
Muramatsu, R., Tanaka, Y. and Miyazaki, H., "Example of increasing productivity and product quality through satisfying the workers' desires and developing the workers' motivation," Proceedings of 1981 Spring Annual Conference and World Productivity Congress, AIIE, Inc. PP.652-660.
Naisbitt, John, (1982). Megatrends, Warner Books. QC Circle Koryo, "General Principles of Quality Circles," JUSE, Tokyo, Japan, 1980.
Robinson, M., (1982). Quality Circle, A Practical Guide, Gower Publishing Co. Ltd., Aldershot, England.
Teaming up for Quality: Quality Excellence Teams Education Guide, IBM Data Systems Division, Poughkeepsie, New York 12602.

Productivity Management Frontiers — I, edited by D.J. Sumanth
Elsevier Science Publishers B.V., Amsterdam, 1987 — Printed in The Netherlands

PRODUCTIVITY GAINS THROUGH CIRCLES

M. GOMEZ
Harris Corporation, Computer Systems Division, 2101 West Cypress Road, Fort Lauderdale, Florida 33309

ABSTRACT

An overview of the implementation of Quality Circles in a high technology development and manufacturing organization. The effort was conducted as a research project with measurements, observations, and maintenance of statistics and records. Concurrently, the participative style of management was adopted. After five years, the effort is still increasing its contributions to productivity gains. Returns from productivity gains exceed all prior investments. Many other benefits comparable to investment returns, accrued from this effort.

1 INTRODUCTION

Most American business executives believe that the solution to lagging productivity lies in technology. That technology can enhance productivity is true. It is a key link in the productivity enhancement chain -- the first link is leadership. Effective leaders, must earn the respect and support of followers. Support and respect is gained by exhibiting knowledge in coordinating human resources and in applying effective management techniques. Followers, even when they support and respect their leaders, must also earn their place as a link in the productivity chain. They need to be trained in, and be effective at, supporting leaders in achieving productivity objectives. Their primary training needs are in creativity and innovation. This, along with education in investigative, analytical, and evaluative processes, applied in a communicative atmosphere, creates a climate for dedication, unbeatable teamwork, and unsurpassed productivity gains.

The aforementioned management and employee characteristics, combined with institutionalized employee participation, using circles' processes, provides the groundwork for developing a work force that can select and implement the most beneficial and productive technology. So, while business executives are correct in their belief that technology will remedy lagging productivity, they fail to see that human resources development is a necessary prerequisite to technology application. It is also prerequisite to improvements in organization, processes, systems and procedures, and areas of opportunity for productivity gains.

2 THE CSD RESEARCH

The aforementioned facts became evident from experimentation and research in participative management, employee participation, and circles. The research was conducted at the Computer Systems Division, of a large corporation. The effort began in mid-1980 and is now institutionalized as a normal organizational operating mode, ligitimized by documented policies and procedures. Computer Systems Division, (hereafter referred to as CSD) employs about 1,000 employees. They develop and manufacture super-minicomputer systems and operating software. Annual revenues substantially exceed $100,000,000.

This paper provides a summary and overview of CSD's employee participation effort, using circles as the coordinating media. Because of modifications and broader scope of functions, circles were named "Improvement Circles," rather than "quality circles." The effort was conducted as a research project, so as to verify and validate effectiveness and determine if results merited sustained management support and continued funding. The first three years of this effort is documented in a research dissertation.[1] By the end of 1982, circles were institutionalized; however, measurements, budgeting, controls, observations, and analyses, initiated for research have continued. This data is a source for valid and factual management reports. The reports, in turn, provide justification for continued funding, operational support and for promotional activity to sustain interest and growth.

As of the end of 1985, and after five years, contributions by participants continued at an accelerating annual rate. Circles continued to increase in numbers. See Figure 1. Membership was more selective (by member action). Most members wanted to work only with productive participants. While participation was voluntary, members who did not produce, were asked to quit the circle by other circle members. This action slowed membership growth, but remaining members were more dedicated and enthusiastic. See Figure 2 returns from circles now surpass all other previous and presently planned investments. Futhermore, the effort remains unmarred despite recent economic and business climate disappointments.

3 CSD CIRCLES PROJECTS

Continued success, growing effectiveness, and favorable productivity gains are best substantiated by relating specific projects and activities. Ensuing paragraphs relate typical prrojects consumated during five years of participation and circles to 1986.

1. See References and Bibliography

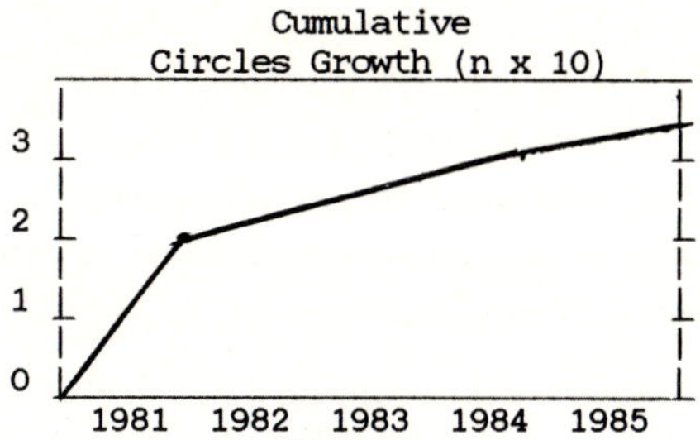

Figure 1.

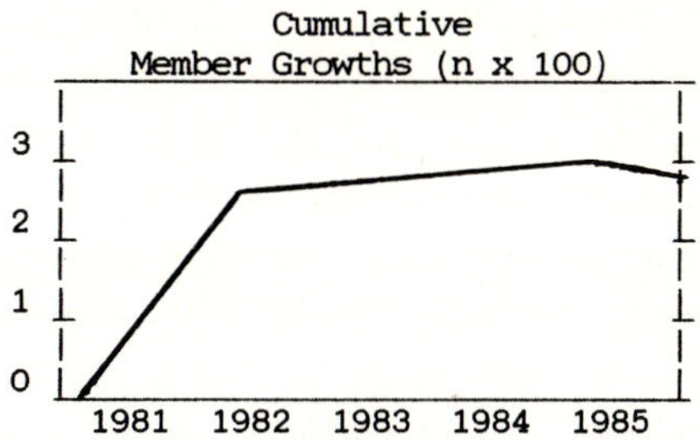

Figure 2.

3.1 Initial Circles Effort (1981 -- 1983)

Initially circles worked to improve their own department. Tasks selected were those where members, or their supervisor, had discretionary authority to effect changes. Typically, circles worked at purging cluttered and disordered work files, including prints, specifications, assembly/test instructions and other work documents. Obsolete and worn documents were replaced, cutting confusion and time lost in job setup. Circles improved handling, packaging, and control of assembled parts and materials during manufacturing, to minimize deterioration, damage, identity loss, and mislocation. A changed issue point of computer accessories saved $25,000 annually in damage, lost or misplaced items, and unneeded reinspection, cleaning, and repackaging of previously QC certified hardware.

Test Technicians had a "fit" problem when inserting Printed Circuit Boards (PCBs) into a backplane connector. Time lost and damage to hardware was most disturbing to Technicians. Process changes by Quality and Manufacturing Engineers had corrected the situation twice in the prior year. Each time the problem returned. The fixes had required complex assembly operations to assure proper mating. Difficulty of assembly frustrated Technicians and procedures were not scrupulously followed. After organizing a circle, the Technicians decided that this costly and frustrating problem needed a design change to simplify assembly. They knew Development Engineering resisted changes because they were pressed for time with new designs. The circle brainstormed for a new design. They discussed and analyzed various ideas. Finally, they invited the Senior Project Engineer to meet with the circle. The engineer responded and was delighted with their ideas. Soon, thereafter, a design implementing the circles' proposal was released. The problem was eliminated.

Incoming Quality Control Circles' statistics revealed that 28 percent of the Computer power supplies were reported to have failed in Systems Integration. When the power supplies were returned to QC they tested good. Replacing, repackaging, handling, processing paperwork, transporting, and rework of these heavy power supplies cost CSD $38,000 annually.

The circle proposed an on-computer test similar to the one used by QC, but using a portable test station. The cost of the test station was $11,000, still gaining a savings of $27,000. Management and Test Engineering gave approval. Project Engineers later found that the test could be accomplished with available equipment. The result? The Circle saved the entire $38,000.

3.2 The Middle Years (1983 -- 1984)

During 1983 and 1984 new circles were formed and circles expanded scope and complexity of projects. Complex projects were longer term projects. Most were improvements not problems. Problems were scarce in mature circles. Without improvement projects, circles would be out of work. Completing significant projects whetted appetities for more complex challenges and greater achievements. Management set no limits or restrictions on circles projects, so long as they contributed to quality, productivity, or quality of work life.

The Accounting Circle wanted to reduce excessive overtime needed for end-of-month closeout. Investigation revealed that computer terminal limitations delayed data entry and created a bottleneck. A Cost/Benefit Analysis disclosed that adding a terminal would save overtime costs that resulted in a payback in two months. Management approved the purchase of another terminal.

Two circle members discussed the problem with friends in Contracts. Contracts also had overtime problems for lack of terminals. Contracts could not do billing and Corporate Reports until Accounting completed entry of end-of-month data. Two Contracts terminals were located near Accounting. Since Contract's terminals were mostly idle while Accounting completed EOM data entries, and Accounting's terminals were mostly idle thereafter, the solution to both departments' terminal problems was obvious. By terminal sharing, Accounting was able to cancel the purchase of a terminal and, thereby, achieved $1,500 in "cost avoidance." Contracts also solved their terminal problem and, consequently, their excess overtime and associated costs.

In Manufacturing, engineering specifications and drawings were not always compatible with processing operations. Screw lengths were too short to meet quality and and Workmanship Standards. PCB and cable markings were mislocated and not visible after assembly (needed by Test Technicians and Field Engineers). Cable lengths, if made to drawing, either would not reach connectors, or excessive lengths caused cosmetic problems.

Manufactured hardware that deviates from drawing or specification results in Quality Control rejection. A QC rejection stops hardware processing until the problem is resolved. Drawing and specification incompatibility caused schedule delays while Manufacturing, Quality Control and Design Engineers expended much high cost labor deciding the action needed. Usually, temporary waiver or deviation documents were issued with a promise of formal drawing or spec change in 30 days. Because of pressure from their own projects, Design Engineering's commitments were usually not met. Hence, another rejection adding to lost time and schedules, ergo added cost. The change commitment backlog grew, as did schedule delays and diverted engineering labor.

Manufacturing Engineering proposed a joint effort by the three Engineering group's circles to find a solution. After many joint meetings, ideas, arguments, and discussions, an acceptable solution evolved. It was agreed that, for items not affecting "form, fit, or function," Manufacturing Engineering would prepare an Engineering Change Package (ECP) and submit immediately to Design Drafting. If Drafting agreed with the "3 F" criteria, it would become effective upon signature of the two engineering groups, subject to later Design Engineering concurrence. Implementing this change immediately cut engineering labor diversions, hardware processing stoppages, and delayed schedules. Productivity gains exceeding $112,000 were achieved the first year through redirecting engineer's labor to their primary tasks.

The Software Development Engineering Circle, of highly skilled programmers, worked many overtime (unpaid) hours to meet project commitments. Programmers would leave projects (programs) running on the computer at the end of the workday. They would return later after a meal with the family. Often, they worked past midnight before returning home. The circle proposed home terminals, so as to monitor projects from home after normal work hours. The General Manager provided fifteen terminals and modems for a trial effort. He wanted to be certain that efficiency and schedules could be maintained by this manner of operating. After six months, home terminals accelerated project completion by over $49,000 in budgeted programming hours. This was in contrast to normal overruns. A female programmer commented, "Now I do my family's ironing while I monitor my project."

3.3 The Fifth Year (1985)

In the fifth year even more complex and sophisticated projects were undertaken by circles members. Scope, effects, and the productivity benefits of projects expanded.

The diagnostics Engineering circle, concerned about criticism of their programs by customers and Field Support Engineers, decided that action was needed. Complaints were that programs were complex and instruction manuals difficult to understand. The circle knew that when computers break down at customer sites, there is intense pressure to restore operations. To help relieve the situation, the circle set the following goals:

- Make all diagnostic programs menu driven.
- Provide easily accessible HELP instructions.
- Provide much more on-screen instructions.
- Standardize instruction manuals and minimize content.

To reduce repair time, a statistical analysis was made of diagnostics programs usage frequency. This resulted in reordering Diagnostics Tapes to load programs in order of most frequent use.

Achieved productivity gains reduced labor costs of Systems Integration and Field Service Engineers by $200,000 annually. These users also reported improvement in quality of work life because of their increased efficiency. Customer goodwill was enhanced by expeditious restoration of computer services. Circles with many active years had few problems. Diligent mental searching for "improvement" projects, was needed for circles to remain productive. Favorite projects involved increasing productivity and quality through standardizing, reorganizing, improving processes, and education and training. Circle members were strongly moved toward improving employees' capability and competence. Many projects of this nature were undertaken. Examples and descriptions of projects that contributed to operational improvement and to job knowledge recently completed, include:

The Drafting Book of Knowledge. This book establishes CSD's standard drafting symbols, terminology, and layout. It includes instructions and application of Design Oriented Graphics Systems (DOGS), plus information for uniformity of drawing and specification structure and process. It also helps new and temporary employees to quickly adapt to CSD drafting standards and procedures.

The Tricks of the Trade. This handbook was developed by assembly circles to provide instructions for assembly operators. It sets rules for use of tools and equipment, jobs set-up, checking work documents, and for verifying product hardware and workmanship quality.

Systems Integrations Group (SIG) Manual and _Peripherals Integration Test (PIT) Manual._ This manual was prepared by circles from Systems and Peripherals Test departments to pool everyone's knowledge so as to standardize effective methods to integrate systems, including power up, preliminary testing, running diagnostics, repairing, revision upgrading, quality checks systems teardown, and prep-for-ship operations. The manual contains instructions for operational testing to assure interface of hardware, software, and test equipment. Also, instructions provide for testing in fixed sequence and in a specific manner. Through uniformity in intergration and test, failures on installation should not occur from technician (tester) oversight or deliberate omission. Field installation failures will, hereafter, have a test added to the manuals to assure that defects are detected and corrected before shipping to customers.

The Hardware Standards Manual. Final assemblers found that specified fastener hardware, and other assembly bulk materials that served the same function on different computers models, were often of different sizes, material content, platings, configuration, etc. The circle requested joint meetings with engineering circles from Quality, Manufacturing, Design Drafting, and Development. Final Assembly recommended use of identical general hardware in all similar computer assemblies, where it served the same purpose. Standardizing, they reported, would reduce part specifications and quantities in assembly bins and stockrooms, thereby, reducing inventory and facilitating assembly. Recommendations were accepted and Final Assembly helped engineers by marking up drawings for revision. The Hardware Standards Manual was published and is now a controlled technical document for current and future hardware applications.

4 ANALYSIS AND RESULTS

Circles generated cost savings provided a valid basis for calculating productivity gains. Managers at CSD discovered that their responsibility, for increasing productivity, was best achieved by circle members' initiated and factually supported cost savings. Prior to circles, CSD had no formal system for problem solving and making improvements. Managers and engineers, overloaded with personnel planning, budgeting, scheduling, and sustaining operational control, had little time to solve problems and less for improvements. Therefore, circles' accomplishments are productivity gains that would not have occurred otherwise. At the end of 1981, verified and validated circles' net savings exceeded \$90,000. By the end of 1985, annual savings grew to over \$1,000,000. See Figures 3. and 4. This is not the full story. CSD rules did not permit credit for downstream beneficiaries. Unless affected circles participate in the action and concur in the Cost/Benefit (C/B) Analysis. Also, savings projections for recurring operations are limited to one year, maximum. Furthermore, most intangible gains cannot be measured in dollars. For example, improving quality, reliability, service, and ease of maintenance and operation sometimes added cost. Yet, these items helped customers, field support, sales, and marketing. Customer good will is difficult to measure. Whether a sale was made because of circle improvements cannot be determined. Also, did simplifying field maintenance raise the quality of work life of technicians and field support personnel, such that they were more enthusiastic and more productive? CSD Management's consensus was that these factors contributed substantial "value added" to the product and, accordingly, productivity gains.

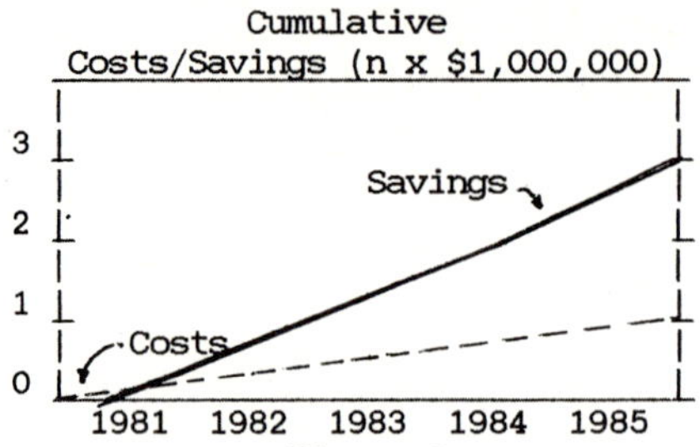

Figure 3.

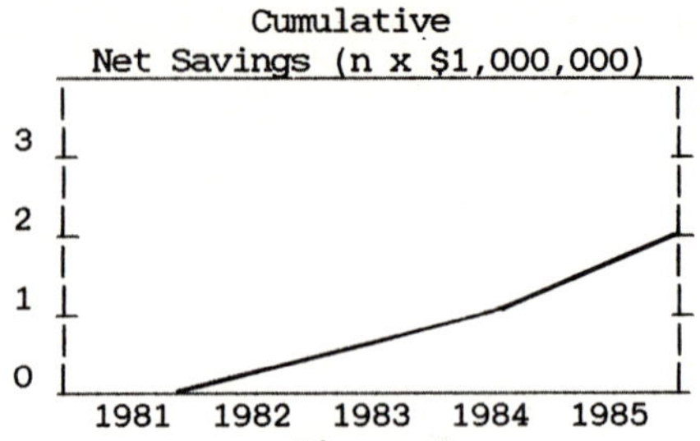

Figure 4.

Productivity improvements are most evident through close analyses of circles accomplishments. For example:

- The $1,000,000 in net circles' savings in 1985 by 265 (average) members amounts to $3,846 per active member.
- Labor cost of member employees averaged $25,000 annually. Therefore, each member earned 15.4 percent of his pay in 2.5 percent of his work hours -- one hour per week of circle time.
- The $1,000,000 in 1985 annual savings was equal to earnings on $14,000,000 of added sales, using a 1985 industry average return on sales of profitable computer companies.
- The industry produces an average annual revenue of $100,000 per employee, or $48.07 per paid hour. Equivalent hourly revenue for CSD employees' during ring circles effort was $1,016.00.

5 CONCLUSIONS

From participation and circles, productivity gains were achieved in every operating area, not just manufacturing. Also, productivity gains were not made exclusively through technology. Extensive gains were made through improvements in administration, organization, and new and better processing, mostly originated through worker ingenuity and persistence.

Many beneficial factors were a fallout of this coordinated effort, including a perception by employees of improved quality of work life. Employees perceived themselves as recognized and valued members of their company.

Management's contributions cannot be overlooked. Without their strong support, participation would not have achieved its remarkable success. Most notable of management's actions, was their receptiveness to employee proposals. Had they not learned to listen and given consideration to ideas, employees might not have lost their inhibitions to boldly speak out and present their projects to managers and engineers. In return, managers discovered that their responsibility, for increasing productivity was best achieved by circle members' initiated and factually supported cost savings. Participation and circles also fostered improved inter-department communications. This is most evident in projects that involved two or more departments to arrive at a solution to a common problem.

Finally, a causative force was the continuing management and employee development, training, and practice to support and improve participative management and circles. This human resources effort contributed to better leadership, more effective participative management, and understanding and effective employees. For example, everyone was trained to support all proposed projects by Cost/Benefit Analysis. Of several hundred projects proposed and supported by C/B Analysis, none were rejected. Training and development resulted in a cohesive organization and a powerful workforce that directed previously unapplied personal human capabilities and resources toward making CSD a more profitable, competitive and sustaining organization.

In the computer industry, foreign and domestic competition has been fierce in the past five years. Many companies have gone belly-up. Others had reduced earnings, concurrent layoffs, and capacity cutbacks. Few have prospered, as is evident by market reports. Yet, CSD has not had a deficit quarter since circles were organized. Two minor layoffs had negligible effect on circles growth or member commitment. The division's profits, while not phenomenal, have been above industry average.

6. REFERENCES AND BIBLIOGRAPHY

Gomez, M. Does Employee Participation Increase Productivity and Perception of Improved Quality of Work Life? Ann Arbor, Michigan: University Microfilms International, 1983.

Hackman, J.R. and Suttle, J.L. Improving Life At Work. Santa Monica, California: Goodyear Publishing Company, Inc., 1977.

Levering, R., Moskowitz, M. and Katz, M. The 100 Best Companies To Work For In America. New York: New American Library, 1985.

Ouchi, W., Theory Z. Reading, Massachusetts: Addison-Wesley Publishing Company, Inc., 1981.

Peters, T.J. A Passion For Excellence. New York: Random House, Inc., 1985

Peters, T.J. In Search of Excellence. New York: Harper and Row Publisher, Inc., 1982.

Productivity Management Frontiers — I, edited by D.J. Sumanth
Elsevier Science Publishers B.V., Amsterdam, 1987 — Printed in The Netherlands

THE 'SHRED COST' PRODUCTIVITY GAINSHARING PLANS

E.M. DAR-EL
Faculty of Industrial Engineering and Management, Technion - Israel Institute of Technology, Haifa 32000 (Israel)

ABSTRACT

Gainsharing is a form of group reward sharing plan. All the "SHRED COST" plans described comprise both Productivity Sharing and Buy-Back (BB) elements. Gainsharing plans may be applied separately to departments, to the entire production facility, or, to the plant as a whole. The Buy-Backs selectively tighten the looser Time Standards - often the result of improvements made by employees themselves. Gainsharing is especially applicable to companies using Quality Circles , Work Improvement Teams, etc. Applied to ten departments, an average of 1 1/2 BB per year (of 10% each) was achieved. It is claimed that Buy-Backs provides the most efficient (and painless) methods known for tightening loose standards.

1 INTRODUCTION

In his book "Productivity Improvement: Employee Involvement and Gainsharing Plans", Dar-El [3] defines Gainsharing Plans as: "Group incentive plans wherein the productivity gains measured in some defined manner are shared between participating employees and the company according to specified formulae".

Gainsharing plans are generally made an integral part of employee involvement programs such as, Scanlon [1], Rucker [2] and SHRED COST [3]; but they can also operate quite satisfactorily without these as in case of Improshare [4]. However, gainsharing provides the logical reward system for use with formal involvement programs that operate for the purpose of improving productivity. Gainsharing plans essentially reflect a developing partnership between employees and the plant's management, where the accent is on improvements and CHANGE rather than on EFFORT, which is the main objective in wage incentive plans.

Improshare is the only one of the well known plans that maintains a strong link between performance and earned bonus. SHRED COST (SHaring REDuctions in COST, first publicized in 1981 [5], is a "take off" from Improshare.

Other literature on Gainsharing Reward Plans include - [6], [7], [8], [9] and [10].

2 GENERAL STRUCTURE OF SHRED COST GAINSHARING PLANS

SHRED COST gainsharing plans contain two elements, "Productivity Sharing" and "BUY-BACKS". Productivity sharing deals with the periodic sharing of productivity gains between the plant and its participating employees, whereas the BUY-BACK (BB) incorporates a process of absorbing the idle time of participating employees when their work performances are shown to consistently operate above a predefined ceiling or target bonus. Two types of gainsharing plans are proposed - Plan A and Plan B.

In Plan A, productivity sharing is permitted upto a predefined ceiling. It requires the mandatory implementation of a BUY-BACK (BB) once the participating group has demonstrated its ability to perform above the ceiling as illustrated in Figure 1. Here, three consecutive points above the ceiling result in a BB. Previously, all productivity gains between the ceiling and the starting bonus (or Base Bonus, B) are shared p:q (p+q=100) between participating employees and the plant. Starting bonus, B, for the gainsharing plan is based on the past average bonus earned by the employees. The BB involves the "purchase" of a selected amount of the group's idle time (e.g. R = 10% in Figure 1) for a one-time cash payment referred to as the BB compensation.

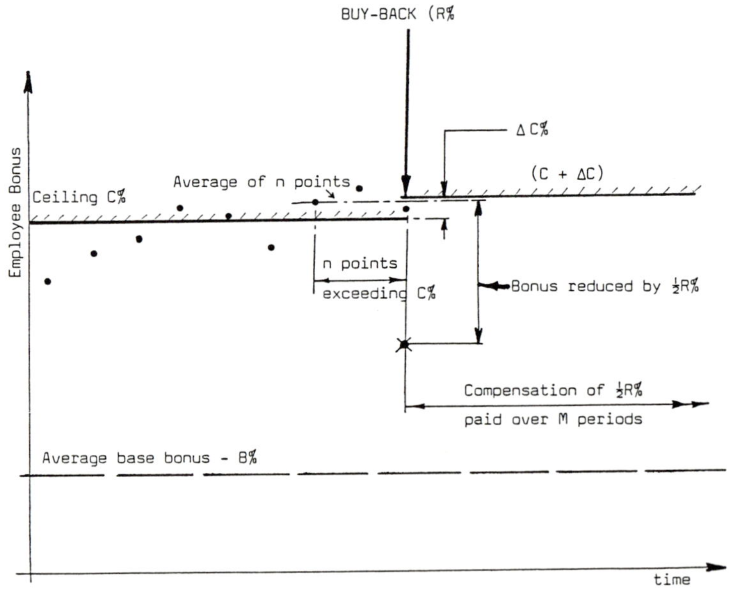

Fig. 1. Illustrating SHRED COST Parameters Associated with BUY-BACK Plan A

Productivity sharing under Plan B is determined by a predefined Target Bonus (TB) whose value reflects the planned long-term bonus earning for the gain-sharing plan. Plan B is based on a periodic evaluation of the group performance in order to implement whatever BUY-BACKS that are possible. BB under Plan B require the tightening of the looser product standards such that the potential bonus earnings at the start of each period is the Target Bonus itself. This is illustrated in Figure 2 which is drawn for an evaluation period of three months. Performances above the Base Bonus B are shared p:q between employees and plant (as in Plan A) but there is no ceiling on the bonus earned. If $\overline{TOTB}(k)$, the average total bonus earned for period k is greater than the Target Bonus, then the BB amount is twice the difference, i.e., $2 \times (\overline{TOTB}(k) - TB)$, since BBs are assumed to be shared equally between employees and plant. Depending on the application, the evaluation period (k) can be three, four or six months.

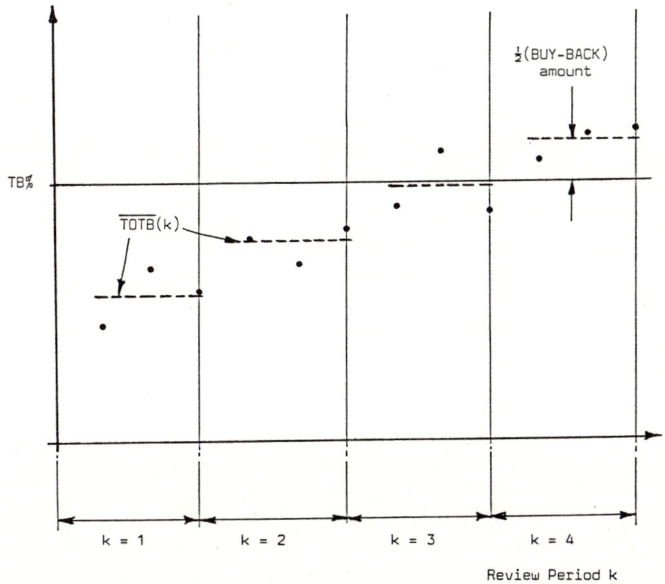

Fig. 2. Illustrating Bonus Earnings Under the Regular Plan B BUY-BACK Operation.

Under both Plan A and Plan B, the BUY-BACK compensation is always taken as one half of the BUY-BACK amount (as a percentage) which is then paid to employees each month over a fixed period such as one year. For example, a 10% BUY-BACK amount means a 5% monthly compensation bonus is paid to each participating employee for the next (say) twelve months. The implication is that appropriate unit labor costs are reduced by 5% during the compensatory bonus payment period, and then by an additional 5% when the compensatory period is concluded. Thus, the entire BUY-BACK benefits eventually go to the plant.

Buy-Backs provide one of the most efficient (and painless) methods for tightening loose work standards known to the writer. SHRED COST applied separately to ten departments for periods ranging from one to three years, averaged one and a half BUY-BACKS (of 10% each) per year per department; the longest running application, taken over a period of three years, experienced a total of five BBs [11].

3 APPROACH TO OPERATING A SHRED COST GAINSHARING PLAN

The following steps should be followed when planning a SHRED COST gainsharing plan:

(a) Select the productivity performance measure(s) to be used

As indicated earlier, SHRED COST is directed towards the productivity of manpower and all that influences it in the organization. Hence, the only meaningful measure must incorporate either, or both, labor productivity and savings in materials, inventory or auxiliaries such as, steam, energy, water, oils, solvents, etc.

Thus, the productivity performance parameters are generally surrogates of overall productivity measures. Depending on the application, these may include some, or combinations of the following types:

(i)	$\frac{\text{SMH Out}}{\text{MH In}}$	where	SMH - Standard Manhours MH - Manhours	
(ii)	$\frac{\text{MH Required}}{\text{"X"}}$	where	"X" is the manufactured products or auxiliary materials used in the production process.	e.g. $\frac{\text{MH}}{\text{\$1000 shipped}}$
(iii)	$\frac{\text{"Y" Output}}{\text{"X"}}$	where	"X" and "Y" are the manufactured products or auxiliary materials used in the production process.	e.g. $\frac{\text{Ounces of Gold}}{\text{Ton of Ore}}$

Selection of the productivity measure(s) must ensure that the participating employees are able to control and/or influence the level of performance.

(b) Establish the past standards of the productivity measures

The past standard(s) is (are) used in determining the Base Ratio(s) of the various measures which are consequently used for calculating the productivity bonuses.

The Base Ratio (BR) is simply the performance level averaged over some stable period in the past, e.g., 6 months to a year. When establishing the BR, one should ensure that during the period of its determination, the participating employees were all working, i.e., no enforced idle time due to business slowdowns, and that the influence of an unusually large number of new employees should be accounted for.

Except for special applications, the BR remains constant in all future productivity sharing calculations. The value of BR is only altered in cases involving changes in the number of non-production employees, and when new norms are established for the consumption of auxilliary materials.

(c) Decide on operational Plan A or Plan B

The writer has used both plans effectively and there are pros and cons with each. Current tendency is to go for Plan B with the periodic review period fixed at 3 months which corresponds to the plant's financial "quarters".

(d) Select the SHRED COST plan parameters

The table below summarizes the parameters used with the various SHRED COST gainsharing plans.

	PLAN A	PLAN B
Productivity Sharing	C, Z, p:q, B	TB, n, p:q, B
BB Application	n, R%, p:q, m,ΔC	k, p:q, m,ΔTB

where:

C - is the ceiling on bonus earned (Plan A).
Typically, 25% to 35%.

Z - is the number of consecutive data points whose average gives the incremental bonus (Plan A). Typically, Z = 3.

p:q - is the sharing bonus ratio between employees and plant.
Typically, 50:50, 60:40 or 40:60.

B - is the past average bonus paid to employees.

TB - is the Target Bonus (Plan B).
Typically, 25% to 35%.

n - is the number of consecutive data points above the ceiling that triggers a BB (Plan A). Typically, n = 3.

R - is the BUY-BACK amount. Typically, 4% to 10%.

ΔC - is the incremental increase in the Ceiling (Plan A).
Typically, 0%, 1% or 2%.

ΔTB - is the incremental increase in the Target Bonus (Plan B).
Typically, 0%, 1% or 2%.

Plan A permits the ceiling to be raised 1% or 2% after a total BB amount of 10% has been implemented. This action is especially important when it is clear that the performance is rapidly achieving an "absolute" value (e.g. the percentage losses in refining edible oils cannot get below a lower limit). The same applies to Plan B where ΔTB, the incremental increase in the Target Bonus, at 1% or 2%, may apply after each BB accumulation of 10% has occurred.

4 THE SHRED COST PRODUCTIVITY SHARING BONUS

The procedure for calculating the SHRED Productivity Sharing bonus is standard for all gainsharing plans. We shall illustrate its application when the performance measure is "SMH shipped/MH invested".

Starting point includes the selection and/or determination of all parameters discussed in the last section that are appropriate to the chosen plan.

The calculations below detail the steps needed for determining the Productivity Sharing bonus.

BR, the Base Ratio has already been obtained.

At the close of each period t:

* Obtain from records -
  (i) SHP(t) - the standard manhours shipped/produced,
  (ii) PH(t) - the payroll hours actually invested.
* Then ALLOW(t), the "allowed" hours associated with PH(t) at which no extra bonus is earned, is given by,

$$ALLOW(t) = PH(t) \cdot BR$$

* The saved hours, SAVE(t) is expressed by,

$$SAVE(t) = SHP(t) - ALLOW(t) \quad (\text{only for } SHP(t) > ALLOW(t))$$

* The incremental bonus, b(t) is then,

$$b(t) = p \cdot \frac{SAVE(t)}{PH(t)}$$

* b(t), averaged over 3 periods is, $\bar{b}(t) = \frac{1}{3} \sum_{j=t-2}^{t} b(j)$
* TOTB(t), the total bonus for each employee is then,

$$TOTB(t) = B + \min\{\bar{b}(t), C\} \ \ldots \ \text{Plan A}$$

or, $TOTB(t) = B + b(t) \ \ldots\ldots\ldots \ \text{Plan B}$

When the productivity measure is some operational parameter such as:

$\frac{\text{Consumption of "X"}}{\text{Output of "Y"}}$, where examples are,

$\frac{\text{Energy Consumption}}{\text{ton steel pipe produced}}$, or, $\frac{H_2 \text{ consumption } (ft^3)}{\text{ton hydrogenated oil}}$, etc.,

we need to calculate the savings SAVE(t), as a "cost" instead of "manhours" (as given earlier). Bonus calculations and productivity sharing then proceed as before.

The following illustrates the procedure for calculating the productivity bonus involving the consumption of some operational parameters:

Let, CONS(t) be the consumption of item X over period t,

PROD(t) is the production of output Y over period t,

P$(t) is the payroll dollars paid for hours actually worked over period t,

u is the unit cost of X,

ECON(t) is the expected consumption of X over period t when the production of Y is PROD(t).

Hence,

$$ECON(t) = PROD(t) \cdot BR$$

If performance is improved (i.e. better than at BR), then the dollar savings in item X, SAVE(t), is expressed as

$$SAVE(t) = u \cdot \{ECON(t) - CONS(t)\}$$

Bonus increment b(t), is given by,

$$b(t) = p \cdot \frac{SAVE(t)}{P\$(t)}$$

and the smoothed value, $\bar{b}(t)$, becomes

$$\bar{b}(t) = \frac{1}{3} \sum_{i=t-2}^{t}$$

If there are m independent performance parameters being considered simultaneously, then TOTB(t), the total bonus paid for period t is,

$$\text{Under Plan A:} \quad TOTB(t) = \min(B + \sum_{i=1}^{m} \{\bar{b}(t)\}_i, C)$$

$$\text{Under Plan B:} \quad TOTB(t) + \sum_{i=1}^{m} \{b(t)\}_i$$

where $\{\bar{b}(t)\}_i$ is the smoothed incremental bonus for operational measure i, $i = 1,2,\ldots,m$.

5 BUY-BACK DETERMINATION

5.1 BB under Plan A

Under Plan A we buy back a fixed amount, R% of the group's work capacity. Recommended value for R = 10%, but even lower values such as R = 4% have been successfully used in conjunction with a lower ceiling. The advantage of a lower R value is that BUY-BACKs are more frequently implemented, which in itself has some motivating elements.

The compensation contains two elements:

(a) The amount of monthly compensation: It is recommended that a 50:50 sharing plan be applied on the BUY-BACK, meaning that the employee gets 1/2R% as a monthly compensation.

(b) The period "M" over which the BB compensation is paid: The value of M times the amount of compensation gives the total amount the plant compensates the employee for giving up some spare (or, idle) time. It is recommended that M be given a value between six to twelve months. The author has used M = 12 months in all his SHRED COST applications.

Let $\bar{U}$ represent the average manpower utilization (expressed as a fraction) taken over the BB period, i.e., usually, the three periods precipitating the BB. Put $\bar{U} = 1$ if its value cannot be readily estimated. If $\overline{PH}$ is the monthly payroll hours, averaged over the last three months and R the percentage of employees' work capacity that is "bought", then the BUY-BACK Hours (BBH) is obtained from,

$$BBH = \frac{R}{100} \cdot \overline{PH} \cdot \overline{U} \ .$$

The quantity BBH must balance out the total savings in processing times obtained when time standards of some of the products (i.e. the ones with loose time standards) are tightened. We achieve this by first identifying which products have "loose" time standards, obtaining a forecast of their future respective annual demands and tightening their time standards by a yet-to-be-determined "reduction factor" and calculating the total expected reduction in processing times in terms of the "reduction factor". This quantity is then equated to BBH in the above equation in order to determine the value of the "reduction factor".

Employees are paid a BUY-BACK compensatory bonus of 1/2R% for the next M periods (usually twelve months). When the value of R is fixed in the order of 10%, one should consider the possibility of raising the ceiling by C (where $C = 1\%$ or 2%) - an action which itself is motivating, since the long-term earning level begins to get some distance away from the original bonus level, B%.

An example for calculating the "reduction factor" in the BUY-BACK process is given below (adapted from [3]).

Consider the plan applied to a production department and that the condition for implementing a BUY-BACK occurs.

Past production data is as follows:

$\overline{PH}$ = 8412 hours (payroll hours for the last 3 months)

$\overline{U}$ (estimate) = 88%,

$R = 10\%$

Then the BUY-BACK Hours, BBH, is calculated as,

$BBH = 0.1 \cdot 0.88 \cdot 8412 = 740.3$ hours.

We now need to define several parameters:

Let $S(ij)$ be the standard time for producing part i with operation j,

$E(ij)$ be the process efficiency, where,

$$E(ij) = \frac{\text{Standard Allowed time for producing batch size (part i,operation j)}}{\text{Actual time for producing the batch quantity}}$$

$Q(i)$ be the expected annual demand for part i ,

L be that lower level of efficiency to which all "loose" standards will be tightened,

$\hat{a}$ be the set of parts with "loose" time standards.

Then EXPSAV(ij), the expected savings (in <u>real</u> time) when tightening the standard for part i from $S(ij)$ to $S(ij) \cdot \frac{L}{E(ij)}$ would be approximately given by:

$$EXPSAV(ij) = \frac{Q(i)}{4.L} \cdot S(ij)\{1 - \frac{L}{E(ij)}\}$$

When L is correctly selected, then the total expected savings must equal the hours involved in the BUY-BACK, i.e.

$$\frac{R}{100} (\overline{PH} \cdot \overline{U}) = \sum^{i\varepsilon\hat{a}} \frac{Q(i)}{4.L} \cdot S(ij)\{1 - \frac{L}{E(ij)}\}$$

The value of L is obtained as follows:

Select a trial value of L, say L'. Then all parts whose operations exhibit efficiencies $\overline{E}(ij) > L'$., are put into set $\hat{a}$.

Calculate L and compare with L'. If L and L' are within 10% of each other, make L = L' and check if set $\hat{a}$ is altered significantly. If set $\hat{a}$ is changed, put L' = L and go through the process of finding L again. One iteration should suffice for most applications.

In the table below, L' is given a value of 1.70. Column (1) lists all the parts in set $\hat{a}$ whose respective efficiency (given in Column (3)) exceeds L'. The expected savings for each part is given in Column (6) whose sum puts the total expected savings at 713.40 hours which is within a few percentage points (3.6%) off the BBH of 740.3 hours. Assuming that $\hat{a}$ remains unchanged, we put L = L' = 1.70 and calculate the new Standard Times given in Column (7).

Department: Automatics　　　Start date: 1st January 1982
BUY-BACK No. 1　　　BUY-BACK date: 31st March 1982

(1) Part Number	(2) Standard Time S(ij)	(3) Efficiency E(ij)	(4) Annual Forecast	(5) New Efficiency (L')	(6) Expected Savings EXPSAV(ij)	(7) New Standard Time
025892	0.0746	2.05	2,500	1.70	4.68	0.0619
142401	0.0277	1.78	16,000	1.70	2.93	0.0265
072650	0.0228	1.90	12,000	1.70	4.24	0.0204
024775	0.0650	2.32	18,000	1.70	45.98	0.0476
163335	0.1156	2.65	18,000	1.70	109.70	0.0742
424238	0.0167	1.92	2,400	1.70	0.68	0.0148
444653	0.0252	1.74	5,000	1.70	0.44	0.0246
025736	0.2450	2.39	8,800	1.70	91.59	0.1743
.	.	.	.	.	.	.
.	.	.	.	.	.	.
.	.	.	.	.	.	.
.	.	.	.	.	.	.
.	.	.	.	.	.	.
426004	0.2310	2.75	1,000	1.70	129.82	0.1428
426422	0.0588	1.96	15,000	1.70	17.21	0.0510
146624	0.0482	1.82	4,500	1.70	2.11	0.0450
023209	0.0851	1.92	8,800	1.70	12.62	0.0753

Total Savings = 713.40 hours

5.2 BB under Plan B

BUY-BACKS under Plan B is illustrated by Figure 2, where the periodic review time is set at 3 months. $\overline{TOTB}(k)$ is the average bonus earned during the review period k and the BB amounts, $\{R(k)\}$, is only initiated when its value exceeds TB, the Target Bonus.

We see from Figure 2 that R(k)%, the percentage that is "bought" is known, i.e., $R(k) = 2\cdot\{\overline{TOTB}(k) - TB\}$ and in a similar manner to Plan A, the BUY-BACK Hours for Plan B, BBH(k), is obtained from, $BBH(k)\cdot\overline{PH}\cdot\overline{U}$, where $\overline{PH}$ and $\overline{U}$ have the same meaning as in Plan A. Compensation and tightening of standards are identical to the method used in Plan A. The Target Bonus could be increased by $\Delta TB(\Delta TB=1\%$ or $2\%)$ when the cumulative value of the BB amounts, $\Sigma R(k)$, equals 10% or so.

The Base Ratio (BR) under both Plans A and B, ordinarily remains constant with respect to time, except when dealing with operational parameters, or, when sizable proportions of non-direct production employees form part of the participating group. In these cases, the BR may need to be modified after each BB. This aspect is fully discussed in [3].

6 SCOPE OF APPLICATIONS

The SHRED COST gainsharing plans may be applied in one, or, more of several ways:

* In Production/Service Departments (parallel applications possible)
* For an entire Division
* For a whole Plant

Care should be exercised in defining the performance measures for service departments that cater to the needs of production - in that these measures should enhance the performance of the production department rather than that of the service departments themselves.

In each situation, the following gainsharing plans may be applied in parallel, so that the employee receives productivity bonuses from several sources: i.e., when performance measures are directed towards -

* Outputs
* Work-in-Process Inventories
* Energy Consumption
* Auxiliaries

In this manner, we can achieve almost the same breath of application as found with Scanlon and Rucker plans, but with the difference that each measure is directly linked to employee/plant productivity.

7 CONCLUDING REMARKS

The SHRED COST gainsharing plans, while applicable as they stand, are nevertheless best used in conjunction with, and in support of employee involvement programs. Only in this way do we achieve enthusiastic cooperation between employees and a true spirit of partnership between employees and the plant's management team.

The payment system is based on scientifically based performance measures and employees actually cooperate in tightening standards - a result that cannot be achieved (consistently) in any other operating system.

Finally, it is claimed that the payment systems described in this article, result in reduced overall unit costs, while at the same time enhances many factors that raise the QWL (Quality of Work Life) of the plant's employees.

8 REFERENCES

1. Lesieur, F.G. (Ed.), 1958. The Scanlon Plan. The MIT Press, Cambridge, Ma.
2. Rucker, A.W., 1952. Progress in Productivity and Pay. The Eddy-Rucker-Nickels Co.
3. Dar-El, E.M., 1986. Productivity Improvement: Employee Involvement and Gainsharing Plans. Elsevier Science Publishers, Amsterdam.
4. Fein, M., 1980. An Alternative to Traditional Managing. Mitchell Fein, Hillsdale, N.J.
5. Financial Reward Systems, Sept. 1981. Workshop organized by the Productivity Evaluation Center, Virginia Tech.
6. United States General Accounting Office, March 1981. Productivity Sharing Programs: Can They Contribute to Productivity Improvement? Document AFMD-81-22, U.S. GAO, Washington, DC.
7. O'Dell, C.S., 1981. Gainsharing: Involvement, Incentives, and Productivity. American Management Association, NY, 1981.
8. Graham-Moore, B.E. and Ross, T.L., 1984. Productivity Gainsharing. Prentice-Hall, Englewood Cliffs, NJ.
9. Freund, W.C. and Epstein, E., 1984. People and Productivity. Dow Jones-Irwin, Homewood, Illinois.
10. Lawler, E.E., 1981. Pay and Organization Development. Addison-Wesley, Reading, Massachusetts.
11. Dar-El, E.M., 1979-1983. Consulting services with an Israeli metals manufacturing company.

Productivity Management Frontiers — I, edited by D.J. Sumanth
Elsevier Science Publishers B.V., Amsterdam, 1987 — Printed in The Netherlands

INCENTIVE SYSTEM FOR MULTI-PLANT, MULTI-PRODUCT INDUSTRIAL FACILITY

Y. Hosni,* H. Taha,** A. Elimam*** and S. Abdel-Bar***

* University of Central Florida, Orlando
** University of Arkansas, Fayetteville
*** Kuwait Institute for Scientific Research, Kuwait

KEYWORDS

Incentive system, profit sharing, multi-product.

OVERVIEW

Incentive plans in production systems, when properly developed, can result in significant increases in productivity [Maynard, 1970]. The main principle behind incentive plans is that "employees are compensated for increased productivity above an acceptable productivity level, at a predetermined participation ratio, in proportion to the increase in productivity or in accordance with an established plan" [Fein, 1973].

For the case of incentive plans of production systems where single product is assumed, productivity is measured against a predetermined rate of production. Labor is compensated based on its share in the production cycle, and the increase over the standard rate. This is computed as straight piece rate or a standard hour, depending on the pay system implemented in the organization. Rates are determined based on time studies. Such plans usually guarantee a percentage (80 to 100 percent) of a normal day's pay to a worker regardless of productivity. Piece rate and standard hour plans are adequate for industrial systems where tasks or operations are independent and well-defined.

Currently, where operations are more complex, and where tasks are accomplished in "cooperation" with automated systems, we are seeing substantial success with group incentives, which are more commonly called "productivity sharing plans" [AFT, 1983]. Industry tends to treat such cases with the use of company-wide incentive plans, such as the Scanlan plan [1,5] and the Lincoln Electric plan [1,5], which include motivation techniques along with the profit sharing. The accounting system which translates the plan into pay is tailored to the industry applying it. It is our opinion that such plans are meant to be applied in large companies with stable product and established records for job specification and production standards.

The case of multi-plant production systems, where production lines may have more than one product and where products differ in their production costs as well as their net revenues to the company, have not been addressed thoroughly from the incentive system point of view. The case of incentive systems on automated lines are rarely treated in the literature. This is partially due to the belief that production lines have a regulated pace which guarantee production line productivity. On the other hand, it was found that automated production lines may vary in their productivity according to the "group" managing the line (including maintenance). It is the combination of a number of factors which contributes to the increase or decrease in productivity regardless of the "automation" element of the line.

Presented here are the details of an accounting incentive system for a multi-plant industrial facility. The system assumes the extreme case of complexity in the facility structure which may affect the incentive plan. It assumes the case of a company composed of several plants, each plant having one or more products in a single production period. The system adjusts itself to the case under study through configuration parameters. Products differ in their production costs as well as their net revenues to the company. The system provides for incentives to company management at different levels, as well as to direct and indirect labor. The governing algorithms are translated into a computerized incentive system, where three levels of reporting are produced: Company, factory, and production line (individual share).

THE SETTING

The production system considered in this paper is a company composed of several plants (factories), each having one or more production line. Figure 1 is the schematic for a system of three factories, each with two to three production lines. The system is characterized by the following:

1. The nature of the products is the same; however, they may differ in their production costs as well as their net revenues to the company. For example, the company may specialize in producing cement products. One factory may be producing cement blocks and slabs while another is producing cement pipes, and a third may be producing ready-mix concrete.

2. A production line within a factory may produce one or more products in a production period. For example, in the cement block factory, a production line may switch from cement blocks to slabs, or from one size of blocks to another. Switching may require setting up the line for the new product.

3. Production orders are periodically issued by the company management to the various factories. An order would specify minimum "quota" required of various products during a particular period. Quantities are determined based on sales and market studies. Factory management makes production assignments to the various lines within the factory, based on compatibility and schedule. However, there is always a "pool" of standby production orders which can be assigned to production lines if factory quotas are met in shorter period, i.e., there is always the chance for labor to earn more through increasing productivity.

4. Production lines can be fully or partilaly automated.

5. A production line is staffed by a production group during a production shift. The group is responsibile for line setups, maintenance and actual production during a production period.

6. A typical production group is composed of a supervisor, indirect labor (cleaning and maintenance), and direct labor involved in the production cycle.

7. Compensation for labor is paid on a monthly basis as fixed rate (guaranteed based on 100% work day); however, incentives should be computed based on company earnings due to extra production above preset standards for each product on each production line. Incentive pay may be paid monthly or quarterly.

The setting of the production system as described is not unusual. The system in its simplest form would be a company made up of one plant which produces a single product on one production line. For such a case, there is also a single management. The extreme case would be a company with multiple factories, each with a number of production lines which can produce more than one product.

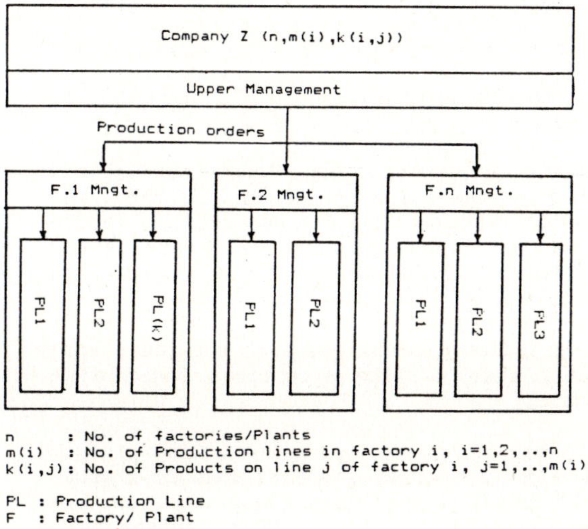

Fig. 1. Production system setting.

THE SYSTEM

The system was designed in a flexible way to accomodate any variations in the company structure. The system size itself is based on a set of parameters which are supplied at the time of system initiation. Information regarding productivity during production periods is considered transactions, and is supplied to the system periodically. Variations regarding the rules of incentive distribution are built into the system; however, it also depends on parameters supplied by the user.

The system is described here in two sections: The rules of eligibility and incentive distribution, and a description of the computer system.

Incentive Eligibility Rule Base

Following are the rules which govern the system logic concerning eligibility for incentives and distribution among concerned parties (company, labor):

1. The extra earnings due to increases in productivity are shared between company and labor (management, direct, and indirect labor) based on a ratio set by the company and labor representatives. It is an input parameter to the system.

2. Upper management can share in the incentive based on a pre-set ratio from the overall labor share (an input parameter to the system).

3. A factory is eligible for incentives if any of its production lines are found to be eligible for incentives. Factory contribution to the overall extra earnings is the accumulation of all its eligible production lines' contributions.

(It is important to differentiate between the terms "contribution" and "share" for the sake of clarity to the rules discussed. Factory/line contribution is the actual dollar value of extra production (above standard) earned by the factory or line. Factory/line share is the incentive amounts distributed to the factory or the line. Contribuion is not necessarily equal to the share.)

4. Factory administration and indirect labor (support services, factory maintenance and cleaning) share in the incentives based on a pre-set ratio from factory share.

5. A production line is eligible if the net value due to its production is above a standard value determined by standard rate of production for each product produced on the line during the produciton period (standard rate and value per unit are input to the system by a pre-built data base).

6. All eligible lines are normalized, so that actual increases in productivity are the bases for distributing the total share among eligible lines. Consequently, line share is not necessarily equal to the line contribution.

7. Once a line is found eligible, all group members (individuals) operating the line during the production period may receive incentives. The system provides for the following individual distribution rules:

7.1 "Across-the-board" (A/B) payment, i.e., an equal amount to be paid for each individual on the line.

7.2 Based on the basic salaries (relative to each other) paid to labor on the line.

7.3 Up to five factors, with a weight ratio associated with each factor ($f(i) \leq 1$ where $f(i)$ is the weight for factor i, $i = 1$ to 5).

7.4 Combinations of the rules 7.1, 7.2, and 7.3. User specifies through input parameters ratios of distribution among 7.1, 7.2, and 7.3.

8. Factors used in computing individual shares (case 7.3) are assigned to meanigful attributes such as effort, degree of involvement (with respect to automation), absenteeism, and degree of responsibility. These attributes are part of the individual record in the system. Through the proper selection of factors and their respective weights, coupled with a combination of A/B estimates, the effect of line automation on the productivity increase can be neutralized, leaving only group effect and individual characteristics as the main factors which affect the incentive base productivity.

As outlined, the rules of distribution are all based on input parameters, which make the system flexible to a greater extent. Still the system is limited in its flexibility on the individual share distribution. Once the individual record is built, it remains applicable for all types of products, which may be unfair in certain cases. To demonstrate this point, assume the case of a Quality Assurance (Q.A.) operator working on the cement block production line. His job is to visually check the blocks coming from the mold press as they pass before him, looking for any out-of-shape blocks. If the press is producing three units per stroke, with a stroke time of 30 seconds, he has 10 seconds to check each block. If the product changes and for a different type of blocks, the press produces six units per stroke, the Q.A. operator has only five seconds to check each block. The fact that an individual's record in the system does not change with the product prevents in some cases the reflection of added responsibility on the part of individuals. To recognize such effects, the system would require a great amount of data.

The Computer System

The system as outlined in the block diagram of Fig. 2 is composed of a data base of three files: Company, individual, and rules. Building the data base through a special component in the system is performed once at the system initiaion (unless updated due to changes). Periodic input to the system includes production orders and productivity figures on the production line level. The system produces four levels of reports: Company, factory, production line, and individual share.

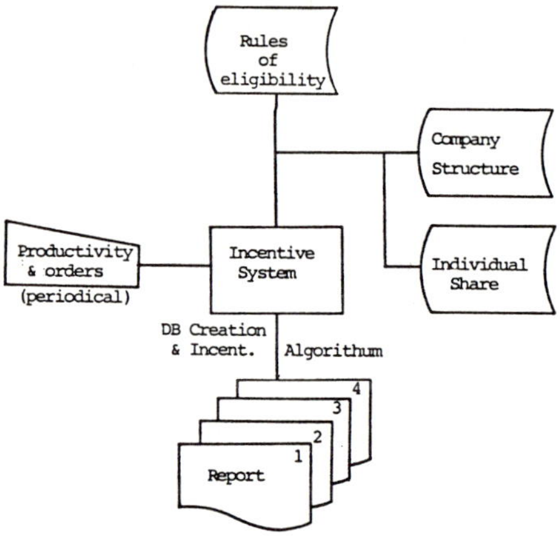

Fig. 2. Incentive system block diagram.

The system data base is composed of files describing the overall structure of the company and is the base on which the system sets itself. It includes the number of factories, the number of production lines in each factory, types of products produced, standard rates, and revenue value for each product. Names and identification codes for the lines and products are also part of this file. The individual share file is classified by group and there is a record for each individual in the group. An individual's record contains his basic salary and the weighting factors used in computing his share as outlined in the section on rules of eligibility. The rules file contains the various ratios and fixed parameters used in determining eligibility and incentive distribution among involved parties.

The system provides the following four levels of reporting:

1. Company-level, in which amounts and shares for the company, upper management, and the total share of the factories (labor) are reported.

2. Factory-level, in which productivity of various lines of production in each factory are detailed, and their status in terms of eligibility and their contribution are reported. Reasons for ineligibility and factory contribution is also part of this report.

3. Production line level, for which the details of income generated by each product is reported along with the percentage increase or decrease above or below the standard.

4. Individual level, for which the details of eligible lines are reported, including actual payments by name to the various workers on the production line. Values computed as across the board, salary base and factor base, as well as the total incentive and its ratio to the worker salary are reported.

SYSTEM PROGRAMS

There are two main programs in the system, the data base (DB) creator, and the incentive algorithm. The DB creator is a data base management program; its function is to create and update the data base file outlined in the system structure. Through an interactive dialog with the user, files are created in a way compatible with the incentive algorithm (a sample of the dialog is in the application example section).

The incentive algorithm is the element of the system which determines the eligibility and computes the incentive amounts for individuals, production lines, factories, and the company. It is made up of a set of sucessive blocks (routines) schematically represented in Fig. 3.

The program first accepts production orders and productivity figures for the period on the production line level. It then accesses the data base, and determines pertinent standards as well as computational parameters. For every factory in the company, and for every production line in the factory, the eligibility for incentive payments is determined. If eligible, the value of extra income is computed and added to a company accumulator, and the factory is marked eligible to receive incentive.

Another component computes the actual increase in productivity over the standard for every product produced, and determines the increase as a percentage. In another pass, company eligibility is checked and if found eligible, both company share and labor share are computed as well as upper management share if pertinent. Eligible factories are scanned, and their administration share is computed before computing individuals' shares in their eligible production lines. Individual shares are computed based on the individual profile record in the individual data base. The four levels of reports outlined in the system structure are produced at different stages of the logic.

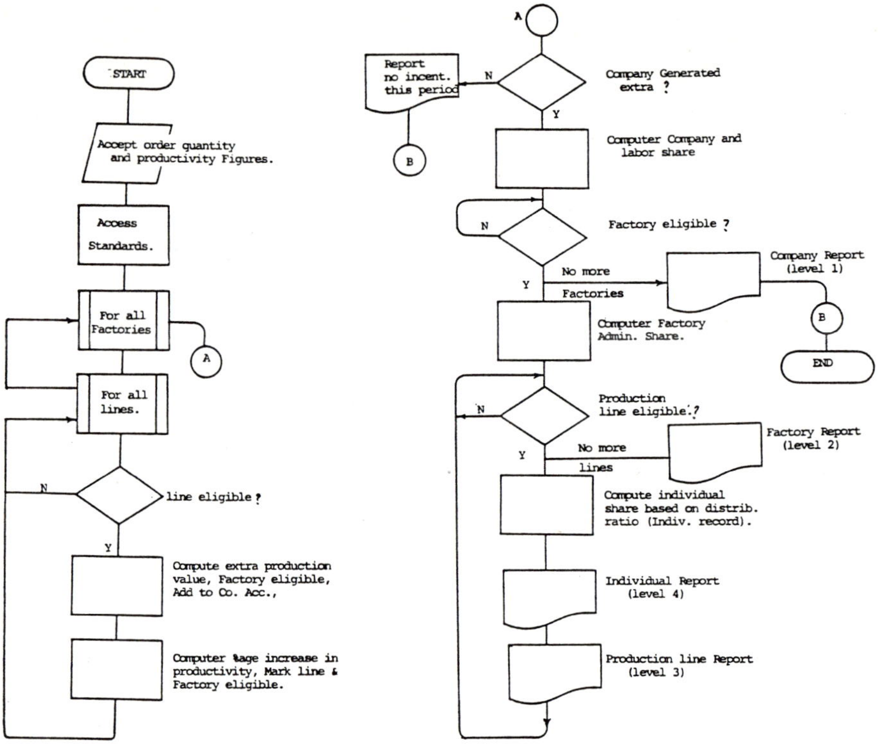

Fig. 3. Incentive algorithm.

CASE APPLICATION

The following is a sample run and demonstration of computing incentives for a case of two factories, each operating two production lines for a given company in a given production period. This version of the system has been modified to demonstrate the building of the data base, and to show the ratios used in the computation.

The system execution starts by displaying the assumptions, the rules of computations, and a set of user's instructions. Next, the system begins a data entry session in which the user supplies answers to questions posed to him/her on the screen. During the data entry session, data regarding governing ratios, actual production, standard rates and product revenues are supplied (a large portion of this data can be input directly from an existing company data base). Once data is entered, the system will perform the necessary computations before starting to report on the various levels of reporting. For individual types of reporting, questions pertinent to ratios of across the board to salary base, to factor distribution are posed and answered. User may perform sensitivity analysis until he/she settles on an acceptable ratio. Detailed reporting of individuals' shares are displayed on the screen for eligible production lines.

In the example that we used, three lines were found eligible for incentives, and their contribution was found to be $1,556 of which the company share was $778 and the factory's share was $778. As shown in the example, each facory received an amount comparable to a percentage increase in productivity after normalization between eligible lines.

As may be observed from the output reports, reasons for ineligibility are reported (case of fourth line of production). Some values of incentives were paid for lines F1L1 and F1L2 ($296.23), however, F1L1 generated $836 and F1L2 generated $420. The reason for equality in the paid share is that each line contributed the same increase in productivity (25%) above standard. Some lines (F2L1) were ineligible since they failed to reach the standard.

```
ENTER COMPANY NAME ? UCF
ENTER NO. OF FACTORIES IN UCF ? 2
ENTER COMPANY SHARE (%) OF ANY EXTRA REVENUES ? 50
ENTER TOP MANAGEMENT RATIO (%) OF INCENTIVE ? 10
ENTER FACTORY  1  NAME ? FACTORY 1
ENTER NO. OF LINES IN FACTORY 1 = ? 2
ENTER FACTORY  2  NAME ? FACTORY 2
ENTER NO. OF LINES IN FACTORY 2 = ? 2
ENTER MANGEMENT SHARE FOR FACTORY 1 FACTORY (%) ? 30
ENTER MANGEMENT SHARE FOR FACTORY 2 FACTORY (%) ? 25
```

```
   ENTER STARTING, ENDING DATES OF PROD. PERIOD (eg.DD/MM/YY) ? 1/1/86,2/1/86
   ENTER TOTAL WORKING HOURS THIS PERIOD ? 200
   ENTER CURRENCY TO BE USED            ? $
HIT ANY KEY TO CONTINUE
         FOR FACTORY FACTORY 1 :
         -------------------------

    ENTER NAME OF LINE  1  ? F1L1
    HOW MANY PRODUCTS PRODUCED ON LINE F1L1 THIS PERIOD ? 2

        ENTER FOR PRODUCT  1  :
                 ENTER NAME : ? F1L1P1
                 ENTER ACTUAL PRODUCTION (# UNITS) : ? 10000
                 ENTER NO. OF HRS. SPENT IN PRODUCTION (HRS) : ? 140
                 ENTER STANDARD RATE FOR THIS PRODUCT (#/HR.) : ? 60
                 ENTER VALUE PER UNIT OF PRODUCT ($ ) : ? .4

        ENTER FOR PRODUCT  2  :
                 ENTER NAME : ? F1L1P2
                 ENTER ACTUAL PRODUCTION (# UNITS) : ? 1000
                 ENTER NO. OF HRS. SPENT IN PRODUCTION (HRS) : ? 60
                 ENTER STANDARD RATE FOR THIS PRODUCT (#/HR.) : ? 12
                 ENTER VALUE PER UNIT OF PRODUCT ($ ) : ? .7
```

Sample Data Entry Dialog

```
                    COMPANY : UCF
TOTAL EXTRA AMOUNT GENERATED : $ 1556
COMPANY SHARE                : $ 778
TOTAL FACTORIES SHARE        : $ 700.2
TOP MANAGEMENT SHARE         : $ 77.8
HIT ANY KEY TO CONTINUE
```

Company Level Report

```
FACTORY NAME : FACTORY 1
TOT. NORMALIZED AMOUNT      : $ 592.4769
FACTORY MNGT. SHARE         : $ 177.7431
FACTORY DIRECT LABOR SHARE  : $ 414.7338


 LN#     LN.NM          % PR.INC.       AMT.TO DISTR.
==========================================================================


 1       F1L1           25              207.3669
 1       F1L2           25              207.3669
                                  FOR FACTORY FACTORY 1
                                  ----------

PERIOD : 1/1/86-2/1/86                TOT. HRS. :  200  HRS.
TOT.EXTRA : $ 1256

 LN# LN.NM    ELIGIB.   GEN.        STAND.  EXTRA     %PR.INC.  COMMENTS
=====================================================================

  1  F1L1      Y         4700        3864   836        25
  2  F1L2      Y         2100        1680   420        25
HIT ANY KEY TO CONTINUE
                                  FOR FACTORY FACTORY 2
                                  ----------

PERIOD : 1/1/86-2/1/86                TOT. HRS. :  200  HRS.
TOT.EXTRA : $ 300

 LN# LN.NM    ELIGIB.   GEN.        STAND.  EXTRA     %PR.INC.  COMMENTS
=====================================================================
  1  F2L1      N         3000        3300   -300      -9.100001 (= OVL.PR. ST.
  2  F2L2      Y         3600        3300    300       9.09
```

Factory / Line Report

```
ENTER ACROSS THE BOARD SHARE FOR LINE (EG.20,30,..ETC.) F1L1 ? 5
 ENTER % OF VALUE TO DISTR. BASED ON SALARY DISTR. ? 50
FACTORY NAME                   : FACTORY 1
LINE NAME                      : F1L1
PRODUCTION PERIOD              : 1/1/86-2/1/86
NO. OF EMPLOYEES OF THE LINE :  6
TOTAL SALARIES OF LINE         : $ 560
TOTAL AMOUNT OF INCENTIVES     : $ 296.2384
INCENTIVE TO SALARIES RATIO  :  .52

SQ.  ID#   NAME    SALARY  A/B    SL.BS.  FCT.BS.   INCENTIVE  RATIO
==========================================================================


 1   1001  NAME1   90      2.468  22.614  24.509    49.592     .55
 2   1002  NAME2   120     2.468  30.152  26.239    58.851     .49
 3   1003  NAME3   100     2.468  25.127  23.932    51.528     .51
 4   1004  NAME4   110     2.468  27.64   23.932    54.041     .49
 5   1005  NAME5   80      2.468  20.101  19.627    42.178     .52
 6   1006  NAME6   60      2.468  15.076  22.491    40.036     .66
```

Individual Share Report

CONCLUSION

Productivity sharing plans are among the most acceptable forms of incentive systems due to increases in productivity. The success of the plan depends on the accompanying accounting system. The system should account for the value of the product, and the actual extra effort demonstrated by labor in attaining the above-standard production. The problem tends to be complex in a multi-plant, multi-product environment, and where automation may be involved in the production cycle. With the advent of powerful microcomputers, better systems of incentive eligibility determination and distribuiton can be created. Presented here is one such system, which produces reports on company, plant, production line, and individual levels. Through system algorithms and manipulation of distribution rules, the degree of automation in the production process can be accomodated and fairness in incentive distribution is guaranteed. Sensitivity analysis can be performed, and through the system reporting component, both company management and labor can trace the source of incentive amounts, thus eliminating any ambiguities.

REFERENCES

[1] Aft, L., Productivity Measurement and Improvement, Reston Publishing Company, Reston, VA, 1983.

[2] Fein, M., "Wage Incentive Plans," Publication 2 in the Monograph Series of the Work Measurement and Methods Engineering Division of the Institute of Industrial Engineers. Copyright in 3rd edition of IE Handbook, McGraw-Hill, New York, 1971.

[3] Hanssen, G. M., "Productivity Gains from the Shop Floor Up," Production Engineering, March 1982.

[4] Henderson, R., Compensation Management, Reston Publishing Company, Reston, VA, 1980.

[5] Sumanith, D. J., Productivity Engineering and Management, McGraw Hill Book Co., 1984.

[6] Taha, H., Y. Hosni, A. Elimam and S. Abdel-Bar., "Productivity Study for Cement Plants," interim report, Kuwait Institute for Scientific Research, Kuwait, 1985.

[7] Watmough, E. B., "The Case Against Incentives," Journal of Industrial Engineering, November-December 1965.

Productivity Management Frontiers — I, edited by D.J. Sumanth
Elsevier Science Publishers B.V., Amsterdam, 1987 — Printed in The Netherlands

MOTIVATING MANAGEMENT TO MOTIVATE: A CONCEPTUAL FRAMEWORK FOR INCREASING EMPLOYEE PRODUCTIVITY

W.B. WERTHER, JR.
Chair Professor of Executive Management, School of Business Administration, University of Miami, 414 Jenkins Building, Coral Gables, Florida 33124

ABSTRACT
Werther, Jr., W.B., 1987. Productivity improvement through focused behavioral reinforcement often fails because managers lack a conceptual understanding of "how" behavioral theory works. This paper offers an understandable model.

Both laboratory and field experiments demonstrate in a rather convincing way that organizational behavior modification does work and can increase employee productivity. The translation of successful experiments into actual use by managers, as a practical matter, seldom occurs. Many managers begin with the "best of intentions." Often, several, sincere applications of behavior reinforcement take place, usually with little or no noticeable results. Even the most dedicated managerial believers who might persist longer than others, in time, become disenchanted and give up. When managers "give-up," they stop using positive reinforcement, sometimes before the behavior becomes established. Now the lack of reinforcement is noticed, leading to questions about the managers' sincerity in the original applications. Not only may the results be confusing for employees, the next attempt at positive reinforcement may engender more resistance and suspicions, further reducing the success of this otherwise useful management tool.

Part--and perhaps a good part--of the problem stems from the managers' lack of motivation to continue with the necessary, positive reinforcement. Three causes for this lack of follow-through are addressed in the paper:

- o Managers may lack the necessary education and training;
- o Appropriate reinforcement from their superiors; or,
- o Immediate reinforcement provided by improved employee performance.

The first part of the problem--insufficient education and training--is a minor one, readily solved through academic behavioral and motivational courses, or in-house training programs.

The second part of the problem--the lack of feedback from superiors--is not easily solved. Although some specific recommendations are offered along the lines of "training the boss" and developing "peer reinforcing networks," the individual manager is limited in his or her ability to ensure reinforcement.

The third part of the problem--a poor understanding of why reinforcement efforts take so long--is addressed by providing a conceptual model of the behavioral reinforcement process. This model relates rewards, needs, self-image, expectancy theory, and effort to explain productivity improvements that result in employee motivation. Then, the problem of long-term reinforcement is addressed within the context of both behavioral and conceptual motivational theory. The paper concludes with a discussion of what managers and their superiors must do to "Motivate Managers to Motivate."

INTRODUCTION

Organizational behavior modification works and can increase employee productivity (Fedor et al., 1981). It is based upon a long line of research that can be traced from Pavlov's experiments (1927) and Thorndike's work (1931) in the last century, to Skinner's (1969) more recent efforts. And yet, managers too seldom use organizational behavior modification, and even more seldom use it correctly (Werther, 1986a).

The research base supporting the use of organizational behavior modification consists of laboratory and field experiments. Although the bulk of the studies reported in the psychology and management literature are laboratory studies, field applications exist that demonstrate the practical efficiency of organizational behavior modification in improving employee productivity in on-going organizations. In short, it works. But even with the evidence of successful laboratory experiments or field applications, few managers apply these powerful behavioral concepts regularly and systematically. Why?

This paper posits an explanation of why managers are often ineffective in their application of organizational behavior modification to improve employee productivity. It discusses the three primary barriers to successful application: managerial understanding; superior reinforcement; and employee reinforcement. The paper concludes with a conceptual model that identifies the salient considerations and inter-relationships underlying the long-range perspective needed for successful applications of organizational behavior modification.

MANAGERIAL UNDERSTANDING

Organizational behavior modification, as used here (Luthans et al., 1975) is a disarmingly simple concept: Through the proper application of behavior reinforcement, managers can modify the behavior of people in organizations. In its most basic form the manager must:

- Identify the desired behavior.

- o Determine the baseline frequency of that behavior.
- o Identify appropriate rewards for the behavior.
- o Apply the rewards on some schedule when the behavior occurs.
- o Monitor the frequency of the desired behavior to assure that the reward/interventions are modifying the requency of the desired behavior.

The theory underlying this procedure is called the Law of Effect. It states that, "Behavior is a function of its consequences." That is, the occurrence of a behavior is influenced by the consequences the person receives for having exhibited the behavior.

Conceptually, Exhibit I depicts the relationship between behavior and its consequences. Once a behavior occurs, it meets with consequences that range from positive to negative. The consequences, in turn, give a positive or negative valence to the antecedents that proceed the behavior. Restated, positive or negative consequences lead to positive or negative attitude associated with the behavior, with a positive consequence leading toward greater frequency of that behavior.

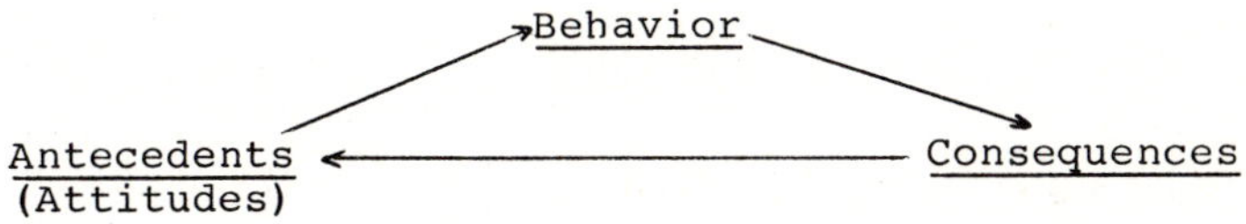

Exhibit I. The Relationship of Behavior-Consequences-and Antecedents.

Although considerable research by Skinner and others (Skinner, 1969 and Luthans et al., 1975) has been devoted to the frequency or schedule of reinforcement, variable or intermittent reinforcement appears to be the most effective schedule, once the behavior is established by using more frequent reinforcement at the beginning of the planned interventions (Skinner, 1969).

Obviously, managers who are unaware of these concepts cannot apply them. Through training and reading, many managers have been exposed to these ideas. For all the laments about a growing flood of MBA's graduating from our universities, it's hard to imagine a business school graduate during the last twenty years being unaware of these concepts. Even among managers who have never been through business curriculum, nor taken a non-credit course in motivation, the common sense notion of rewarding good behavior should be familiar, given how it is embedded in the myth and reality of organizations run as meritocracies. Still, one shortfall in the application of organizational behavior modification may simply be a lack of understanding about the concept, its applications, and the application of positive reinforcement.

Overcoming the barriers therefore, requires a widespread education of new managers as they move into management from other disciplines or up into management from the ranks of workers. Although educational institutions can provide this needed education, in-house supervisory and management training programs offer a more realistic and immediate solution.

SUPERIOR REINFORCEMENT

Education and training alone are not enough to close a widespread application of organization behavior modification. Although they may reflect a necessary condition, alone they are insufficient. If education and training were all that were needed, then the hundreds of thousands of managers who have been exposed to these concepts would be a vanguard of users slowly but steadily spreading the concepts across industry and throughout their individual organizations. Yet, no such groundswell seems evident (Werther, 1986a).

When asked, "How many of you get enough recognition for the job you do?" less than twenty per cent of the managers surveyed believe they get enough recognition. And, the survey is based on more than three thousand managers, in one hundred groups from virtually every industry in the United States and Canada (Werther, 1986a). Without the advantage of a superior who serves as a role model and a coach, it is unlikely that any amount of education and training would cause managers to use a concept not practiced on them. Not only would they lack the encouragement to use the positive reinforcement, but even if they used it, managers would be unlikely to receive positive feedback for their efforts. The underlying irony is that managers do not use organizational behavior modification on their employees because it is not used on them. And, even when it is used on an unplanned basis, the lack of reinforcement they receive is unlikely to create an established pattern of behavior among these managers.

Further still, managers are unlikely to use the concept without reinforcement from their superiors because of a threshold effect. As Exhibit II illustrates, most managers set a minimum performance standard. Although the standard varies from employee to employee, and varies over time for each employee, managers expect some minimal level of effort or performance. When an employee performs below that level, managers properly intervene with discipline, which is shown as a minus sign (-) in Exhibit II and which may range from a brief discussion to more severe interventions.

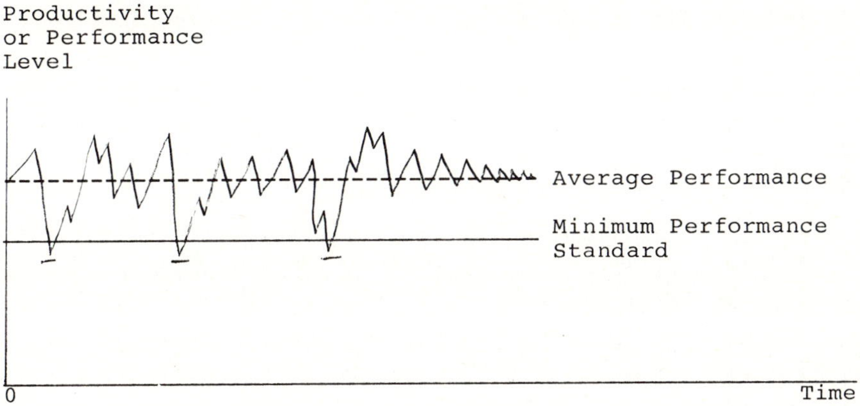

Legend: + Application of Positive Reinforcement
- Application of Discipline

Exhibit II. Short-Term Results of Positive Reinforcement and Discipline.

Even though the employee may perform below the standard at a later time, the disciplinary intervention usually works. (And, when it fails, usually after being progressively more severe, it may result in the manager either lowering the standard or separating the employee from the work group through transfer, discharge, or other means.) Since discipline is known to work by the manager, the use of positive reinforcement seems counter-intuitive.

Further confounding the use of positive reinforcement is also evidenced in Exhibit II. The use of positive reinforcement should occur when the employee's performance is above the average, as illustrated by the dash line in the exhibit. However, when positive reinforcement is applied for exceptional performance, subsequent performance, by definition, is not likely to be exceptional. Instead, it is likely to migrate back toward the average. After several applications of positive reinforcement, even the most dedicated manager's apt to abandon the use of positive reinforcement. And, why not? From an individual manager's perspective, his or her application of positive reinforcement probably lead to a <u>decline</u> in performance (from the peak level) back toward the average. But discipline often led to an <u>improvement</u> toward the average. The likely conclusion is that discipline works and positive reinforcement does not.

Should these managers become the recipients of positive reinforcement as a consequence of their behavior for any reason, they would experience the benefits. Having experienced these benefits, they would be more receptive to the idea of using positive reinforcement. They would be more likely to give this positive reinforcement to ensure that the desired behaviors of their employees met with positive consequences. And, perhaps most importantly, they would be receiving positive reinforcement for their efforts to reinforce employees, thus lengthening the managers' use of positive reinforcement before they abandon it.

EMPLOYEE REINFORCEMENT

Under the idealized conditions of an educated managerial work force that was reinforced properly by senior managers, the use of positive reinforcement would be more likely to occur. However, managerial use of positive reinforcement would probably decline and even extinguish without the reinforcement of employee behavior. Unless managers see some primary benefit in the form of improved employee productivity, their continued use of this approach is unlikely.

Managers receive reinforcement for their use of positive reinforcement when employee performance improves. However, that improvement is likely to be both erratic, and take some time before a positive pattern of improvement emerges. Contrast, for example, Exhibit II and Exhibit III. As evident, average performance improves as shown in Exhibit III. But the improvement is not immediately evident in the short run, as depicted in Exhibit II.

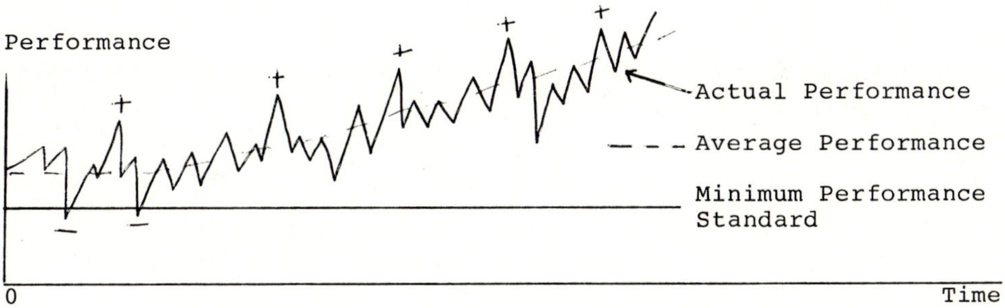

Legend: + Application of Positive Reinforcement.
- Application of Discipline

Exhibit III. Long-Term Results of Positive Reinforcement and Discipline.

CONCEPTUAL MODEL OF BEHAVIORAL REINFORCEMENT

Managers are concerned with productivity improvement through people. Although productivity through people engulfs a broad domain (Werther et al., 1986b), employee performance is the central issue. Sophisticated technology, expensive capital outlays, and innovative concepts all contribute to productivity, but these are only tools (Sumanth, 1984). More directly impacting productivity is the motivation and commitment of the people in the organization. Above all else, it is people who apply the sophisticated technology, extract output from expensive capital, and generate the innovative concepts that lead to competitive advantage.

Productivity--commonly defined as the ratio of output to input--is depicted as the "bottom line" in Exhibit IV. To achieve a high level of productivity requires effort. Effort, however, will be forthcoming only if there is an expectation that the effort will lead to desired results (Vroom, 1964). Whether one has such an expectation depends largely on one's self-image about his or her ability to perform. In fact, an a priori argument exists that actual productivity will be approximately equal to the employee's self-image of himself or herself as a worker.

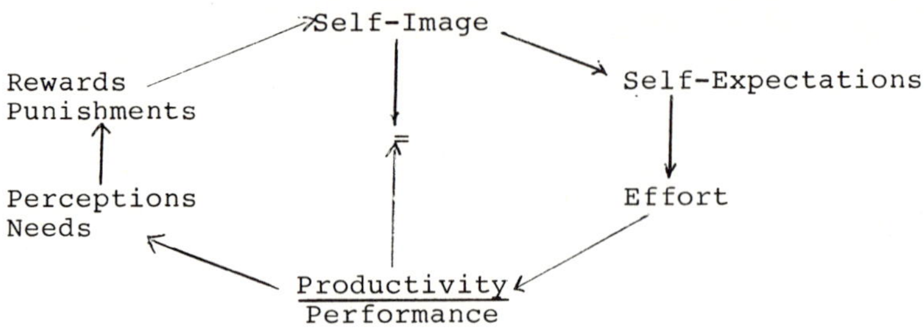

Exhibit IV. A Conceptual Model of Behavioral Reinforcement

The self-image is shaped through the feedback one receives from the environment (Bandura, 1977), the rewards and punishments (or consequences) associated with performance (or productivity). What determines whether a particular stimulus is a reward or a punishment is the employee's perceptions and needs.

Therefore, if the results of an employee's efforts (his or her productivity) are met with rewards (positive reinforcement) as perceived by the employee's needs, then the self-image of the individual is likely to be enhanced. A higher self-image leads to higher self-expectations and those expectations become the basis for higher effort and improved productivity.

Why positive reinforcement works, in part, comes from the self-image enhancement of positive reinforcement. When only discipline is used, or is the primary "motivator," the self-image can deteriorate with negative consequences for self-expectations, effort, and ultimately productivity. What is needed, simply put, is more positive reinforcement, more recognition for good performance. Moreover, that reinforcement must be continued long enough to positively impact the employees' self-image before the average performance in Exhibit III begins to trend upward.

REFERENCES

Bandura, A., 1977. Social Learning Theory, Englewood Cliffs, N.J.: Prentice Hall.

Fedor, D.B. and Ferris, G.R., 1981. Integrating OB Mod With Cognitive Approaches to Motivation. Academy of Management Review, Volume 6, No. 1, 115-25 pp.

Luthans, F. and Kreitner, R., 1975. Organizational Behavior Modification, Glenview Illinois: Scott, Foresman.

Pavlov I., 1927. Lectures on Conditional Reflexes, New York: International.
Skinner, B.F., 1969. Contingencies of Reinforcement, New York: Appleton-Century-Crofts.
Sumanth, D.A., 1984. Productivity Engineering and Management, New York: McGraw-Hill Book Company.
Thorndike, E.L., 1931. The Psychology of Learning, New York: Teachers College.
Vroom, V.H., 1964. Work and Motivation, New York: Wiley & Sons.
Werther, Jr., W.B., 1986a. A Point of View. National Productivity Review, Volume 5, No. 1, 1 p.
Werther, Jr., W.B., Ruch, W.A., and McClure, L., 1986b. Productivity Through People, St. Paul: West Publishing Company.

Productivity Management Frontiers — I, edited by D.J. Sumanth
Elsevier Science Publishers B.V., Amsterdam, 1987 — Printed in The Netherlands

PRODUCTIVITY IMPROVEMENT IN A MULTIBRANCH DISTRIBUTION COMPANY

R. E. CRANDALL
College of Business, Appalachian State University, Walker Hall,
Boone, NC 28608

ABSTRACT
Crandall, R. E., 1987. Productivity Improvement in a Multibranch Distribution Company.

This is a case study about a productivity improvement program in a wholesale distribution company. The program involved the following stages:

(a) Measure the present level of productivity.
(b) Analyze to determine the potential for improvement.
(c) Involve managers in planning for improvement.
(d) Design and implement the improvement program.
(e) Measure results and start the next improvement cycle.

As a result of the program, productivity was improved in a number of areas, including: sales per employee, sales per vehicle, sales per square foot, and annual inventory turns. Improvements were achieved in all branches with the magnitude of the improvements directly related to the effectiveness of the managers.

1 INTRODUCTION

This is a case study about a productivity improvement program in a wholesale distribution company with seven branch operations, ranging in distance from the company headquarters of 80 to 350 miles.

The company had been highly centralized when it was operated as a major division of a public corporation. The branches were structured to function as sales and distribution centers but many operating decisions such as inventory planning, purchasing, and product mix selection were made at the corporate level. Although the company had computerized systems, they were primarily accounting-oriented and did not provide for effective analysis of branch operations. As a result, while the branches had profit potential, the individual operations managers were not informed about their overall branch performance.

The operations managers were, without exception, individuals who had "come up through the ranks" and who possessed considerable product knowledge because of their experience. The normal progression was to start as a driver-warehouseman, move to inside (counter) sales, and finally to branch operations manager. None were college graduates and their only formal management training came from in-house programs provided by the parent company. The branch managers reported to a Vice-President, Operations. In several branches, there were salespersons who had a responsibility for branch sales and who reported to a Vice-President, Marketing A. The company had a second major product line in three of the branches where there were additional salespersons for the second product line, and who reported to the Vice-President, Marketing B. At some branches, therefore, there could be as many as three persons reporting to three different individuals at the corporate offices. Needless to say, no single individual felt responsible for the total profitability of the branch.

Over a five-year period, the company had experienced steadily declining earnings and finally started losing money. The parent company decided to appoint a new President who instituted the productivity improvement program which is described in the remainder of this paper.

This program was implemented during the same period of time when the company was phasing out one of their major product lines, which had been a major contributor to the company's unprofitability. As a direct result of the program, even though sales decreased by 12% from 1982 to 1983, the major productivity measures increased from 11% to 29%. (See Exhibit C)

2 APPROACH

The productivity improvement program consisted of the following major steps:

(a) Revise the organization structure to establish each branch as a profit center under the supervision of a branch manager.

(b) Develop a management reporting system for use in planning and control at the branch level.

(c) Train the branch managers in the analysis and use of the reports to identify problems and take actions to improve.

(d) Involve the branch managers in the planning process, including both a profit plan and a productivity improvement plan.

(e) Refine the reporting system to compare the actual results with plan and identify action to be taken.

The newly-appointed President had direct involvement in all aspects of the program.

3 ORGANIZATION

The first step involved changing the reporting responsibilities at the branch level. This involved selecting an individual at each branch to be

Branch Manager with responsibility for all branch operations, reporting to the President. At the corporate level, responsibilities which had previously been divided among the Vice-President, Operations; Vice-President, Marketing A; and Vice-President, Marketing B were realigned so that with two positions were eliminated and the remaining position was that of Vice-President, Marketing.

The purpose of the reorganization was to establish a clear-cut profit center responsibility for each Branch Manager. In order to move quickly, the President chose the best person available at each branch who in practically all cases was the previous operations manager. While they were experienced and dependable, it turned out that some were not sufficiently oriented to sales to be completely effective. It should be noted that the company had previously experienced high turnover of employees and many of the most capable individual managers had left during the five-year decline of profitability.

This change was implemented within a month after the President was appointed and was received very well at the branch level as a positive move. The President also established a regular monthly visitation schedule to review the operating results with each branch and discuss actions needed to improve the operation. This was a new experience for the branch employees who had not seen the previous President, except on rare occasions.

4 MANAGEMENT REPORTING SYSTEM

With the new organization structure in place, the next major step was to develop a management reporting system which could be used at the branch level. It was decided that the number of reports should be kept to a minimum until the branch managers could be trained in their use.

One of the first reports to be developed was an Income & Activity Report. (See Exhibit A). This report was, in part, an Income Statement, because it provided sales, gross profit, expense, and profit contribution at the branch level, on a monthly and year-to-date basis. The remainder of the report was designed to provide information about the utilization of the major resources in each branch--employees, inventory, delivery vehicles, and space.

The reports were prepared manually at first by extracting information from the computerized accounting reports. Later, the reports were designed into a spreadsheet format on a microcomputer for easier updating and processing.

The President reviewed the results of the previous month with each branch manager, and key employees, during his regular monthly visits to the branches. The initial reaction by the branch managers was one of interest; however, when confronted with the need to improve productivity, most felt they were already operating as well as they could and did not have much to suggest in the way of improvements.

On a weekly basis, a Key Indicator Report was used. (See Exhibit B).This provided each branch with information about how they were doing both absolutely and in comparison with other branches. This report was eagerly awaited each week by the branches and provided a great deal of competitiveness among branches to see who could outperform the other. Although sales were always shown, the Key Indicator Report had a flexible format which could be changed, as desired, to focus on various areas of the business such as slow-moving inventory, % gross profit, sales of promotion items and the like.

Other information, such as sales to key customers by month, sales by product groups, and sales of products from each vendor were developed to help direct the emphasis of the sales programs.

5 TRAINING THE MANAGERS

It was obvious that the Branch Managers needed additional training in the interpretation and use of the management reports if they were to take corrective and improving actions. The President had already told them that they would be fully involved in the planning process for the next year.

In addition to the monthly review sessions with the President, the branch managers participated in a regular Branch Manager´s Meeting, which provided a means of discussing current topics of interest such as new products, competitive pricing, and the like. A portion of each meeting was also used for training and discussion of some aspect of the operating results. As time went on, increased attention was focused on the following areas of productivity:

(a) Sales per employee
(b) Inventory turns
(c) Sales per delivery and sales vehicle
(d) Sales per square foot of space

By comparing the results from the branches, it was determined that there were vast differences in productivity levels. It was also established that the seasonality of the business affected some branches more than others because of more northerly locations but that all had to cope with slow seasons which could affect productivity significantly.

6 PLANNING PROCESS

Analyzing actual results and taking actions to improve set the stage for the Branch Managers to become involved in the annual planning process. Although they were all willing, the rigors of planning proved difficult for them at first. As with most planning programs, the first plan was just a beginning. Each year the plan became better and the ability of the managers to integrate their own thinking into the plan became greater. The planning process was kept flexible and relatively simple with an absence of numbers-crunching at the branch level. As with the management reports, most of the planning documents were prepared on a microcomputer, using spreadsheet software, with the same format as the monthly reports.

Some managers responded well and it was clearly evident that the ones who did the best job of planning had the branches who achieved the greatest gains in productivity and profitability.

7 SYSTEM REFINEMENTS

As the participation of the managers increased, refinements were made in report formats and planning steps; however, the basic system remained essentially the same as originally installed.

8 RESULTS

As a result of the program outlined above, productivity increases were realized in all branches. The results for the total company during the first full year of the program are summarized below and the results for each branch are presented in more detail in the Comparison of Productivity Indicators (Exhibit C).

Sales volume, in dollars, was used as a reference to measure the total level of activity in the company. While not completely satisfactory, it was considered the best indicator available. Inasmuch as selling prices remained stable for some products and declined for others, the actual level of activity was slightly higher in the second year than the sales volume would indicate.

(a) Sales decreased by $1.1 million from 1982 to 1983, or 12%, as the company phased out a major unprofitable product line.

(b) The average number of employees decreased from 47 to 35, thereby resulting in an increase in sales per employee from $202 thousand per year to $239 thousand per year, an increase of 18%.

(c) The average number of vehicles declined from 30 to 24, with a corresponding increase in sales per vehicle from $318 thousand per year to $354 thousand, an increase of 11%.

(d) The average number of square feet of space decreased from 76,800 to 56,600, while sales per square foot increased from $123 per year to $148 per year, an increase of 20%.

(e) The average inventory decreased from $2.1 million to $1.4 million, and annual inventory turns increased from 3.6 to 4.7, an increase of 29%.

There was a wide range in productivity increases among branches; however, each branch had increases in every productivity measure except Branch 3 where the sales decrease was very high at 39%, and Branch 6, with a sales decrease of

29%. In Branch 3, while there were decreases in sales per vehicle and sales per square foot, both longer-term in response time, there were increases in sales per employee and inventory turns of 6% and 25%, respectively. In Branch 6, there was a decline in sales per square foot but increases in the other three measures.

The overall conclusions were that there were across-the-board improvements which could be attributed to the system changes which had been implemented. In addition, some managers demonstrated the ability to manage better than others within the guidelines provided by the productivity improvement program. These results only confirm that good systems must be put in the hands of good managers for best results.

8.1 Sales per Employee

In a business which averaged 20% gross profit, payroll was a major expense. In order to be profitable, it was necessary for the payroll to average 8.0% of sales at the branch level. In a typical branch, there were 4 to 6 employees and, even in a seasonal business, it was necessary to keep employees on the payroll year-round to assure having experienced workers during the busy season.

The following actions were taken during the study period to improve the productivity of the branch employees:

(a) Eliminate the practice of scheduling all employees for 42.5 hours each week. This level of scheduling was obviously not needed during the slow season and not needed for all employees during the busy season. While this action did not show as improvement in the sales per employee ratio, it did reduce payroll costs as a percent of sales.

(b) Plan and schedule deliveries to customers better so that the driver could get the deliveries completed faster and have more time available to work in warehouse activities. The resultant timely and consistent deliveries provided for more improved customer service.

(c) Increase the emphasis on employee selection and training. The job scope was enlarged so that each employee could be used in a greater variety of assignments, thereby providing increased flexibility in matching availability of service to the customer demand. While the wage rate increased slightly, the added versatility and utilization of the workers more than compensated for the higher rate.

(d) Increase the branch manager's effectiveness by providing him with better information about the customers. He could then do a better job of allocating his selling efforts among customers and, by better time management, still be available at the branch when needed.

The net effect was a reduction in number of employees in almost all branches, even those with increasing sales, so that the increase in sales per employee ranged from 6% in Branch 3 to 37% in Branch 2.

8.2 Sales per Vehicle

Delivery and sales vehicle expenses initially represented 3-5% of the sales dollar in most branches. The eventual target was to reduce that to 2-3%. Several actions were taken to improve productivity in this area:

(a) One of the major steps was to decide which types of vehicles were best suited to the need. In the past, the choice of vehicles had been largely left to the branch manager and the selection had been too often based on the "best deal" at the local car dealer rather than a study of the actual requirements. After some analysis of length and frequency of trips, size of loads, and ease of loading and unloading, it was decided that a panel van worked well for the smaller branches and a 16-foot step van for the larger branches. Requirements of the branch managers or salespersons were best satisfied by medium-size station wagons which were economical, yet large enough to carry sales samples and demo units.

(b) The improved scheduling of deliveries previously mentioned also increased sales per vehicle by making it possible to reduce the number of vehicles required at the larger branches.

(c) In the past, the lack of good inventory management had resulted in many rush trips to the supplier to pick up an out-of-stock item. Better planning largely eliminated the need for these trips and reduced the cost of operating the vehicles and made them more available for customer deliveries.

(d) All vehicles were leased from a national truck leasing company, a program which assured good maintenance procedures and reduced the major repairs which had previously been experienced. It also made financial planning easier and eased the cash flow requirements for capital expenditures.

The net effect of these actions was an overall increase in annual sales per vehicle of 11%, with the range being from a decrease of 6% in Branches 3 and 6 to a high of 79% in Branch 7.

8.3 Sales per Square Foot

The initial reaction by most branch managers was that there was little they could do about sales per square foot. As it turned out, a number of things could be done, with the major ones listed below:

(a) The improvement in inventory management was the major change. As inventory turns increased, less inventory (and space) was required for a given sales volume or conversely, increased sales could be handled with the original inventory load, and floor space.

(b) At least three branches were located in buildings considerably larger than needed because they had originally been the only operations in the company. As branches were added to better service the customer, the original ones required less space. To reduce costs, they were relocated to smaller, more functional buildings.

(c) The branch manager did a better job of warehouse layout which resulted in better space utilization and thereby made it possible to generate more sales from a given amount of space.

(d) Studies indicated that a branch of approximately 6,000 square feet could be considered the "standard" size because it represented a branch size conducive to good customer service and management.

These changes resulted in a net reduction of 20,200 square feet of space, and a corresponding increase in sales per square foot of 20%, ranging from a low of -14% in Branch 3 to a high of 76% in Branch 5.

8.4 Inventory Turns

The next major area to be considered was utilization of inventory, or inventory turns. This program has been described separately in detail but the major actions are summarized below. (1)

(a) It was found that as much as 20% of the inventory in the branches had been there for more than one year and needed to be classified as slow-moving. Once identified, it became possible to steadily reduce the slow-moving inventory and free up funds for products that turned faster.

(b) Through a classification of inventory items, it was found that 35% of the items carried in the computer product file and in various levels of inventory, represented less than 3% of the sales. Although not all items could be eliminated, some could and the result was not only less inventory but also less space requirements and less time required to manage the inventory.

(c) A great number of purchase orders were being written, particularly for small items, because each branch was placing individual purchase orders. A central distribution center was established so that each branch could obtain many items, especially replacement parts, more easily. This also reduced the average cost per unit because of the benefit of centralized purchasing.

(d) Provide more meaningful sales and inventory information on a regular basis so that, as the branch managers became more aware of their inventory needs, they were able to adjust not only the frequency of ordering but also the product mix of their orders to provide both better availability of product and reduced average inventory levels.

The inventory management program was the productivity measure which appeared to best demonstrate the difference in the branch managers' abilities. The better managers averaged six or more turns per year while less effective managers had turns of four to five times. The total increase in average inventory turns for the company was 29%, with the range from 7% in Branch 5 to 45% in Branch 4.

OTHER BENEFITS

There were other less tangible benefits from the program. The most important was that two of the branch managers proved very capable managers and were promoted within the company.

Another benefit was the increased flexibility within the company to adopt new programs and eliminate out-dated ones. The organization, and its managers, were able to respond more quickly to needs and opportunities.

REFERENCES:

1. Crandall, Richard E., "Human Factors, Computer System are Key Elements in Inventory Management Program", Industrial Engineering, October, 1985, pp. 64-70.

EXHIBIT A

INCOME & ACTIVITY REPORT
OR THE 12 MONTHS ENDING OCTOBER 30, 1983
DOLLARS IN THOUSANDS

PREP BY:
DATE:

BRANCH: 4

INCFTP83	NOV	DEC	JAN	FEB	MAR	APR	MAY	JUN	JUL	AUG	SEP	OCT	TOTAL	% SALES
SALES	61.1	50.0	38.0	45.0	70.2	88.0	75.9	114.6	86.6	94.1	81.9	91.1	896.5	100.0%
COST OF SALES	48.5	39.9	30.3	36.1	56.3	70.5	60.1	91.0	68.7	75.0	66.0	73.0	715.4	79.8%
GROSS PROFIT	12.6	10.1	7.7	8.9	13.9	17.5	15.8	23.6	17.9	19.1	15.9	18.1	181.1	20.2%
PAYROLL	5.4	5.0	5.3	4.5	6.0	6.3	6.5	6.5	6.3	6.2	5.1	4.9	68.0	7.6%
VEHICLE	1.8	1.0	0.7	3.0	2.2	1.5	1.6	1.7	2.2	1.8	1.4	2.5	21.4	2.4%
RENT & UTILITIES	1.2	1.2	1.2	1.1	1.2	1.5	1.2	1.4	1.2	1.3	1.3	1.1	14.9	1.7%
TELEPHONE	0.8	0.7	0.4	0.8	0.8	0.9	0.8	1.0	0.8	0.8	0.5	0.7	9.0	1.0%
INTERWHSE FRT	0.3	0.4	0.4	0.3	0.4	0.2	0.5	0.1	0.4	0.5	0.7	0.6	4.8	0.5%
BAD DEBTS	0.1	0.2	0.2	0.4	0.7	0.9	0.7	1.2	0.9	0.9	0.8	0.9	7.9	0.9%
OTHER	0.5	0.2	0.3	0.4	0.2	0.2	0.2	0.4	0.2	0.4	0.3	0.3	3.6	0.4%
TOTAL EXPENSES	10.1	8.7	8.5	10.5	11.5	11.5	11.5	12.3	12.0	11.9	10.1	11.0	129.6	14.5%
PROFIT CONTRIBUTION	2.5	1.4	-0.8	-1.6	2.4	6.0	4.3	11.3	5.9	7.2	5.8	7.1	51.5	5.7%
% GROSS PROF-MONTHLY	20.6%	20.2%	20.3%	19.8%	19.8%	19.9%	20.8%	20.6%	20.7%	20.3%	19.4%	19.9%	20.2%	
ACTIVITY REPORT														
SALES - LAST YEAR	28	62	51	53	64	77	61	61	85	94	75	63	774	
SALES-% LAST YEAR	218%	81%	75%	85%	110%	114%	124%	188%	102%	100%	109%	145%	116%	
NO OF EMPLOYEES	3	3	3	3	3	4	4	4	4	4	3	3	3.4	
ANNUAL SALES/EMP	244	200	152	180	281	264	228	344	260	282	328	364	262	
NO OF VEHICLES	2	2	2	2	2	2	2	2	2	2	2	2	2.0	
ANNUAL SALES/VEH	367	300	228	270	421	528	455	688	520	565	491	547	448	
NO OF SQ FT	6.0	6.0	6.0	6.0	6.0	6.0	6.0	6.0	6.0	6.0	6.0	6.0	6.0	
ANNUAL SALES/SQ FT	122	100	76	90	140	176	152	229	173	188	164	182	149	
AVERAGE INVENTORY	110	103	97	117	103	113	116	109	129	121	121	98	111	
AVG DAYS OF INVEN	68	77	96	97	55	48	58	36	56	48	55	40	56	
AVG ANNUAL TURNS	5.3	4.6	3.7	3.7	6.6	7.5	6.2	10.0	6.4	7.4	6.5	8.9	6.4	

EXHIBIT B

KEY INDICATOR REPORT

PREP BY: SEPT, 1982

DATE:

DOLLARS IN THOUSANDS

BRANCH	NO OF EMPL	% GR PR	WK 1	WK 2	WK 3	WK 4	WK 5	TOTAL
			|---------- SALES - BY BRANCH --------|					
BRANCH 1	11	20%	109	50	45	42		246
BRANCH 2	5	19%	7	29	20	18		74
BRANCH 3	5	19%	25	18	19	23		85
BRANCH 4	4	17%	27	13	16	20		76
BRANCH 5	7	20%	32	16	35	31		114
BRANCH 6	7	18%	30	17	26	22		95
BRANCH 7	4	21%	23	21	14	14		72
TOTAL SALES	43		253	164	175	170	0	762
TOTAL GROSS PROFIT		19%	48	32	34	34		148
NEW ACCOUNTS OPENED								
BRANCH 1			2	3	2	1		8
BRANCH 2			3	1	5	2		11
BRANCH 3			2	2	3	1		8
BRANCH 4			2	3	1	4		10
BRANCH 5			3	1	4	2		10
BRANCH 6			1	2	3	1		7
BRANCH 7			2	4	6	3		15
TOTAL - ALL BRANCHES			15	16	24	14	0	69
SALES (SELECTED VENDORS) IN THOUSANDS OF DOLLARS								
VENDOR A			6	7	11	8		32
VENDOR B			34		4	1		39
VENDOR C			1	2	2	6		11
VENDOR D			12	10	9	5		36
VENDOR E			9	2	10	8		29
VENDOR F			15	12	13	11		51
VENDOR G			4		5	7		16

EXHIBIT C

COMPARISON OF PRODUCTIVITY INDICATORS
FOR THE YEARS OF 1982 VS 1983
DOLLARS IN THOUSANDS

PREP BY: REC
DATE: 9-15-86

UMPRODC1	BR 1	BR 2	BR 3	BR 4	BR 5	BR 6	BR 7	TOTAL
FOR THE YEAR 1982								
SALES	3037	1025	1297	774	1260	1276	775	9444
COST OF SALES	2404	841	1031	626	1019	1018	605	7544
AVG NO OF EMPLOYEES	12.0	7.3	7.5	3.4	5.7	7.1	3.8	46.8
*ANNUAL SALES/EMPLOYEE	253	141	173	228	221	180	204	202
AVG NO OF VEHICLES	6.0	4.8	4.1	2.0	4.6	5.4	2.8	29.7
*ANNUAL SALES/VEHICLE	506	214	316	387	274	236	277	318
AVG NO OF SQ FT-M´S	20.5	15.0	8.5	6.0	12.0	8.8	6.0	76.8
*ANNUAL SALES/SQ FT	148	68	153	129	105	145	129	123
AVG INVENTORY	634	322	375	141	189	279	127	2067
*AVG INVENTORY TURNS	3.8	2.6	2.7	4.4	5.4	3.6	4.8	3.6
FOR THE YEAR 1983								
SALES	2321	1335	790	897	1111	908	992	8354
COST OF SALES	1852	1078	637	715	896	716	761	6655
AVG NO OF EMPLOYEES	8.1	6.9	4.3	3.4	3.8	4.8	3.7	35.0
*ANNUAL SALES/EMPLOYEE	287	193	184	264	292	189	268	239
AVG NO OF VEHICLES	4.4	5.3	2.7	2.0	3.1	4.1	2.0	23.6
*ANNUAL SALES/VEHICLE	528	252	296	449	358	221	496	354
AVG NO OF SQ FT-M´S	14.6	12.0	6.0	6.0	6.0	6.0	6.0	56.6
*ANNUAL SALES/SQ FT	159	111	132	150	185	151	165	148
AVG INVENTORY	363	320	185	111	155	167	116	1417
*AVG INVENTORY TURNS	5.1	3.4	3.4	6.4	5.8	4.3	6.6	4.7
INC(DEC) - 1983 VS 1982								
SALES	-716	310	-507	123	-149	-368	217	-1090
% INCREASE (DECREASE)	-24%	30%	-39%	16%	-12%	-29%	28%	-12%
AVG NO OF EMPLOYEES	-3.9	-0.3	-3.2	0.0	-1.9	-2.3	-0.1	-11.8
*ANNUAL SALES/EMPLOYEE	33	52	11	36	71	9	64	37
% INCREASE (DECREASE)	13%	37%	6%	16%	32%	5%	31%	18%
AVG NO OF VEHICLES	-1.6	0.5	-1.4	0.0	-1.5	-1.3	-0.8	-6.1
*ANNUAL SALES/VEHICLE	21	38	-20	62	84	-15	219	36
% INCREASE (DECREASE)	4%	18%	-6%	16%	31%	-6%	79%	11%
AVG NO OF SQ FT-M´S	-5.9	-3.0	-2.5	0.0	-6.0	-2.8	0.0	-20.2
*ANNUAL SALES/SQ FT	11	43	-21	21	80	6	36	25
% INCREASE (DECREASE)	7%	63%	-14%	16%	76%	4%	28%	20%
AVG INVENTORY	-271	-2	-190	-30	-34	-112	-11	-650
*AVG INVENTORY TURNS	1.3	0.8	0.7	2.0	0.4	0.6	1.8	1.0
% INCREASE (DECREASE)	35%	29%	25%	45%	7%	18%	38%	29%

Productivity Management Frontiers — I, edited by D.J. Sumanth
Elsevier Science Publishers B.V., Amsterdam, 1987 — Printed in The Netherlands

PERFORMANCE ACTION TEAMS: CASE STUDY

D. SCOTT SINK and LARRY SHETZER
Virginia Productivity Center, Virginia Tech
Blacksburg, VA 24061 U.S.A.

DEAN MARION
Burlington Industries
Radford, VA 24141 U.S.A.

ABSTRACT
Sink, D.S., Shetzer, L., and, Marion, D., 1987, Performance Action Teams: Case Study.

The Performance Action Team Process is a continually evolving, comprehensive, and top-down implemented partcipative management process. Its purpose is to improve performance by fostering planning, decision making, and problem solving at the lowest appropriate level in the organization. In mid-1985 implementation of the PAT process began at Burlington Industries' New River Textiles Plant. The implementation process is described. Early results are reported. Also, the quantifiable costs/benefits of the PAT process are presented.

1 INTRODUCTION

The Performance Action Team Process is a continually developing and evolving total participative management process. It has been designed on the premise that you don't change organizational culture or organizational performance by talking about or just by creating awareness about the issues. The PAT process is designed on the premise that complex organizations modify/evolve their culture(s) and their levels of performance by modifying the management process. The underlying theoretic questions are: (1) Does one change the culture to change the process? or, (2) Does one change the process to change the culture? Anyone close to the literature and major proactive company interventions today realizes the implications of these two questions. We believe that culture is an outcome that results from management practices and processes. "We are more likely to act our way into a new way of thinking than to think our way into a new way of acting." Therefore, our belief is that the challenge facing many, if not most, American managers is to design, develop and effectively implement new management processes that get people behaving in ways that will support the levels of performance necessary to be competitive in the 80's and 90's. We believe the PAT process is one that has been designed to:

(1) Make <u>everyone</u> in an organization more proactive relative to performance improvement.

(2) Move planning, problem-solving, and decision-making to the lowest appropriate levels in the organization.

Accomplishing these two major objectives moves an organization closer towards "total participative management."

The Performance Action Team Process is a continually developing and evolving, comprehensive and top-down implemented participative management process. It's development began in 1976 at the Ohio State University (Sink, 1978). It obtained its name/title using the following logic. First, what American organizations and managers should be striving for most is improved performance, not just quality or efficiency or productivity. The performance of an organizational system (work group, plant, firm, function, etc.) is a function of seven interdependent and interrelated criteria. Those criteria are: effectiveness, efficiency, quality, productivity, quality of work life, innov(tion, and profitability/budgetability. For detailed definitions of each of these criteria and for discussion as to their interrelationship see <u>Productivity Management</u> by Sink (1985). Managers, staff, unions, and line employees will embrace the concept that performance is multidimensional and will be skeptical and less supportive of "programs" with a narrower focus (i.e., quality circles). Hence, we needed the word performance in our generic title for this process.

Second, what we lack in America is <u>not</u> an awareness of what the problems are or even an understanding of the solutions. We lack the commitment and discipline neccesary to effectively <u>implement</u> known solutions. We lack mechanisms for generating consensus and support for perceptions of problems and designs for solutions. This observation inspired Peter Drucker's high impact quote, "What we need to learn from Japan is not what to do, but to do it." We felt we needed the word action in our generic title.

Finally, any management process must be compatable with the existing organizational culture. One aspect of the American culture that we believe is important, but often untapped in American organizations is that of team effort. American's don't do things in circles, they do things in teams. Therefore, the final word in our generic title was team.

American quality management, participative management, and productivity management programs are by and large extremely poor examples of the true potential these efforts can and do have. As an example compare the results of a typical Small Group Activity process described by Davidson (1982) with typical results of an American quality circles program (see Table 1). The comparisons are dramatic and reveal what we believe is a fundamental deficiency/finding in how American management plans for, designs, develops, and implements productivity, quality, and participative management efforts.

TABLE 1

Japanese Small Group activity compared to U.S. Quality Circles

	Small Group Activity	U.S. Quality Circles
Development Time:	5 years	< 5 months
Scope of Involvement: (Definition of Employee)	Everyone	Line supervisors, Blue-collar workers clerical staff
Focus of Improvement:	Performance	Quality
Proposals Developed per group per year:	100-600	1-10
Proposals completed per month per group	30-50	< 1
Percent Implemented:	80-90	10-50
Evaluation Criteria	Constant improvement Implementation Effectiveness Total involvement Quality of process	Benefit to cost ratios Short term payoff

2 BACKGROUND

There is increasing realization that if the American organization is to reverse the trend towards declining productivity, a radical re-orientation of management practices is in order. Data from Nominal Group Technique sessions held with over 1,000 managers and professionals from a wide variety of organizations (public, private; large, small; hi-tech, low-tech) suggests that there is a recurring and common thread to perceived roadblocks to productivity in the American organization. Consistently identified and top ranked roadblocks to productivity in American organizations are:

- insufficient and ineffective training;
- ineffective, poorly communicated and coordinated planning at all levels;
- lack of cooperation and coordination within the organizations;
- resistance to change;
- lack of an incentive to improve;
- poor labor management cooperation;
- ineffective leadership skills;
- lack of implementation effectiveness;
- lack of an awareness, understanding, and commitment to productivity on the part of top management.

One strategy for inducing gains in worker productivity is to improve the quality of management practices and processes through increasing the level of employee participation in the planning, problem-solving, and decision-making functions of the organization. Employee participation in management has become a major political, social and economic issue throughout the world, particularly in Europe. A number of forms of participative management have taken permanent root in many parts of the world: The German program of co-determination and the Israeli kibbutz are prominent examples which have been in operation over twenty-five years (Strauss, 1984). In Europe, Strauss notes, participative management was designed to change the predominantly adversarial character of labor relations, a strategy which has seen considerable success, as evidenced by Germany's postwar record of labor peace.

Of course, no introduction to participation is complete without mention of Japanese Quality Circles, small group activity, semi-autonomous workgroups and autonomous work groups. Group level problem solving designed primarily to improve performance and remove roadblocks is prevalent and in certain Japanese organizations highly developed. While the U.S. at first rushed to adopt this Japanese import (Cole, 1983), Quality Circles have failed in this country at a rate of approximately seven to one (Swim, DeVries, and Sink, 1984). It has become increasingly clear that while the style of participation in QC's is congruent with Japanese culture and expectations, it cannot be effective in an American organization without major modification to either the process or the culture in which they are implemented. As Kerr (1984) points out, participation needs are satisfied very differently in the U.S. and Japan. The participation needs of Japanese employees are met when subordinates have had the honor of presenting their views to organizational superiors. In the U.S., on the other hand, employees are likely to feel that genuine participation has not occured without evidence that eventual decisions reflect their input.

3 THE PERFORMANCE ACTION TEAM PROCESS

The Performance Action Team Process is a comprehensive, top-down implemented, participative management process. The process has two major components. The first is a performance improvement planning process that focuses on development of consensus, 2-5 year developmental goals, and on specific action programs that must be implemented in order to move successfully toward those goals. This process component, as you will see, is implemented in the upper and middle-levels of the organization (company, plant, division, etc.). We have found that this process supplements and enhances typically existing strategic planning activities in areas such as business planning, capital budgets planning, facilities planning, etc. Figure 1 illustrates the interrelationship between these two planning efforts.

The second major component of the PAT process is a participative problem-solving process. This process component is implemented with staff, line supervisors, line workers, and clerical personnel (i.e., first level

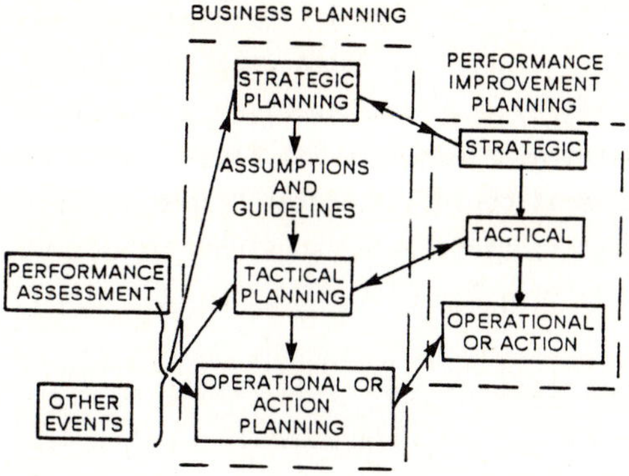

Figure 1. Business planning and performance improvement planning.

management and below). We will review each of these components and then illustrate how the two components are brought together in an integrated fashion.

3.1 Performance Improvement Planning Component

One of the common critical and recurring problems associated with application of participative management theory and technique is the lack of understanding of, involvement in, real commitment to and support for the "employee" involvement efforts. We recognized the need to have a process achieve more than just "lip service" acknowledgement of top management legitimization and support. However, we also recognized the demands placed upon top-management in most organizations. Our design problem then was to satisfy at least two critical objectives or constraints. We have developed a process that involves management in a relevant and substantive activity while simultaneously exposing them to and allowing them to feel structured team processes. As a result management is not only cognitively involved in the process but also feel the process and can react accordingly.

Participative management efforts don't fail because employees can't make them work. They normally fail because once into the effort managers in the organization don't understand the process, feel threatened by the process, or just plain don't agree with the process. Participative management programs fail at a very high rate not because they can't work and not because their potential payoffs are not high, but rather because we fail to make them an integral part of the management culture. Successful participative management efforts are developed in a top down fashion. The question we struggled with was how to do this effectively and efficiently.

There is an imbalance between strategy and policy formulation, long-range/strategic planning, and incremental performance improvement planning. Our effort in the PAT process has been devoted to trying to improve the effectiveness and efficiency of the overall planning process in an organization. We do this by establishing a participative performance improvement planning process that integrates with existing strategic business planning processes. We believe this process begins to address some of the deficiencies in American strategic planning identified by Hayes and others in the recent literature.

3.2 Performance Improvement Planning Sub-Process

The basic steps in the performance improvement planning process are detailed elsewhere (Sink 1985a, 1985b). We provide an outline of the steps in the process in Table 2.

A manager once commented that trying to understand this process by reading about it or hearing a lecture on it is analogous to trying to read your favorite record album. Her point was that the process has a dimension to it that can only be appreciated fully by experiencing it. Some of the most critical aspects of the process are very difficult to capture on paper.

3.3 Participative Problem-Solving Component

Once the performance improvement planning process component has been implemented through all layers of management down to the line managers, the PAT process transitions from a planning and action team orientation to a partcipative problem-solving process involving line managers and their work team. The early stages of the PAT process involves natural work units (at all levels; white collar, staff, as well as line). Note that to this point in the PAT process the focus of improvement planning has been on the organization, not the work group. At this stage in the process this orientaion shifts and we begin to focus improvement efforts at the work group, department, or function itself. Table 3 shows the steps involved in the partcipative problem-solving process.

3.4 The "Grand Strategy" -- Integrating the Two Components

Figure 2 is an example of a top down implementation sequence for the PAT process. The figure demonstrates how the process structures a flow of information top-down, but also bottom-up during the performance improvement planning process and the participative problem-solving process. Several specific points are worthy of slightly more detailed discussion.

First, certain necessary preconditions must be satisfied in order for a PAT process implementation to be successful. We present them in their order of importance. Upper management, the key decision makers in the organizational system targeted for PAT process implementation must be involved in the initial design and evolving design and development of the process.

TABLE 2

Performance improvement planning steps

Steps		How to execute the step	Avg. Time
Pre-condition: Awareness of long-range planning, mission of organization, culture, strategic business plan, business conditions, etc.		Prime the group with data.	
STEP 1	External Strategic Analysis-- Identify and discuss those external factors that should be considered in development of the PI plan.	Nominal Group Technique Steps 1 & 2 only, Discussion.	1-1/2 hrs.
STEP 2	Internal Strategic Analysis-- Identify and discuss those internal factors that should be considered in development of the PI plan.	NGT Steps 1 & 2 only, Discussion.	3/4 hrs.
STEP 3	Performance Improvement Plan Assumptions-- From Steps 1&2 develop a list of specific assumptions upon which plan will be based. Each assumption is then evaluated as to its importance to the plan and the perceived validity of the assumption.	NGT Steps 1 & 2, Importance-Certainty Grid Analysis, Discussion.	1 hr.
STEP 4	2-5 Year Performance Improvement Goals-- Prioritized goals for performance improvement for the organization are developed.	NGT Steps 1-4, Discussion of results, repeat Step 4 (voting & ranking) if necessary.	1 hr.
STEP 5	Performance Measures-- Development of measures of performance for the organization.	NGT Steps 1&2 or Steps 1-4, Discussion, Refinement & Operationalization.	3/4 hr.
STEP 6	Action Programs-- Prioritized objectives we should devote resources to in the next year in order to move toward our 2-5 year goals. Ideas for performance improvement.	NGT Steps 1-4, Discussion of results, formation of Action Teams in Step 7.	1-1/2 hrs.
STEP 7	Action Teams-- Formation of Action Teams (small groups of 2-5 persons) to develop proposals for top priority action programs.	Participants in the planning session volunteer for at least one action team assignment.	1/2 hr.
STEP 8	Proposal Review, Implementation, Evaluation	Accountability chart is important. Project management and coordination is critical.	

TABLE 3
The partcipative problem solving process

PIP Interfacece	Step	How to accomplish	Outcome
Step 6&7	STEP 1: Form Performance Action Group, Ideas for Performance Improvement. Form Performance Action Teams	Nominal Group Technique Steps 1-4, Discussion	Consensus, Prioritized Ideas for Performance Improvement, Action Teams
	STEP 2: PAT's develop proposals (scoping proposals)	Provide PAT with outline of what a scoping proposal should contain, establish deadline, coach & counsel if necessary, educational intervention if necessary	Proposals detailing how to implement idea (problem-solving has occured)
	STEP 3: PAT presents proposal to PAG	The PAT's each present their proposals to the PAG for review/ critique	Proposals are improved or approved by PAG
Step 8	STEP 4: Review Session	The PAT's present their proposals to a "review committee" for decision	Proposals are approved, modified, or rejected
	STEP 5: Feedback Session	The "review committee" meets with the PAT's and communicates the decision within one week after proposal presented	Decision Communicated, Action initiated
	STEP 6: Implementation, Project Management, Accountability	The project plan outlined in the proposal is executed	Action
	STEP 7: Measurement & Evaluation, Feedback, Charting	Visibility room(s) is/are developed to track implementation progress and effectiveness	Follow-through, feedback, visibility
	STEP 8: Maintenance Mode for PAG's	Once proposals have been implemented, groups meet bi-weekly in visibility room(s) to discuss issues	Continued communication and coordination
	STEP 9: Recycle annually, integrate with budget planning process, continued evolution of process		

They must be completely committed to and knowledgeable about the process. We recommend that any organization not able to achieve this level of commitment and involvement not pursue such a process. The second major precondition is the selection and placement of a champion, coordinator, manager for the process. We have developed a set of specifications we require an individual to meet. This person/position will make or break the process. Finally, we believe that the PAT process must be viewed as a long-term process. We feel that the process must be seen as a way of doing business. The process must be understood as an integral part of our overall productivity management process.

Second, the process begins at the "top" of the organizational pyramid with the eight-step performance improvement planning process. As the process moves downward in the organization (firm, function, division, plant, etc.) information from the higher levels is passed downward in a "lagged" fashion. Lagging the data ensures that lower levels of management are not biased by upper level views. We can then compare and contrast views and identify areas of goal and action program inconsistency, incongruity, etc. This is an extremely valuable and important aspect to the process.

4 THE PAT PROCESS AT BURLINGTON INDUSTRIES

We would now like to present a case study of a comprehensive pilot study of the PAT Process. The pilot study got started by a briefing on the PAT process to upper-and middle-management at Burlington Industries. After that briefing, corporate management and the Greige Sales Division management decided to undertake a pilot test of the process in a specific plant within the Division. The management briefing was made in early 1985 and our pilot study began June 1, 1985. The Virginia Productivity Center was provided with a gift/grant to support their research effort with the pilot study. A PAT process coordinator and manager was selected from within Burlington. Dr. Sink, Director of the VPC, and staff were to guide the New River Textiles Plant through a 12-month pilot test of the process while continuing research and development on the process itself. Mr. Dean Marion, Employee Involvement Manager at New River Textiles was to learn the process, coordinate the pilot study, and develop necessary skills to maintain, expand and develop the process upon completion of the VPC pilot test. The plan was for the VPC, under the guidance of Dr. Sink, to complete one cycle of the process down through and including several wage employee work groups. Mr. Marion observed the leadership and facilitation and studied the background theory and techniques during that first top-down cycle. Upon completion of the first cycle, Mr. Marion then changed roles with Dr. Sink and he became the process leader and facilitator.

4.1 The Company

Textiles is considered to be the nation's oldest industry. Burlington Industries has been a part of this heritage since 1923, when J. Spencer Love established the first plant in Burlington, NC. Burlington has since grown to become the nation's largest and most diversified manufacturer of textiles and related products for apparel, the home, and industry.

4.2 The Approach

The Performance Action Teams Process at New River Textiles has been defined as "a process by which staff, management, line employees - everyone - can become more involved in planning, problem-solving, and decision-making." This structured process is an attempt to modify the management process. We view it as an attempt to move from a control oriented strategy to a committment oriented strategy (see Walton, 1985). The process gets people actively involved by generating consensus regarding ideas for performance improvement. It provides a mechanism for them to identify and deal with nagging problems that we all knew existed but never did anything about.

We have successfully executed one cycle of the PAT process and are now beginning our second cycle. We know that modifications will need to be made as we continue to develop the process in the Plant. The past year has been very rewarding, but not without predictable difficulties. The process is not a panacea and is not easy to implement. It has required persistence, patience, and skill. Despite the costs, we are confident that the benefits are significant. We would like to review activities and accomplishments to date.

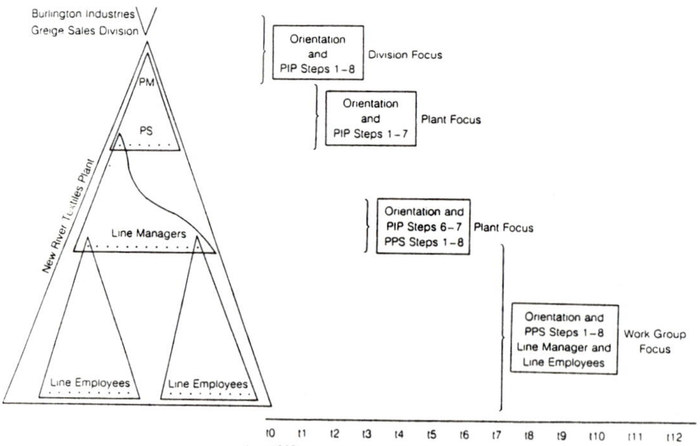

Figure 2. PAT process top-down implementation

4.3 Major Milestones

Figure 2 presents the project plan that was followed in the case study. The study began with a strategic planning session with management of the Greige Sales Division on June 6, 1985. This was followed by a four-hour corporate management briefing on June 20. These two sessions helped finalize the approach taken at the plant. From July 1985 to March 1986 the PAT process was implemented at the plant. The process was gradually moved from senior management, down to line managers, and then down to the work-group level.

In March 1986 the second cycle of the process began. This cycle was started at the "bottom" of the organization. The second cycle has just been completed, and participants are pleased with the progress that has been made.

5 RESULTS

The data we present in this section of the paper represents results accrued from June 1, 1985 to March, 1986. Table 4 shows the extent of invlovement in PAT's and PAG's achieved. Comparisons with other types of employee involvement efforts in the U.S. suggest that these levels are impressive. Table 5 shows the number of recommendations/ideas generated by groups at each level. 81% of the recommendations have already been implemented. Table 6 shows the costs/benefits of the process to date.

TABLE 4

Employee involvement achieved

Group	PAG's	PAT's	% Involved
Wage	11	36	30
Clerical	2	10	93
Line Mgmt.	3	9	96
Plant Mgmt.	1	4	100
Division Mgmt. and Appropriate Corporate Staff	1	4	100
TOTAL	18	63	40

TABLE 5
PAT process results

Group	# of recoms.	# of ideas
Wage	56	258
Clerical	(in process)	53
Line Mgmt.	30	99
Plant Mgmt.	28	31
Division Mgmt. and Appropriate Corporate Staff	20	22
TOTAL	134	453

TABLE 6

Quantifiable costs/benefits of PAT process

COSTS		BENEFITS	
Labor (involvement time)	$7,766	Materials, Reduced Waste	$100,000
Materials	3,000	Labor Savings	100,000
Overhead Costs	40,000		
Pilot Study Contract Costs:	75,000		
Total	$125,766		$200,000

6 CONCLUSIONS

Plans have been formulated to have total plant involvement by year end 1986. Is PAT essential in order for New River Textiles to survive? Management feels that it is an integral part of the plant's success. The PAT process has been and will be the device that will guide our employees to seek and act on productivity and quality improvements. It is management's responsibility to be supportive and nurture this process. We believe we are well on our way to becoming our vision of the "organization of the future," and the PAT process is the mechanism that is ensuring we make this progress as a team.

7 REFERENCES

Cole, R.E., 1983. A Japanese Mnagement Import Comes Full Circle, Technology Review, May/June.

Davidson, W.H., 1982. Small group Actvity at Musashi Semiconductor Works, Sloan Mnagement Review, 23(3), Spring.

Kerr, S., 1984. Leadership and Partcipation, in Productivity Res. in the Behaviorial and Social Sciences, edited by A.Brief, Praeger, New York.

Sink, D.S., 1978, unpublished doctoral dissertation, Ohio State University.

Sink, D.S., and Kiser, K.J., 1985. Performance Action Teams: Update on Continuing Developments, Procs. Annual IIE Intl. Conf., Norcross.

Sink, D.S., 1985. Productivity Management: Planning, Measurement & Evaluation, Control & Improvement, John Wiley, New York.

A version of this paper originally appeared in the National Prod. Review

Productivity Management Frontiers — I, edited by D.J. Sumanth
Elsevier Science Publishers B.V., Amsterdam, 1987 — Printed in The Netherlands

TOTAL EMPLOYEE INVOLVEMENT IN PRODUCTIVITY IMPROVEMENT: A CASE STUDY

Howard W. Oden
Worcester Polytechnic Institute
Worcester, MA, USA, 01609

ABSTRACT
Oden, Howard W., 1987, Total Employee Involvement in Productivity Improvement: A Case Study

This paper describes an innovative and comprehensive program for involving all employees, and particularly middle management and non-manufacturing personnel, in improving productitvity. Jamesbury had implemented a number of changes to their manufacturing system which greatly improved their productivity; however, these changes only involved top management and manufacturing personel. To further improve productivity, Jamesbury decided to implement a comprehensive program that would involve all employees in a meaningful way. This program not only achieved significant productivity improvements in its own right, but changed the attitude of employees so they were willing to support productivity improvements from all sources.

1 INTRODUCTION

Although Jamesbury was a very successful company with a long history of worker involvement in productivity improvement, increased competition indicated that additional improvement was needed if Jamesbury was to remain a leader in the valve industry. Jamesbury had implemented a number of productivity improvements to their manufacturing system which put them well on their way towards obtaining a fully integrated capability, these productivity improvements were carried out primarily by top management and manufacturing with little involvment by middle management or personnel outside of manufacturing.

To ensure that Jamesbury would continue to make productivity improvements with the full acceptance and support of all workers and supervisors, it was decided to implement a comprehensive productivity improvement program that would involve all employees in a meaningful way. In the following paragraphs we will describe this program, including the situation that spawned it, the initial concept of the program, how it was implemented, and how it is currently being managed. We will conclude by outlining the future directions of the program and by providing some guidelines for implementing a similar program in other companies.

2 BACKGROUND

2.1 Jamesbury's Situation

Jamesbury got its start in 1954 with the introduction of a radically new design for the ball valve. In 1967, the wafer-sphere, a new concept in high performance butterfly valves, was added to the product line.

By the mid-seventies, Jamesbury had become an acknowledged leader in the ball and butterfly valve industry. Their growth continued throughout the seventies and by the end of the decade they employed more than a thousand people, with most of them located in the four plants in Worcester, MA.

During this early period of innovation and rapid growth, Jamesbury based their strategy on their engineering capability. Production costs were important, but they were not the major factor for Jamesbury, since their design and quality were far superior to that of their competition.

However the early eighties found the company being challenged for leadership in the standard ball and butterfly valve market by a number of low cost producers. Jamesbury continued to be a market leader for high quality valves for special applications, but the situation was far different for the standard valves. Other producers had been able to essentially duplicate Jamesbury's designs, and were able to produce the valves at a cost much lower than Jamesbury could.

To reduce production costs, Jamesbury value engineered their valve designs and implemented a number of improvements to their manufacturing system. They completely redesigned their small ball valve line and designed and installed a new flexible production line to reduce the cost of production and product changeover. They also made major simplifications in the design of their wafersphere valve, and installed a highly flexible semiautomated system for fabricating and assembling it. These technological changes were expected to yield significant increases in productivity, but if Jamesbury was to retain its lead in the extremely competitive standard valve market, it was apparent that further increases in productivity had to be made. But how?

Jamesbury has a long history of worker involvement in productivity improvement. A company-wide Quality Circle program had been in operation since 1980. A Rucker Productivity Sharing program had been implemented in the early seventies. These programs and a number of others had been instrumental in improving the productivity and employee participation at Jamesbury in the past, but were currently not yielding significant results. Also these programs were not well coordinated and were often competing for resources of personnel, money, and time.

2.2 Concept of the Program

It was at this point that the idea for a comprehensive Productivity Improvement Program was born (Oden and Greene, 1984). The intent of this comprehensive program was to revitalize existing programs, initiate new improvement programs, and coordinate the management of all programs to provide the most effective utilization of scarce resources of time, talent, and money.

The general concept of the program was to provide an umbrella under which all the separate efforts could be coordinated into a single coherent program. By creating a master program that would include and coordinate all the other programs, the most effective use of available resources could be made.

At the same time it was desired that every employee be involved in improving the company's productivity. Jamesbury realized, as had many other companies (Sink, 1985), that all available resources must be utilized to their fullest if they were to remain competitive. The active involvement of all employees would not only generate valuable productivity improvements, but would also foster an attitude that would make productivity improvments generated by other means, namely automation and technological change, much more acceptable.

3 INITIAL IMPLEMENTATION OF THE INVOLVEMENT PROGRAM

3.1 Planning Phase

Document Corporate Philosophy. The planning phase encompassed all those activities that should occur before the program can be formally initiatied. One of the first activities undertaken was that of documenting the corporate philosophy. It was not that Jamesbury had no corporate philosphy, but rather that they had never taken the time to document it. If Jamesbury was to get everyone involved in improving the productivity of the company, it was imperative that all employees understand and accept the corporate philosphy.

Promulgate Improvement Policies. Another early activity was to establish and clearly state that no one would lose their job as a result of a productivity improvement. If some improvement did abolish jobs, the people affected would be retrained and relocated but not fired. If a company expects the whole-hearted committment of every employee to productivity improvement, it must promise they will not lose their jobs as a result of their efforts. Layoffs may result from business downturns and slumping sales, but not from productivity improvements.

At the same time, we clearly stated that the only way that Jamesbury could expect to maintain current employment levels in the face of productivity improvements, was to sell more products. To sell more products, we must improve our productivity faster than the competition does. Thus to maintain our current level of employment, we must improve our productivity faster than the competition, which means improving productivity by every means available.

Establish Permanant Organization. If the program was to be successful, it was essential that the necessary resources be applied to make it happen. As a minimum this meant that the program should have a full time manager. Finding a person qualified for the position and making him available took some time, but it was finally accomplished.

To plan and direct the program, inputs from all parts of the company was deemed necessary. To accomplish this, a steering committee was set up composed of representives from all levels and all functions at Jamesbury. This advisory committee meets regularly and at least monthly to establish policy for the program manager and to approve plans before they are implemented.

Develop Program Plan. This program was established on the belief that every Jamesbury employee wants to be involved in what his company is doing. He wants to feel that he belongs to company that is doing something worthwhile and that he is contributing to this effort. He wants to improve himself and the company in which he works.

However, he has to be provided with a program that will enable him to become involved in a meaningful and concrete way. Participation simply for the sake of participating is hollow and counter-productive. We didn't want to generate a lot of enthusiasm for improvement and then not have any means by which every employee could contribute in meaningful way. To avoid this situation, we developed a specific plan of action, the details of which are outlined in the following paragraphs.

3.2 Initiation Phase

The first visible step in the program was to emphasize the importance of the program and demonstrate top management's involvement. A mass meeting of all employees was considered, but was discarded in favor of smaller meetings in each of the manufacturing plants attended by the President and Vice Presidents of the company, the Productivity Improvement Manager and several members of the steering committee.

The initiation meetings held in January 1984 stressed the need for improving productivity at Jamesbury and the part that each employee could play in this improvement. Copies of the corporate philoshpy were handed out and discussed by the President. The concept and plans for the program were explained by the Program Manager. Members of the steering committee were introduced and employees were encouraged to send their suggestions and ideas about the program to the committee.

3.3 General Education Phase

Immediately after the initiation meeting, a series of general education meetings were scheduled. The purpose of these meetings were to provide general information about Jamesbury, its products, its goals, and its competition. It was considered that a better informed employee will be a better motivated and committed employee.

For these meetings, the employees were divided into approximately fifty groups of about twenty employees each. Meetings for each small group was scheduled every two weeks. The central part of each meeting was a videotape prepared in-house by in-house personnel. The first twenty minutes conisted of viewing a videotape, with the next fifteen minutes utilized for a small group discussion of the tape.

The first series of meeting centered on the subject of competition to give all employees an understanding of the competitive pressures being felt by Jamesbury. In a videotape the Vice President of Manufacturing described the companies and products that were in competition with Jamesbury. Many employees were surprised to learn that several companies were producing valves with essentially the same quality as Jamesbury, but at much lower costs.

As an additional means to provide general information and to keep everyone posted on the progress of the program, a newsletter was started. The newsletter was published bimonthly and served as a means to recognize and publicize those people who made a significant contribution to the program, as well as provide general information to supplement the video tape program. For information of a more transitory nature, special bulletin boards were established and used extensively.

4 MANAGING THE ON-GOING INVOLVEMENT PROGRAM

4.1 Continuing Education and Training

A number of mechanisms were utilized for general education in addition to the meetings describes above. One of the most useful was the Update sessions where all employees met in small groups with the President and one of the Vice Presidents over a one year period. Invitations to the sessions were issued alphabetically by shift, with groups of approximately forty people from different plants doing different jobs attending each session. Top manage-ement shared information on how Jamesbury was doing and listened to the concerns and improvement ideas from the employees.

4.2 Selecting and Implementing New Improvement Efforts

The Steering Committee which consists of the President, all the Vice Presidents, and a representative from all levels and all functional areas of the company, is the primary decision making body that decides which and when new productivity improvement efforts will be introduced. Since the beginning of the program two new major efforts have been initiated.

MRP II Program. In order to improve customer service while reducing the inventories carried, it was necessary to improve the production planning and control system. After looking at a number of alternatives, Jamesbury decided that implementing the MRP II program developed by Oliver Wight (1981), represented the most cost-effective solution to the problem.

Quality Improvment Program. The next program that was initiated was the Quality Improvement Program to implement some of the quality concepts that have proven so effective for the Japanese. The program was based on the technique developed by Juran (Juran and Gryna, 1980) and consisted of a sequence of structured projects based on the premise that 80% of quality problems can only be solved by management. The program also include training in blueprint reading, geometric dimensioning and tolerancing, statistical process control, and other aspects of quality assurance, to provide every employee with the necessary skills to produce high quality products.

4.3 Measuring and Evaluating Improvement

The following overall measures of corporate performance are being measured and posted each month: orders as percent of budget, shipments as percent of budget, and percent of on-time shipments. Total measures of productivity in terms of dollars worth of product produced versus the dollars worth of labor input have been developed for all manufacturying groups and departments and are being tracked and posted regularly. For other aspects of the program, a number of less formal measures are utilized, such as the number of suggestions made in the productivity sharing program, the number of presentation made to management by Circles, etc..

4.4 Rewarding Involvement and Improvement

To motivate employees to make improvements there must be an effective system to reward them for their efforts. Although all share in the company's overall improvement through the productivity sharing program, an employee's individual effort has little effect on the reward received. A more immediate and visible reward was needed.

To provide such an immediate reward system, a point system was developed to reward employees for such improvement activities as: participating in quality circle activities, participating in ad hoc problem solving groups, making suggestions, etc. Points can be exchanged for more tangible rewards such as a company jacket, a trip to Bermuda, or cash. Since the points awarded are widely publicized, the points themselves serve as a reward by giving recognition for a person's contributions.

5 FUTURE DIRECTION OF THE PROGRAM

5.1 Evolutionary Nature of the Program

The Productivity Improvement Program at Jamesbury is an on-going continuing program that will adapt to the changing needs of the company. Although the area of emphasis and the techniques for implementing improvements may change with time, the never relenting striving for excellence through continuous improvement will not change.

5.2 In the Near Future

The current emphasis is on getting middle and lower management, and personnel outside manufacturing, more involved in the productivity improvment program (Oden and Bibeau). Prior emphasis had been upon worker groups in manufacturing, with little involvment of supervisors and middle managers in these efforts. With manufacturing becoming more automated, the major improvements through employee involvement will come in the areas of management and support efforts, such as marketing and engineering.

As an aid to getting middle and lower management more involved, a management newsletter, the Jamesbury Outlook, was initiated in October 1985. The Outlook provides middle and lower management with information on what and how well Jamesbury is doing in a form that can be shared and/or discussed with their subordinates. In addition to being a source of very useful information, the Outlook will also serve as a vehicle for encouraging discussions of productivity improvement at all levels throughout the company.

5.3 A Longer View

The Productivity Improvement Program is still very much in its forma tive stage. It would be foolhardy to attempt to predict its exact future form. However, we do have a vision of what we expect in the future.

We expect the improvement process to evolve until the following philosophies and practices become commonplace in the company.

* Every department and individual will consider it of utmost importance to continually be searching for ways to improve departmental or individual work with the idea that we will strive for, but never reach, perfection.

* Everyone in the corporation will want and be able to participate in the planning, improving, and correcting of his/her work. Employees will be able to do this through formal and informal mechanisms. Employees will be able to, and be encouraged to, move in and out of problem solving teams.
* Employees at all levels will determine significant measures and actively track them with the intent to continually improve productivity as indicated by these measures.
* Training will become a "way of life" and the skills of all will be continually upgraded by formal and informal training.

6 APPLICATION TO OTHER COMPANIES

The Productivity Improvement Program has answered a real need at Jamesbury by providing a comprehensive program in which all employees could become involved in productivity improvement. It is enabling Jamesbury to make effective use of its most valuable resource, its people. Although many of the details of the program are specific to Jamesbury, the general approach could be applied to any company.

6.1 Should Jamesbury's Approach be Applied in Other Companies?

The answer to this question depends on whether the other company is in the same situation as Jamesbury. Is it facing stiff price competition for a standardized product from its competitors? If it is, then it must do everything in its power to improve productivity. This means mobilizing its entire work force behind the strategic objective of improving productivity and reducing costs. This can best be done with a total involvement program similar to that at Jamesbury.

Like Jamesbury, most companies have many parts of the program already in place. Which means that they don't have to start from from ground zero, but can add to what they already have. Thus the logical first step is one of integrating the existing parts into a coordinated and effective whole. With the program started by coordinating and rejuvenating existing efforts, then one can begin to add efforts which will provide the greatest benefits.

6.2 Guidelines for Implementing a Productivity Improvement Program

Since other companies will be starting from different points and will be adding different efforts to the program, the detailed approach of each company will tend to be unique. However, Jamesbury has learned a number of lessons in implementing their program which are applicable to any program. These guidelines are outlined in the following paragraphs.

Establish a formal continuing program. A productivity improvment program is more "a way of life" than a one-time program (Sumanth, 1984). What one is doing is literally changing the corporate culture, and such an effort can not be achieved by a one-time effort. To ensure that it continues, it must be formally planned and have a permanent management organization. Without a plan and a permanent management structure it is apt to falter when confronted with a slump in business or other crises.

Ensure Top Management Involvement. For a productivity program to be successful, top management must be more than committed to the program, they must be actively involved. The employees must see them actually participating. Not only must they participate but their participation must be seen as sincere. The participation of Jamesbury's President and other members of top management in the general education phase of the program was instrumental in convincing the employees that top management was really interested in the contribution of each individual.

Make everyone aware of the need. Everyone must be made aware of the need to improve productivity before any meaningful involvement can be expected. This can best be done by an honest, factual presentation of the current state of competition with concrete examples of competing products and prices. The primary purpose of Jamesbury's initial general education program, was to make all employees aware of the need to improve productivity.

Have a plan for involvement. With everyone aware of the need and enthused to do something, there must be some way that all this motivation can be channeled toward some concrete improvement activity. Don't formally initiate the program until you have plan that provides an opportunity for every employee to contribute. If a quality circle program is not in place, this should be one of the first new efforts to initiate. Active involvement in an improvement effort not only results in improvement, but also reinforces and strengthens the desire to make more productivity improvements.

Include all levels of management and staff. If a company

starts with top management and then moves directly to involving workers, the worker improvement groups will not have access to the information and assistance they need from staff groups and middle management to make intelligent recommendations. It is very demoralizing for a group to work on a project for six months, formulate a very constructive improvement proposal, and then find out their proposal will not be acted upon because the product is being phased out. Worker improvement groups must have access to this type of information, and it will normally not be forthcoming until middle management and staff groups are actively involved in the improvement program.

Emphasize continuous improvement. Throughout all facets of the program, it should be emphasized that the program is one of continuous improvement and not a one-time effort. Throughout Jamesbury's program, we have emphasized that we are seeking "improvement" and not "change". Although improvement can not be achieved without change, we find that people are much more receptive to "improving" than they are to "changing". Emphasizing improvement causes people to focus on the end result to be achieved rather than on the mechanism for achieving it.

Be flexible in your program planning. Have a plan, but don't hesitate to change the plan if the situation warrants. If some person or group is eager and ready to initiate a new improvement effort in their area, incorporate that effort into the plan.

7 CONCLUSIONS

The program outlined in the preceding paragraphs certainly should not be construed as a substitute for implementing the technical elements needed to achieve improved productivity. It is, however, a powerful complementary program that not only achieves productivity improvements in its own right, but changes the attitude of everyone in the company so they are willing to accept the the necessary technological changes to achieve improved productivity.

With every employee imbued with the idea of continuously seeking productivity improvements in their own sphere of influence, a corporate culture evolves which is willing to accept and support productivity improvements wherever and however they may occur.

8 REFERENCES

Juran, J.M. and F.M. Gryna, 1980, Quality Planning and Control, McGraw-Hill

Oden, Howard W. and Marshall W. Greene, 1984, "Performance Improvement Through Worker Involvement", Synergy-84 Conference Proceedings, APICS/SME, pp. 134-136

Oden, Howard W. and J. Michael Bibeau, 1986, "Involving All Employees in Computer Integrated Manufacturing", CIMTECH Conference Proceedings, CASA/SME, pp. 10.39-10.52

Sink, D. Scott, 1985, Productivity Management: Planning, Measurement and Evaluation, Control and Improvement, John Wiley & Sons, New York, p. 434

Sumanth, David J., 1984, Productivity Engineeriing and Management, McGraw-Hill, New York, p. 477

Wight, Oliver W., 1981, MRP II: Unlocking America's Productivity Potential, CBI Publishing Co., Boston

H. IMPLEMENTATION OF PRODUCTIVITY MANAGEMENT

Productivity Management Frontiers — I, edited by D.J. Sumanth
Elsevier Science Publishers B.V., Amsterdam, 1987 — Printed in The Netherlands

PRODUCTIVITY AND QUALITY ENHANCEMENT - AN EXPERIMENT

O. W. MILLER, M. S. LEONARD, L. G. DAVID
Industrial Engineering Department, College of Engineering, University of Missouri-Columbia, Columbia, Missouri 65211, (USA)
JAY GOLDMAN,
Dean, School of Engineering, University of Alabama-Birmingham, Birmingham, Alabama 35294, (USA)

ABSTRACT
Miller, O. W., Leonard, M. S., David, L. G., and Goldman, J., 1987. Productivity and Quality Enhancement - An Experiment.

There is emphasis today on encouraging the active involvement of government, universities and private industries in sharing their resources to improve productivity and enhance the quality of the outputs of their respective organizations. In the State of Missouri, an experiment was conducted wherein the state government, two campuses of the state university system, and twelve private manufacturing companies participated in the development and operation of a productivity and quality enhancement program. The goal of this program was to provide an approach and an associated implementation methodology that the participating organizations could use to increase their productivity and improve the quality of their goods and services. This paper describes the results of this initial experiment and how these results are being used to make a productivity and quality enhancement training program available to a large number of Missouri firms.

1. TEAM PROGRAM

True and long-term improvement in the competitive position of a company is likely to come only when every company employee takes active roles in the production of quality products and the enhancement of manufacturing productivity. Company management, from top to bottom, must assume responsibility for the direction and support of these quality and productivity enhancement efforts.

An ever growing number of academics, oonsultants, and practitioners speak to the ills of American corporations and support the use of enhancement programs in their articles, books, videotapes, and seminars. Yet it is still surprising to observe the positive results that can come when a quality and productivity enhancement philosophy becomes a central element in the management and operation of a company. Several members of our UMC Department of Industrial Engineering faculty have had the opportunity to see firsthand these positive effects.

These UMC faculty members helped to develop and test a productivity and quality enhancement program for Missouri manufacturing and service industries.

This program is called the Team Effort Advances Missouri (or TEAM) program and it has been funded by the Missouri Department of Economic Development. The TEAM program represents a cooperative effort among state government employees, business leaders throughout the state, and faculty from the Rolla and Columbia campuses of the University of Missouri. This cooperative effort is directed toward increasing employment opportunities for Missouri citizens and advancing the economic development of the state.

The basic premise underlying the TEAM program philosophy is that an organization will make most effective use of its resources when it produces only quality goods and services and when it constantly strives to improve its productivity. For this to occur, productivity and quality enhancement activities must become integral parts of the firm's strategic plan, priority structure, and business routine. Specific enhancement plans must be developed and implemented. Concern for achieving these enhancements must be as real as the concern for meeting other corporate objectives, such as sales and profit levels.

The TEAM program is built upon the cooperative efforts of all company employees. Every individual in the organization must share in productivity and quality enhancement efforts. Subordinates watch managers to see what their managers consider most important. Our TEAM program experience has demonstrated that if managers simply announce the importance of productivity and quality but then constantly give their attention to production schedules, profit margins, and sales quotas, their subordinates will recognize the relative unimportance of productivity and quality. Managers must demonstrate commitment to productivity and quality enhancement by being active participants in enhancement activities, discussing the subject regularly, and giving it equal priority with other organizational activities and functions. Enhancement related efforts must become an important element of everyone's regular job duties.

Establishing specific improvement targets or goals is the first step in attaining this level of company-wide involvement. For the companies that participate in the TEAM program, the top management team sets overall corporate enhancement targets, and each organizational unit within the company develops its own goals and plans in support of these corporate objectives. Enhancement goals are specified in clear objective terms, using measures of productivity and quality selected to be meaningful with respect to the goods and services produced by each organizational unit. Typically these productivity and quality performance measures will be different for each unit. Graphic plots of the performance measure values should be maintained and prominently displayed so that all members of each organizational unit know how their unit is performing with respect to quality and productivity enhancement and whether or not their unit is meeting its targets.

Before they adopted the TEAM program philosophy, the managers of many of our participating Missouri companies believed that high productivity and high quality were incompatible; that one could be attained only at the expense of the other. This attitude appeared to be the result of the managers' past approaches to achieving quality. Most of them tried to "inspect in" quality. In order to have higher quality, they inspected more. This inspection approach to quality enhancement adds to costs and reduces productivity. The TEAM program answer is not to inspect more but to prevent the defects from occurring in the first place. Doing the job right the first time reduces the costs associated with scrap, rework, and inspection. High quality and high productivity go hand in hand. High quality does not cost more money, low quality does. Instead of spending time compensating for and correcting defects, time should be taken to prevent the defects from occurring in the first place.

The development and testing of the TEAM approach was conducted in a variety of Missouri-based manufacturing firms. Companies were selected to participate in the program on the basis of their size and types of products. The experiences of one of the TEAM companies is discussed here to illustrate some of the positive effects that can come from participating in this type of program.

The company to be discussed is a manufacturer of electrical metering and distribution enclosures. This company is one of the four largest producers of electrical enclosures in the nation. Company employees are not represented by a union.

The company began its participation in the TEAM program when several company executives, including the president, attended a set of educational sessions describing the philosophy and operation of the program. Subsequently, the company adopted the TEAM concept for quality and productivity enhancement at each of the four organizational levels within the firm. Each organizational level has identified its own quality and productivity measures, and performance data related to the measures is reported upward through the organization on a weekly basis. Plots of company quality and productivity measure values are displayed prominently on the wall of the company conference room.

The quality of company products has improved significantly over the twenty-four month period that the company has been participating in the TEAM program. At the same time company productivity has been enhanced dramatically. For comparable five-week calendar periods in 1984 and 1985, production has increased by 34 percent with 15 percent fewer employees. In another comparison across the same time periods, payroll cost has increased an average of 14 percent as a result of salary raises yet total labor cost as a percentage of sales revenues has remained virtually constant.

2. A MODEL FOR IMPLEMENTING A QUALITY AND PRODUCTIVITY PROGRAM

1. Identify and select output units.
2. Identify customer(s).
3. Determine customer(s) specifications.
4. Identify and select your input resources.
5. Document current quality in terms of time period in which quality units are delivered (i.e. time period during which output units meet customer specifications).
6. Document current productivity in terms of the value of the ratio of output units meeting current inspection procedure to input resources.
7. Provide for the continuing display of quality and productivity measures over time.
8. Establish initial quality enhancement target with your manager.
9. Establish the initial productivity enhancement target with your manager.
10. Identify the barriers to achieving the quality and productivity targets.
11. Prioritize the barriers.
12. Reduce or eliminate barriers in priority order.
13. Review performance in relation to targets.
14. Review quality and productivity enhancement targets for possible modification.
15. Return to Step 10.

Steps 10, 11, and 12 usually require educational sessions on the use of specific tools and concepts, such as

- Team Building
- Communication
- Statistical Quality Control
- Statistical Process Control
- Brainstorming
- Cause/Effect Analysis
- Pareto Analysis
- Process Charts/Flow Diagrams
- Methods Analysis/Work Enhancement Checklists
- Economic Analysis/Cost Benefit Analysis

3. ONE COMPANY'S EXPERIENCES WITH THIS MODEL (24-MONTH PERIOD)

The company president and several of his department heads attended the initial TEAM Program presentation. Company teams at four organizational levels were then assigned the responsibility for implementing the program.

The first echelon consisted of department heads of manufacturing, sales, engineering, office and quality control and inspection. These were called the "A" Team. Their charge was to develop quality and productivity measures for the entire plant, to set quality and productivity enhancement targets and to implement the program at the "second and third echelons."

The second echelon consisted of these same department heads plus members of their respective departments. They in turn, were to develop and implement quality and productivity measures for their respective departments.

The third echelon ("C" Teams) consisted of manufacturing foremen who were to establish quality and productivity measures commensurate with those of the manufacturing department heads.

The fourth echelon ("D" Teams) consisted of manufacturing hourly employees who were assigned to a team made up of members of their work group and their foreman. These groups were also asked to develop productivity and quality measures and to identify problems and barriers to the solution to those problems.

Almost 6 months were required for the "A" team to identify and agree to quality and productivity measures. Somewhat shorter periods of time were required for "B", "C", and "D" Teams.

1. Statistical Process Control/Statistical Quality Control

A manual was recently developed in which quality specifications and measurements are defined for each of the manufactured products. In addition, statistical acceptance plans were developed for both incoming raw materials and outgoing finished products. Procedures for the specific use of these plans are outlined in the manual.

2. Program Monitoring

The top management (first echelon-"A" Team) meets monthly to monitor and plan future activities in the program.

A typical agenda for these meetings is

1. Review of written summary report on the previous monthly meeting
2. Quality target review
3. Cost of quality review
4. Productivity target review
5. Problem list review, update
6. Goal setting review, update
7. Report on second ("B" Team), third ("C" Team), and fourth ("D" Team) echelon teams' progress with respect to quality, productivity, and problem lists.

The company reports the following benefits from their two-year involvement in the TEAM Program:

1. The overall productivity (based on the number of units per man-hour) improved by 16.3% (latest fiscal year).
2. In the last quarter of 1985, the greatest improvement was achieved: 28.5% increase in productivity. This was achieved by producing over 18% more units with 8% less man-hours than the same period the previous year. If the company had continued to produce at the same productivity rate as it did the previous year, it would have cost the company an additional $300,000 to get the same output.
3. The work-in-process inventory has been reduced by approximately 40% compared to the prior year-end fiscal inventory.
4. In February 1985, the total time required to fabricate, weld, paint and assemble a typical production order averaged 15 working days. By February 1986, that average had been reduced to 10 working days.

5. As a result of the productivity improvements, existing safety stocks were either completely eliminated or reduced, and production runs of fabricated parts were reduced to exactly equal the quantity required for a one-week period.
6. A "Just-in-time" final assembly station was developed by several plant employees. Initially, the employees were skeptical and not inclined to work together in a new environment. But after several weeks they became enthusiastic and designed a workplace station consisting of five operators who assemble, test and package the final product.

3. Conclusion

At first it seemed to the faculty members participating in this experiment that the "wheels of progress" do indeed turn slowly. As teachers, we expect our students to learn techniques quickly and apply them correctly to textbook problems. We knew, and this experiment confirmed that persons must be convinced that it is to their benefit to modify their work habits and approaches to solving problems. This could not have been a successful experiment without top management's absolute commitment to improving its competitive position by applying quality and productivity enhancement procedures that have been around a long time.

4. REFERENCES

Crosby, Philip B., 1980, Quality is Free, Mentor Executive Library, New York, N.Y.

Team Effort Advances Missouri, Books One and Two, State of Missouri, Department of Economic Development, Jefferson City, MO 65102.

Quality and Productivity Enhancement Video Tapes:

Roadmap for Change (Deming), Pontiac Motor Division, General Motors Corporation, Detroit, Michigan.

Introduction to the Deming Management Philosophy, (Dr. Myron Tribus), Massachusetts Institute of Technology, Center for Advanced Engineering Study.

Team Building, CRM McGraw-Hill Films, Del Mar, CA 92014.

The Japanese Approach to Productivity, Automotive Industry Action Group, Boulevard Center Building, Detroit, MI 48202.

Productivity Management Frontiers — I, edited by D.J. Sumanth
Elsevier Science Publishers B.V., Amsterdam 1987 — Printed in The Netherlands

PRODUCTIVITY MANAGEMENT THROUGH IMPLEMENTATION OF A QUALITY IMPROVEMENT SYSTEM

John S. W. Fargher, Jr.
Management Controls Department Head and Comptroller
Naval Air Rework Facility, Cherry Point, NC 28533-5030

BACKGROUND

The six Naval Air Rework Facilities (NARFs) and the Naval Aviation Logistics Center (NALC) are undergoing a significant change in management philosophy and worker participation with a phased implementation of the statistical quality control approach. Statistical quality control/quality improvement methods are being applied for process measurement and management action to obtain improvements in quality and reliability, reductions in reprocessing, and resultant increases in total productivity. The purpose is to (1) detect out-of-control conditions, (2) structure for problem detection and analysis, and (3) cause resolution of the root causes through management action and worker commitment. The total Quality Management Team approach is applied multi-department, multi-disciplinary involvement, focusing on areas which provide either plant-wide support or have high visibility because of defect problems. Analysis is structured around specific production processes.

The paper will outline the phased plan of action and milestones for implementation of the Quality Improvement System as identified from reported research, from visits to private sector companies that have successfully implemented quality management approaches, and from experience at the six NARFs and NALC. The experience of NARF Cherry Point on the application of the statistical quality control/quality improvement methodology and "lessons learned" are also related.

1 INTRODUCTION

A recent New York Times Sunday Magazine cover story, "The Danger from Japan", described Japan as "...dismantling American industry." While most manufacturing plants in Japan are going full blast, U.S. factories are closing. Since 1979, in machinery production, the U.S. has lost 288,000 jobs; in primary metalmaking, we have lost 439,000 jobs; in fabricated metals, 241,000 jobs; in motor vehicle production, 118,000 jobs; and in textiles, 186,000 jobs. All in all, in the old shops, America has lost 1,834,000 jobs since 1979.

The Japanese have moved from a demolished country 40 years ago to a major world force by restructuring. Part of Japan's transformation is managerial and part of it is technological. In March of 1948, Japan's scientists and engineers formed an organization to meet this challenge. The organization, called the Union of Japanese Scientists and Engineers, became aware of an Edwards Deming, who had made several visits to the general headquarters of the supreme commander of Allied Powers. The Union was hard at work pioneering services in statistical quality control and, to support its aims, decided to sponsor Dr. Deming in a series of seminars which took place in the early 1950s. The organization also published a booklet of Deming transcripts and created the now-famous Deming prize. This prize, regarded with high esteem world-wide as well as in Japan, is awarded to Japanese individuals and companies for achievements in research and application of quality control.

Quality control based on statistical methods became a subject for mass education throughout Japan's entire nation. From the mid-50s to the mid-60s, Dr. J. M. Juran, a leading authority on the subject, became famous for his lecture tours in total quality control. His principle is that all workers have exposure to statistical quality control knowledge and jointly participate in company-wide quality control practice. As a result, it was necessary to plan for improving quality at the first line supervisory level. And this is how the seeds of the quality circle movement came to be planted.

Japan's managerial climate emphasizes:

o An obsession with quality;

o Achieving a feeling among employees that they are part of one big team; and

o Identifying and correcting problems with data, not opinions or emotion.

Japan, of course, is presenting a technological challenge, too--by being adept borrowers of our technology. America must also restructure to compete--and time is short.

2 BENEFITS AND EXAMPLES

The application of statistical quality control relies on the operators to judge the quality of their own products and an awareness that each worker has to control quality. The operator must be taught to plot results, seeing patterns of delineation and know precisely what needs to be done if there is a deviating trend which crosses the margin of quality. Total Quality Control (TQC) is defined as an effective system for integrating the development, maintenance, and quality improvement efforts of the various groups in an organization to enable production and service to operate at the most economical and efficient levels to satisfy the customer's product quality requirements. All groups within the organization must ultimately become involved and committed to meet the customer's quality expectations. The TQC approach involves commitment through education in statistical process control.

The advantages of the TQC approach are many:

o It provides time quality assurance through process control with the ultimate goal of achieving a 100 percent defect-free product by building in quality at every step of the production process. By determining the causes which create defects and flaws, TQC and process control enable the employees to identify and correct these causes.

o It opens channels of communication within an organization based upon measurements and data rather than opinion, relying on knowledge, identification of common problems, as well as special causes, and solution of problems and causes before they become detrimental to quality.

Companies in Japan that have implemented the TQC approach include these firms that have won the Deming prize: Fuji Xerox (office copiers), Nippon Kokan KK (steel), Kajima Company (construction), Rhythm Watch Company (small watch company), Komatsu (tractors and earthmovers), and Yokogawa Hewlett-Packard (electronics). Many companies in the U.S. and Canada have also initiated the TQC approach. These include: Ford Motor Company, Nashua Corporation, General Motors, General Electric, and Windsor Export Supply. The TQC approach has also been transplanted from Japan and Korea to U.S. subsidiaries: Toyota, Mazda, Honda, Nissan, Mitsubishi, and Hyundai Motors of Korea. Even in Eastern North Carolina, other firms besides the NARF are applying TQC - Bosch and Yale.

Is TQC a cult? Such concepts as Just-in-time (JIT), a concept whereby an assembler receives shipments of parts as needed for his operations so that he doesn't have to carry any inventory can be viewed as cultist, in my opinion. Only if the economics are there based upon traditional queueing theory and inventory models, should JIT be employed. But then the theory and models would have pointed the way to JIT if the conditions were proper, anyway. TQC is based upon statistical quality control which has been with us since the 1920s. The statistical control chart as well as various statistical analysis methods are most powerful and robust tools for control of processes as well as product performance. Most businesses utilize control charts to track product quality; however, control charts also have a more fundamental use, that of providing a timely method of assessing process stability (variability) and providing an accurate assessment of the capability of the process to meet end product specifications. Businesses using control charts report direct productivity/quality improvements. Figure 1 illustrates the effect of application of TQC.

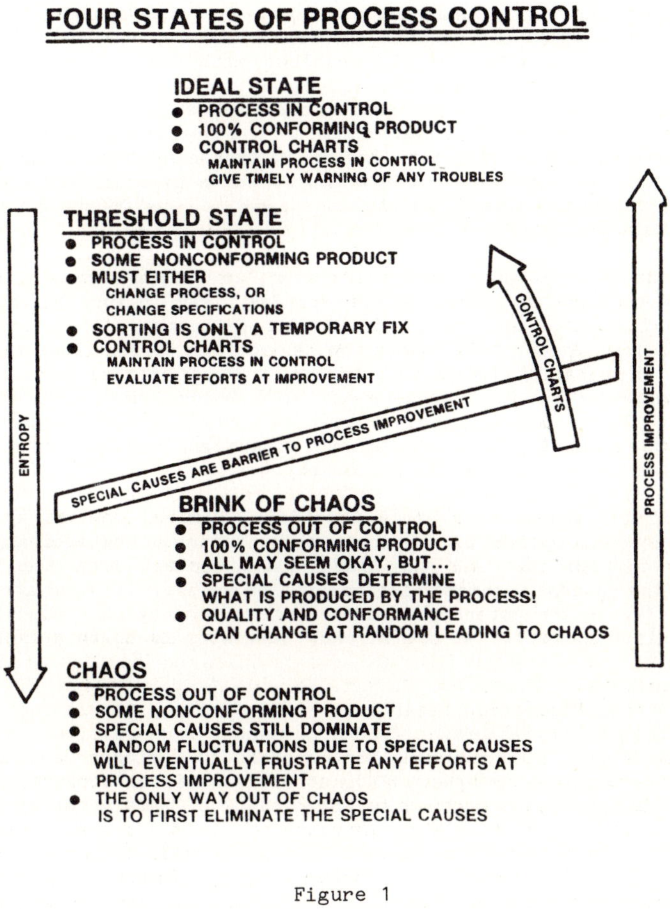

Figure 1

The critical factor in TQC is the application of statistical quality control to the manufacturing and service system and initiating the cycle of continual improvement. The TQC approach advocates change the management control concept from "detection" to one of "prevention". While end-item inspection may not be totally eliminated, measurements are taken on a sampling basis at critical operations, to provide real-time measurements of the process. Adjustments can then be made to the process to greatly reduce and ultimately eliminate scrap and waste. Statistical methods are utilized at this point to logically and systematically evaluate process stability and capability to hold tolerances, and isolate the causes of the problem. Although the application to manufacturing systems is intuitively obvious, service industry also has similar, if paperwork, processes that are also measurable from error rates, etc.

3 STARTING THE PROCESS

Approximately two years ago, several NARF North Island employees became interested and committed to the TQC approach. These individuals were successful in recruiting the incumbent Commanding Officer to their group of believers, which culminated in efforts to implement the philosophy at North Island. Implementation began with awareness training for top managers followed by a "Building 472 Prototype" approach in the Production Department. Initial success in discovering and solving quality problems with the overplating of an H-53 helicopter rotor piston and H-46 helicopter blade fold pin spurred on the efforts. On the H-46 helicopter blade fold pin, a discrepancy in the plating process at NARF North Island resulted in a 70 percent reject rate and schedule delays. After collecting the data and fully examining the process, the plating and grinding process team made process improvements that allowed the reject rate to drop to 2 percent within 7 weeks, saved $52,000 per year, and improved H-46 rotor head productivity.

In addition to the North Island endeavor, these same individuals recruited business leaders from the private sector of San Diego to form an organized group to further and enhance the use of the Deming philosophy. Key individuals have formed a User's Group that meets on a monthly basis and has been very instrumental in bringing Dr. W. E. Deming (several times); Dr. Myron Tribus, MIT; and recently, Dr. W. Scherkenback, Ford Motor Company, to San Diego and North Island. These renowned individuals have provided North Island and the San Diego business community with valuable training and insight into the pros and cons of implementation of the TQC philosophy.

On one of Dr. Deming's trips to San Diego, Admiral John Kirkpatrick of the NALC was invited to attend Dr. Deming's presentation. The Admiral attended and subsequently decided that all the NARFs would benefit from the TQC approach. Thus began the journey toward implementation. Admiral Kirkpatrick recently stated that, "We in the government must change our management approach to embody the total quality concept. We are already beginning to do this in the Naval Air Rework Facilities."

Initially, the Production Department staff, Quality Circle facilitators, and Quality and Reliability Assurance (Q&RA) personnel were involved in what and how of the TQC approach. Top managers were exposed to Deming's philosophy during a four-day seminar held in February 1986 at Jacksonville, Florida. At this time, it became obvious that to succeed in an endeavor of this magnitude, a full-time leader was necessary. Each NARF has appointed a leader for implementation.

Total Quality Control (TQC) has been introduced to all the NARFs and NALC, and it is here to stay. The goal is to achieve a major cultural change involving organizational values, individual attitudes, and business practices. There is evidence of the facilities' determination to ensure that quality has top priority and permeates every action and decision. It is recognized that the total effort will involve every employee. Dedication, patience, and total commitment are required to meet this monumental challenge. By proceeding deliberately and slowly with positive actions, success will be achieved. The other NARFs have had similar experiences in implementing improved quality and productivity through the continuous improvement team process. Besides the examples shown above, NARF North Island has applied the improvement process to "stop gap" manufacturing of H-53 spindle nuts, producing all nuts one month ahead of schedule and at $42,000 less than the lowest commercial vendor with quality superior to the vendor, resulting in returning all USN and USAF H-53 helicopters to flight operations. North Island has also applied the proce-s to the F-14 Standard Depot Level Maintenance process with equally successful results. NARF Jacksonville has applied the process to the A-7 Standard Depot Level Maintenance program and expects to apply it shortly to the P-3 aircraft, avionics, hydraulics, and engine components. NARF Pensacola has cured an excessive turnaround problem in the H-3 main gear box input housing rework process and saved $160,000 in annual costs. NARF Cherry Point efforts will be detailed later in the paper.

Plans of action and milestones have been developed for each NARF and NALC with the assistance of the Naval Personnel Research and Development Center (NPRDC) in San Diego. The actual quality improvement management implementation process varies slightly at each facility; however, the basic process and guidelines as detailed in this paper serve as a guide and "benchmark" against which each facility can measure progress. The process and guidelines were endorsed by Dr. William Scherkenback, Director of Statistical Methods for Ford Motor Company, and conform to the quality improvement management implementation process at Ford. The sites visited by the NPRDC/NARF/NALC team to establish the approach, guidelines, and steps included naval organizations, Portsmouth shipyard, Mare Island shipyard, Long Beach shipyard, Pearl Harbor shipyard, and North Island NARF and private industry, including Pontiac Motor Company, Island Motors, Ford Motor Company, Hughes Aircraft, Dysan Corporation, Mattel Electronics, NCR Corporation, Tektronics, and Polaroid. NPRDC developed material and has made expert personnel available for the NARFs and NALC for quality improvement management implementation. Each NARF has developed the essential elements of the quality improvement management implementation process:

o Executive Steering Committee (called the Quality and Productivity Action Group at NARF Cherry Point).

o Corporate Quality Philosophy Statement

o Implementation Plan

o Training Capability

4 GUIDELINES FOR IMPLEMENTING A QUALITY IMPROVEMENT SYSTEM

The specific guidelines for implementing a quality improvement system fall into six major categories. These categories were identified from reported research, visits to private sector companies that have successfully implemented quality management approaches, and from experience at the six NARFs and NALC. The six major categories are:

- o Top Management Orientation
- o Management and Administrative Structure
- o Training of Quality Management Boards and Project Teams
- o Quality Improvement Actions
- o Expansion of the System
- o Support Activities

A set of 35 specific activities has been identified which falls within these 6 major areas. Many of the specific activities associated with each of these major areas will be run concurrently. The entire range of activities, once top management within the organization has decided to go ahead with the plan, should take about two and one-half years. The chart (figure 2) depicts the relation between the major categories and the sequence of activities. The arrangement of the various steps within the hexagons shown in figure 2 indicates that the steps will occur approximately at the same points in time. The exact order between adjacent steps may change depending on specific organizational needs or concerns; however, the grouped steps indicate their temporal relationship.

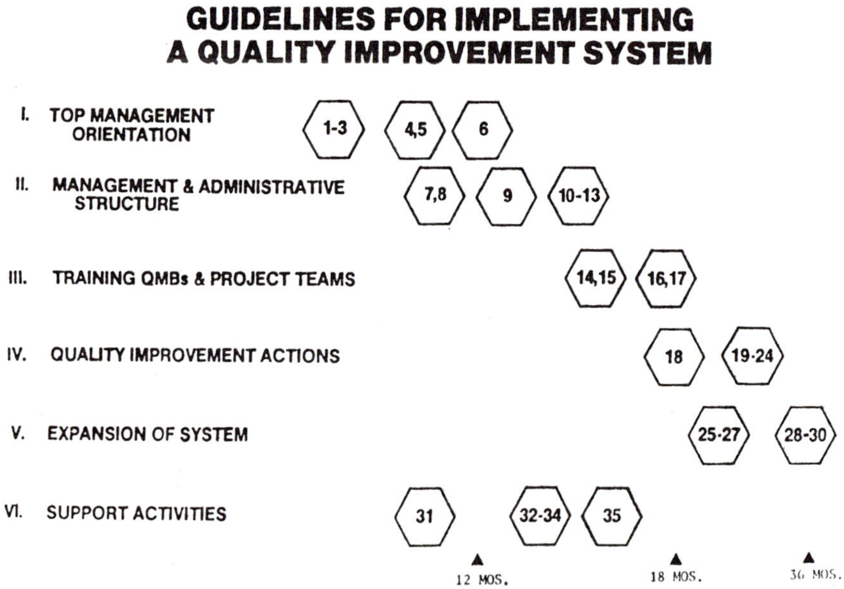

Figure 2

4.1 Top Management Orientation

The initial phase of the implementation process requires the development of an awareness of and commitment to the need for quality improvement. Awareness is relatively easy to achieve because most people will support the need to improve. Commitment, however, is harder to achieve because it requires some activity or action such as the allocation of resources (people, time, and funding).

Step 1 involves several meetings among quality management experts, the top management official (such as the NARF CO), and selected top management (such as the executive committee) to discuss the basic concepts of quality improvement and their applicability to the organization. This step doesn't require working out any details, but rather a discussion of the basic need for such an approach. This step assumes there is some familiarity with the basic quality improvement orientation, such as some members having attended a statistical quality control/quality improvement seminar or discussions with those who have. The objective of these meetings is to achieve an understanding of the basic requirements of quality management.

Step 2 consists of discussions of quality management principles and applications expanded to include top level management such as department heads, division directors, branch chiefs, and top military personnel, if applicable. Meetings should be held between executive committee members (or those identified in Step 1), and small groups of management. Reservations and questions as to the appropriateness of implementing quality management principles should be discussed at these meetings. These discussions should be informal in scope and include an overview of the concepts. Participation and interaction with a statistical quality control/quality improvement expert (guru) to facilitate or provide needed expertise should also be considered a part of this procedure. The purpose of these informal discussions is to provide some familiarity with quality management principles and to stimulate thinking about the expected costs and benefits for the organization to adopt such an approach.

Step 3 is the first "formal" and detailed presentation of quality management principles and tools. Management is provided more detailed and intensive information on concepts, organizational implications, and examples from private and public sectors. This training consists of the "Statistical Quality Control/Quality Improvement Seminar" for all top management (military and civilian). Middle management can attend a formalized presentation called "Level 1", which presents and discusses the theory and applicability of quality control.

Step 4 consists of the activities and processes which occur after formal training. Discussions between the executive committees and top management need to occur once again. These activities are a follow-up to the formal training where many of the ideas can be addressed, questions answered, and perceived values (costs and benefits) reassessed. This is a period of gestation, where ideas have a chance to be evaluated and discussed, both formally and informally. The objective is to determine methods for achieving acceptance and success.

Also, during this phase active support and commitment from top management needs to be established, and further interactions between these managers need to occur. The format for these interactions could include a variety of activities, such as a series of meetings, workshops, management retreats, or similar situations where open exchanges would occur. Whatever types of sessions are used, these interactions are vital because they will determine the success of the implementation efforts.

An important product at this stage is the development of a quality philosophy statement. Every organization operates in accord with particular philosophical principles. Top managers need to develop a statement of their quality management philosophy. This statement will provide a road map of how the organization will be operated. The philosophy statement embodies goals and values which must be communicated to members of the organization.

Another major purpose of this interaction between top management is to formulate a specific organization-wide strategic implementation plan. Consideration needs to be given to the format for implementation, the types and amounts of training that may be required, and how processes will be identified, monitored, and documented.

Step 5 consists of a workshop and/or meetings with middle level management. A similar format to that described above should be followed with this group. The results and important points raised in the workshop described in Step 4 need to be communicated to this group. The philosophy statement and the goals and values decided upon by top management must be discussed. The strategic implementation plan should be presented and middle management's level of involvement must be clearly stated and accepted.

Step 6 is not really an action or activity but rather reflects a point in time indicating a reasonable period for Level I quality management training to have been completed for top management. This time period should be no longer than six months after Step 2.

4.2 Management and Administrative Structure

After awareness and commitment have been achieved, the difficult process of organizational change begins. Both the culture and structure of the organization must change for implementation to have lasting effects. While cultural changes take time to become part of the organization's way of life, structural changes can occur relatively quickly. The structural changes begin with systemic changes which facilitate implementing process control procedures within the organization. As will be described below, these changes emphasize understanding work flow processes and increasing communication and interaction within the organization.

Step 7 begins the structural change activities by the formation of an Executive Steering Committee (ESC). This committee is composed of top management which includes the top management official and several other key managers. The ultimate support, authority, and decision-making responsibility for the quality improvement system rests with this body. This group oversees the implementation process and should receive periodic updates on activities, progress, problems, and evaluations of the effects of the quality improvement system on the organizaiton. One of the tasks for which this group is responsible is to change policies if required to meet implementation goals. Another task is the selection and assignment of members of the Key Implementation Team (KIT).

Step 8 begins the activities towards the actual implementation itself. Members of the KIT are chosen by the ESC. In addition to several decision-making members from the ESC, this group should consist of people who will facilitate the implementation. This group should consist of at least the following members: one person designated as the process control coordinator, one person knowledgeable about the training requirements, an organization development person, a documentation/measurement individual, and a person who understands the interactive relationship between the different departments. The same person could serve in more than one of these functions.

The process control coordinator functions to monitor, plan, collect information about progress, assist with administrative arrangements, or whatever else may be needed to ensure that implementation activities continue. This person is the link to the lower level groups and will provide whatever administrative assistance may be required to facilitate awareness, training, data collection, and evaluation of the activities. As the implementation expands to the entire organization, the duties of this person become more crucial from both a corporate knowledge and a coordinating perspective.

The training person (who may be the person described above) needs to be aware of the training requirements, the training resources available, the required skill levels, and the effectiveness of the training program. This person should be aware of what the various training levels consist of, who will do the training, the abilities of those receiving training, and how this training will be evaluated. This person will also be responsible for maintaining the skill levels of the in-house trainers if they are charged with training the people within the organization.

The organizational development person is responsible for ensuring that the various groups such as the quality management boards and the project teams (to be described below) understand and follow the prescribed set of tasks they are assigned. Not only will the roles of the various groups need to be defined for them so they will clearly know what they are expected to accomplish, but someone will need to ensure that the barriers to communication that sometimes exist between members of different organizational units are reduced or eliminated. In other words, someone needs to be present at the meetings, at least initially, to ensure that threats to one's ego, units, or members of one's unit, are minimized. This person would facilitate (or make sure that others were facilitating) group interaction towards specific tasks and agendas. This person would provide direction or structure as needed.

The documents/measurements person would provide guidance to the groups on how to record group activities, action items, difficulties, successes, and measurement techniques and/or procedures to collect data. This person should play an integral part in the design of forms with which to gather and record data. This person should also be the organizational consultant on statistics or statistical procedures. The duties assigned to this person could be combined with those of others described above.

Step 9 consists of the establishment of Quality Management Boards (QMBs). These boards provide the organizational structure that will reduce friction between various organizational units (e.g., departments) and improve the communication flow up, down, and horizontally, because they are comprised of members from all of the relevant departments. The QMB structure was adapted from Russell Ackoff (1981), who described it in his book, Creating the Corporate Future. QMBs exist at all levels within the organization; at the directorate level, the department level, the division level, the branch level, and the section level. Each board has members primarily from that hierarchical level with a representative from the next level above and below. In other words, if the division QMB is our focus using the NARF structure, membership would consist of individuals from relevant departments with decision-making authority. A representative from the department and the branch would also be present at the division QMB meetings. These two individuals serve as the "linking pins" to provide information up and down the organizational structure. These individuals would report the order of business from the division QMB to their respective QMBs. The same type of organizational structure would be maintained at each QMB level.

The shop level QMB is called a project team. The team is mainly comprised of artisans and/or foremen who are working on a specific issue, problem, or process. The teams are specifically formed to address a particular concern. After that concern has been adequately dealt with, the team is generally disbanded. Multiple teams may be formed within the same shop and may even contain many of the same members. As teams begin to form and work together in the initial stages of the program, problem identification will need to be provided from a higher level board. However, as teams get established and gain experience working together and see the changes that they have recommended being implemented, they may begin identifying and working on problems they themselves have chosen to address. This is a goal to shoot for. The results of the team's efforts would be reported to the next highest level QMB, which would in turn report the activities to the next level, and so on.

Steps 10 through 13 focus on administrative tasks associated with the formation of the initial project teams and the scheduling of training for team members (Levels I and II) and Level II training for management.

After the QMBs are formed it is important for them to identify project team members in order for training to be arranged for them. If the organization will use its own in-house "trainers", it is also necessary to at least begin discussions of who these individuals will be. Some of the initial training may be performed by an external agency, but ideally the host organization should begin the process of selecting and developing its own training arm. The selected individuals should attend the training provided by the outside agency. Step 10 is the identification of the project team members and in-house trainers. During this step initial project(s) should also be identified. Step 11 consists of the scheduling of Level I training (introduction to the philosophy of quality management). Step 12 is the scheduling of Level II training for management. Level II consists of an introduction to group problem-solving (working as part of a multidepartmental team), the use of the Shewhart cycle, and the seven basic graphic methods to depict data and make decisions, and how these procedures will be implemented in the organization. Step 13 consists of the scheduling of Level II training for project team members. The KIT process control coordinator will coordinate the scheduling for the training of the various groups. This person will also monitor the training and subsequent efforts and assist in monitoring the activities of the project teams and the various QMBs.

4.3 Training of Quality Management Boards and Project Teams

This phase of the implementation process consists of the actual training of all relevant participants. The specific composition of the training curricula are described in another source (Hulton, 1985). A brief description of training was also provided above. In addition to lecture/discussion presentations, these training sessions can be combined with invited speakers, use of video tapes, lists of readings, panel discussions, and other media techniques.

Step 14 is Level II training for ESC members, QMB members, KIT, and Trainers. They should all have received Level I training at an earlier date.

Step 15 is Level I training for project team members. This training should be conducted by some combination of ESC, QMB, and KIT members. The reactive roles each of these groups plays should also be discussed and reported to the project team members.

Steps 16 and 17 consist of Level II training for project team members. Part of this training consists of teaching department members ways to work cooperatively on specific problems. The key component of this process is to minimize "finger-pointing" and attributions of faultfinding with particular individuals or departments. Techniques such as process identification through the use of sociotechnical analyses and/or nominal groups have been successfully used in the past (Step 16). Step 17 is the actual Level II training for project team members. As training continues throughout the organization, fewer management people will be involved due to the fact that many of them will have received training at an earlier date. Subsequent training will be, (a) problem generated, and (b) offered only to those who will be putting it to immediate use.

4.4 Quality Improvement Activities

This phase begins the process of actually working on identified problems. Activities include discussions of the processes that will be investigated, the data collection procedures, and the methods of documentation.

Step 18 is the final selection of the problems, projects, or processes that will be investigated. The activities selected will have been discussed previously, because of the prior selection of the project team members. At this point, the final selection from among the possible choices will be made.

At Step 19 the project teams begin their investigation of specific processes. These activities include a thorough discussion and description of the problem/process. Assignments will be given to team members to identify the appropriate data to collect, how it will be collected, who will collect it, and how it will be reported to the group. After the information has been collected and the appropriate action has been determined, if it is within the authority of the project team, then they will have the responsibility to make some changes. If a decision is required from a higher level within the organization, then the proposed action will be relayed to the next level QMB and so on until the appropriate decision-making board has been reached.

Steps 20 and 21 consist of ESC and QMB support and review of project team activities, respectively. The process control coordinator will provide much of the liaison between these groups. While the support will consist of allowing members time to participate in team activities and in answering questions relevant to the activities, a primary focus will be to emphasize that resolution occur at the lowest possible level. The areas of responsibility will have to be clearly defined for project team and QMB members. If decisions can be made at the project team level, members will need to be encouraged and supported to make such decisions. Accordingly, many initial decisions will need to be monitored and reviewed with the project teams in order to clarify areas of responsibility and feasible actions.

Steps 22, 23, and 24 all consist of documentation activities. Appropriate procedures to record the actions and activities of the ESC, the QMBs, and the project teams, respectively, need to become an integral part of those groups' methods of operating. In order to ascertain the success of the implementation, documentation must be maintained. Adequate comparisons, such as before and after, or overtime, cannot be made unless a record is kept of what has been and is occurring. Special forms can be developed on which activities, notes, and action items are recorded. Whatever method is used, it is important that a record is maintained to document procedures. This information subsequently can be used for archival, evaluation, and/or descriptive purposes.

4.5 Expansion of Quality Implementation Activities

As processes and problems are addressed and resolved through the project teams and the QMBs, actions should be taken to institutionalize the process improvement activities which were successful. The efforts that were particularly useful and those that were not need to be identified so that future activities may be based on these learnings from prior experience. As the quality management effort begins to reduce nonproductive work behaviors and improve the indicators of effective performance, not only the ESC, but other organizational areas will benefit form this management approach. Now the activities associated with expansion begin. These activities will need to be closely coordinated in order to obtain maximum benefit from the previous actions.

Step 25 is the point at which the ESC institutionalizes successful quality management efforts. This action includes the documentation of formal policies or instructions and the procedures followed, the ways in which problems were identified, how teams were able to begin working together to "solve" problems rather than finding fault with others (individuals and/or departments), training procedures, documentation tips, and so on.

Step 26 is closely tied with the previous action, and consists of the KIT developing a specific implementation plan. Certainly the activities contained in the present plan may be used. But each organization is unique and certain aspects may be more effective in one organization and not in another. Thus, one of the tasks of the KIT is to incorporate what has been found to be most effective, and use that information in updating the present plan and modifying it to best use the resources of the host organization.

Step 27 consists of the selection of additional sites (shops, departments, divisions) which will serve as the next locations for the implementation of the quality management efforts. Initial selection of sites should be done very carefully to ensure maximum benefit for the host organization.

As new sites are selected, new project teams and team members must be identified. This is Step 28. Also included during this time period is the scheduling and implementation of training (Levels I and II) for these uninitiated individuals. The actions associated with this step is similar to the actions described in phase 3.

As the success of the quality management strategies becomes clear, efforts should be initiated to develop and implement the activities for the entire organization. This requires the utmost care in coordination and execution. Step 29 is the point at which the ESC begins to plan and follow this organization-wide strategy. It is also at this point that individuals with previous QMB or project team experience can assist in the organization-wide implementation. This is Step 30. Some of the activities for these individuals include translating quality management principles into organization-relevant actions, developing ways to best implement quality management principles, identifying impediments and solutions to quality management activities, performing technical analyses of future work processes, and developing and collecting baseline process measures for future sites.

4.6 Support Activities

In addition to the direct actions required to establish and implement organizational change, specific activities are also needed in order to maintain and support the new system. The quality management orientation discussed previously requires the development of and access to a variety of resources. These resources should include human contacts, individuals who are easily approached and able to provide answers to questions in a non-threatening way. The other type of resource should allow individuals to absorb material at their own pace and should include a variety of media presentations in the form of library materials, videotapes, and additional training through attendance at seminars, conferences, and meetings.

Step 31 consists of establishing a library comprised of quality management materials. Someone within the organization, such as a training coordinator, should be assigned the task of maintaining this library. The actions required would be to gather and maintain relevant articles, books, audio and video tapes, and paper presentations. Additionally, maintaining lists and descriptions of conferences and meetings would also be an important function. A further activity would be to establish contacts with other organizations which engage in similar management approaches in order to maintain links and open channels of communications.

Steps 32, 33, and 34 have all been discussed previously. They consist of selecting in-service trainers, providing them with a thorough understanding of the quality management and problem-solving approaches, and selecting an in-service statistician who will be able to direct and advise in the data gathering and analysis phases as well as in providing advanced statistical training.

Step 35 consists of developing an informational or public relations capability throughout the entire organization. While it is listed as Step 35, this activity should begin early in the effort. A plan for disseminating information about the implementation activities to the entire organization should be developed early in the process. In-house publications, announcements, and general discussions can serve in this capacity. A method of reporting activities and results needs to be developed. A specific individual should be selected and given the responsibility for developing such a program.

5 TRAINING PROGRAM FOR NARF TRAINERS

A three-level training program (see Figure 3) has been developed for NARF trainers. Level 1 provides an awareness about and an overview of the quality management approach, and is comprised of two parts: (1) an introductory overview, and (2) a four-day seminar. Level 2 provides guidelines for implementation and shows how to put the TQC principles to work using the Plan-Do-Check-Act cycle and the seven basic graphic techniques. Level 2 is tricomponential: (1) a workshop on guidelines for implementing a quality improvement system, (2) a video course presented by Dr. Myron Tribus on action plans for implementation, and (3) a classroom course on structured problem solving and the basic graphic methods. Level 3 covers advanced problem solving and analytic methods.

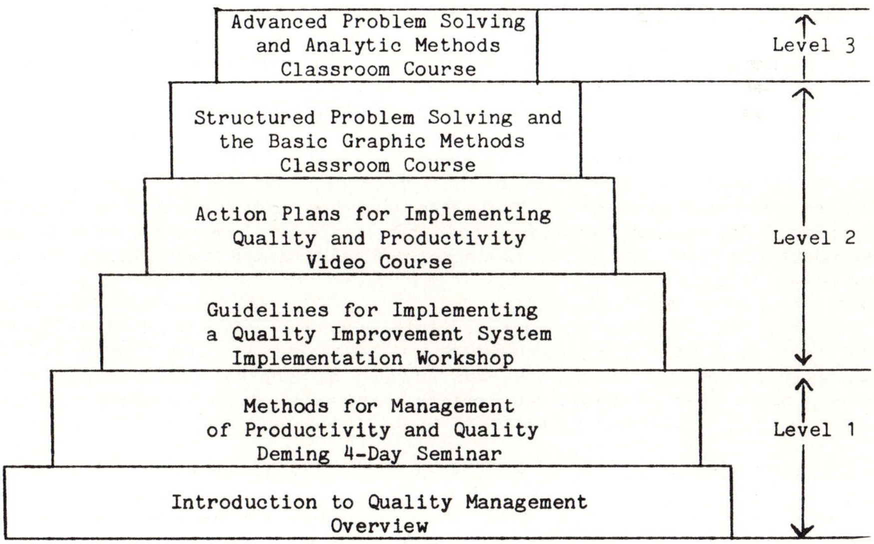

Figure 3. Training Program for Trainers

A booklet of selected readings has also been developed for the trainers. The booklet is divided into three parts. Section 1 readings supplement the information given in the Level 1 courses. The articles in Section 2 deal with implementation issues and the fundamental principles and techniques of statistical process control. Section 3 is comprised of additional articles that are particularly useful for trainers. A description of the training program courses follow:

Level 1 is an overview and seminar:

o Overview. This 3-4 hour session provides an introduction to Dr. W. Edwards Deming - the man and his role in the quality revolution - and the key concepts of his approach to quality management.

o Four-day Seminar. The aim of this seminar, Methods for Management of Productivity and Quality, presented by Dr. Deming, is to teach what top management must do to improve quality and productivity. Dr. Deming discusses his Fourteen Points and provides many examples and practical applications drawn from over fifty years of experience. Seminar participants receive a copy of Dr. Deming's book, Quality, Productivity, and Competitive Position.

Level 2 involves an implementation workshop. The following topics are covered in this one-day workshop, Guidelines for Implementing a Quality Improvement System: Top management orientation, management and administrative structure, training of quality management boards and project teams, quality improvement actions, expansion of the system, and support activities.

o Video Course. This MIT-produced video course, Action Plans for Implementing Quality and Productivity, shows how to put TQC principles to work. In the tapes, Dr. Myron Tribus, Director of the Center for Advanced Engineering Study, MIT, discusses such topics as: Your commitment to stay in business, how to get quality without inspection, and pinpointing your problems with statistical methods. The course consists of three 30-minute videotapes presented in the tutored video instruction mode.

o Classroom Course. The thrust of this ten session classroom course, entitled Structured Problem Solving and the Basic Graphic Methods, is two-fold: (1) The Plan-Do-Check-Act cycle is described and set forth as the foundation for a structured problem solving approach. (2) The seven basic graphic techniques are introduced, utilized, and applied within the Plan-Do-Check-Act framework. The first part of each session is a lecture; the second part is a hands-on problem solving laboratory. Many NARF examples and practical applications are provided. Additional sessions for the trainers are provided as necessary dependent on the trainer's teaching and statistics background, skills, and experience.

Level 3 is a classroom course. In this course, Advanced Problem Solving and Analytic Methods, the requirements, application, and interpretation of statistical control charts will be covered in depth. Participants will also become acquainted with the fundamental concepts of experimental design. The textbook for this course is the AT&T Statistical Quality Control Handbook.

6 LESSONS LEARNED

Even though the NARFs and NALC are only one year into a many-year effort to install the TQC approach, it is already evident that improved quality has been gained in targeted areas through better quality management, use of statistical quality control techniques to discover the root causes of quality problems, and involvement of workers and managers to solve these basic problems using their collective wisdom and experience. Both workers and managers must be taught statistical quality control techniques and simple, powerful statistical methods. We have seen that the diligent application of the TQC approach through basic statistical methods such as both attribute techniques--percent or number nonconforming and control charts for percent defective--and variable techniques --X bar and range charting, tests for stability and evidence of control, frequency distribution, and estimation of the probability distribution function (PDF), and resolution of problems to bring the process into control and within specification--have been very powerful in detecting and analyzing specific problems as well as monitoring the process as solutions are applied as feedback data to the group.

To enhance effective communications and stimulate quality and productivity, an action group made up of selected members of the top management team was formed. Known as the NARF Cherry Point Quality and Productivity Action Group (QPAG), it has the responsibility for implementation of the TQC program from the top. The function of the QPAG is the establishment of TQC implementation policy and oversight of the TQC implementation process. Additionally, the QPAG forms the top layer of a chain of interlocking Quality Management Boards (QMBs) which extend from the top to the bottom of the organization. A chain of interlocking working groups, involving three levels of management down to the shop floor, provides a responsive, participative, multidisciplinary, multidepartmental organization to allow improved operations changes around the NARF. The various subgroups are formed to study problem areas and submit recommendations for improvements to the next higher group if the decision authority is beyond their level of responsibility or implement solutions if it is within their collective areas of responsibility, which is the normal mode. This structure, overlying the existing chain of command, significantly improves internal communications in both vertical and horizontal directions. The QPAG has selected the initial cadre of TQC facilitators and key implementation team members. A complete training plan has been developed and is being implemented. The QMBs are currently working towards the definition and localization of the problems in their areas and will shortly begin the appointment of members to the Process Improvement Teams (PITs) to study the problem process.

Most NARF Cherry Point managers and supervisors have had some training in graphical and statistical methods and a course has been developed at NARF request by Craven Community College in New Bern, North Carolina for those that have not. As discussed earlier, 85 NARF managers and supervisors attended the four-day Deming seminar in February 1986 in Jacksonville, Florida. Training for all managers and supervisors, in the form of short seminars, will have been conducted by the Naval Personnel Research and Development Center (NPRDC) with assistance from the NALC, and fully supported shop floor level QMBs will be in place by 19 December 1986. This date is not the end of the implementation phase, but rather the end of the new <u>beginning</u>.

Several QMBs are already operating at the shop floor level, including the secretarial Employee Participation Group (EPG) and the Turbine Engine Blade/Vane Rework Shop. These groups, and others, are endeavoring to improve their processes and some successes are evident. The following is a discussion of the successes of the secretarial EPG.

7 EXAMPLE OF THE TQC PROCESS

7.1 Naval Messages

Naval messages are a highly responsive means to officially communicate with other organizations. NARF Cherry Point, in order to alleviate problems with delays in the processing of outgoing messages, frustration, possible damage to equipment, and decreased productivity, and to protect the NARF's reputation for quality and accomplishment of its mission, selected outgoing message quality as an area of concern for the secretarial EPG. The secretarial EPG listed the following highest error rate message error problems:

o Addressee unidentifiable or invalid; unauthorized plain language address (Type 4).

o Classification line incorrect (Type 19).

o References incorrectly typed (Type 17).

o Unauthorized or invalid character in heading or text; use of lower case letters; unacceptable/intermixed typewriter font; text single spaced (Type 5).

o Crooked message forms (Type 21).

o Lack of proper equipment to type messages (Type 20), and

o Inappropriate/inferior brands of ribbons, printwheels, fonts, etc., for equipment used to type messages.

These were then ranked by type of error versus the percentage of total errors. The Type 4 error represented 47 percent of all errors. Others ranged from 10 percent (Type 19) to 5 percent (Types 20 and 21). Solutions, therefore, concentrated on Type 4 errors. The causes found by the secretarial EPG included:

o Lack of current USN Plain Language Address Directory (PLAD) and Routing Indicator File Data Management Report.

o Lack of formal in-house training for all drafters and typists plus the high turnover of personnel.

o No central point of review prior to releasing messages from NARF to the Cherry Point Communications Center.

o Lack of sufficient/quality message drafting tools.

o Possible lack of overall pride and interest in outgoing messages.

The solutions recommended were:

o Request that the NARF be included on automatic distribution for current printouts of the Routing Indicator File Data Management Report directly from the Marine Telecommunications Center at Camp LeJeune. Also request assignment to direct distribution lists for Army and Air Force directories. Request a program be written for Office Automation (OA) equipment for message preparation. Establish control file on-line for correct addresses.

o Establish a mandatory formal in-house training program.

o Appoint liaison between NARF and Cherry Point Communications Center and develop tracking system to track error rates (Figure 4 illustrates this tracking).

o Demand high quality tools to accomplish the job. Report unacceptable forms (printed/cut/crooked) to the Government Printing Office. Print messages only on tractor-fed forms. Print messages only on plain white paper cut to correct size.

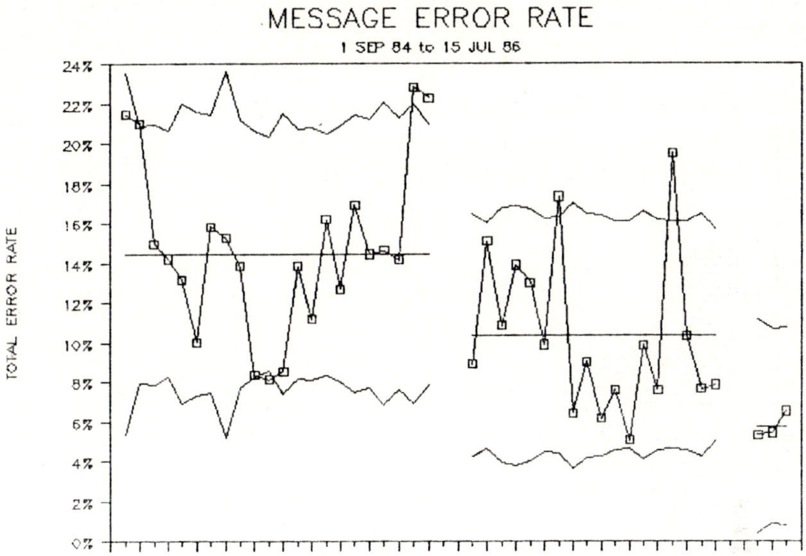

Figure 4. Message Error Rate Control Chart

o Stress responsibility of the drafters to use current references for correct addresses, etc. Establish/tighten departmental procedures and reemphasize duties of department secretaries concerning message preparation/outgoing correspondence.

o Establish preprinted message worksheets for facility use.

o Reduce the number of messages; use speedletters, letters, telephone.

Obviously, the secretarial EPG is well on the way to solving message error problems with assistance from management. Current problems include purchase requisitions, personnel forms, training forms, and travel request forms. Training is being emphasized to keep solved problems from becoming problems again. Office Automation with the establishment of on-line programs with help screens, and an in-house formal OA user training program was also identified as a long-term solution and is in place.

7.2 H-46 Aft Transmission Reliability Improvements

The H-46 aft transmission has been characterized as having low reliability, a high number of Quality Deficiency Reports (QDRs) from the fleet, and general user dissatisfaction with the product as received from overhaul. An analysis of the failure data (see Figure 5) confirmed a low mean time between failure (MTBF) with the main causes identified as high vibration and water contamination. The water contamination was identified as a fleet problem caused by the method used to wash the aircraft. Instructions on proper washing procedures were issued to the fleet by NARF Cherry Point to eliminate the problem.

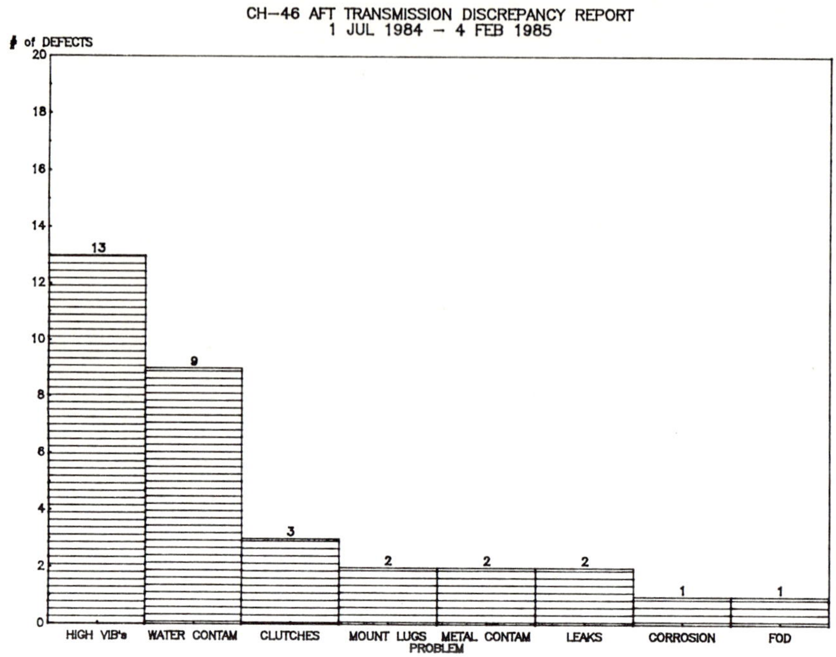

Figure 5

A study of the high vibration levels revealed two causes: input idler out of balance and misalignment of mix box bores. Further study revealed that the problem of misaligned mix box bores was not related to any one cause for the high vibrations. The primary concern, however, was the specification on the parallelism between the gears. At some stage in the past, the manufacturer, Boeing-Vertol, had increased the bore placement tolerances as these tolerances could not be held by that generation of machine tools. New generation numerically controlled machines and inspection equipment, however, can meet the original specifications. All transmission cases are now being reworked and inspected to this original specification, producing a significantly improved transmission. The study of the out of balance idler hubs revealed that a significant number of idler hubs were outside of the specification limits of 0.52 grams (see Figure 6). What was even more significant was that the readings were not repeatable, with variances between the production run and the repeatability run (as shown in Figure 7) averaging 0.17 grams difference, with one sample even exceeding the specification value. An improvement to the process was obviously required. After a change to the process, maintenance of the balance equipment and training for the operators was accomplished, a dramatic improvement in the process capability was achieved, with all idler hubs now being significantly within the out of balance specification (see Figure 8) and much greater repeatability, averaging 0.10 grams difference, between the production and repeatability runs (see Figure 9). Study of this and other problems affecting H-46 aft transmission reliability are continuing to improve product quality.

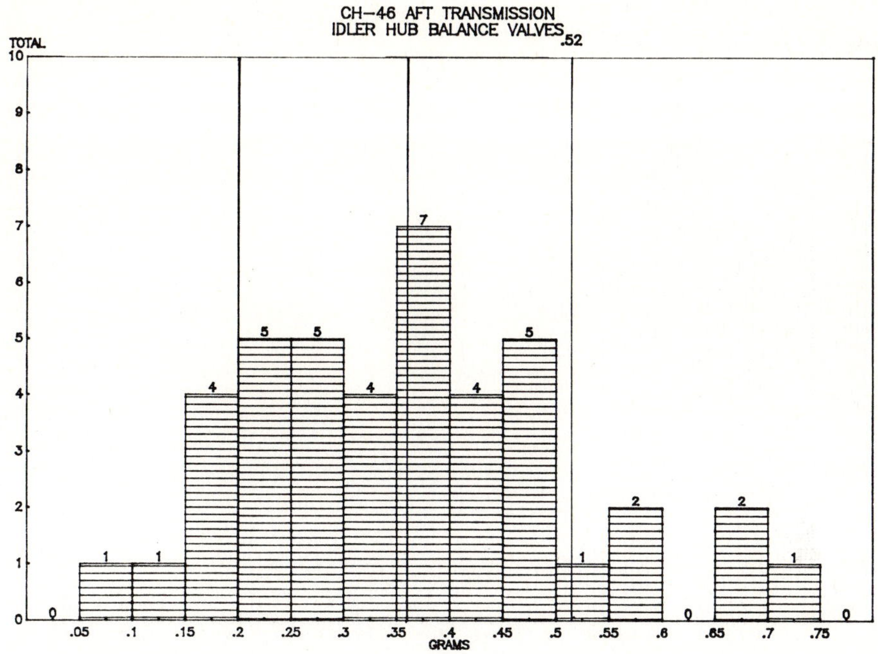

Figure 6

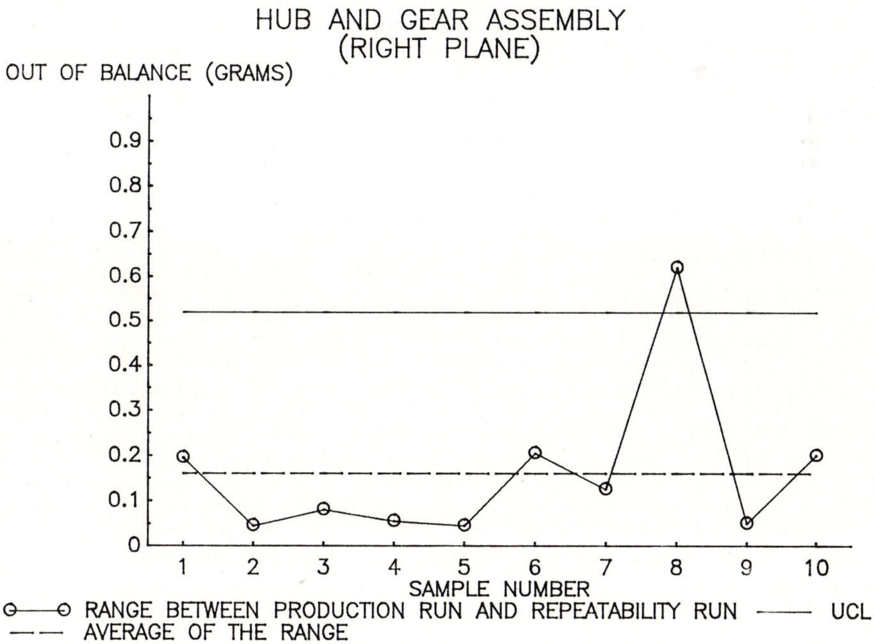

Figure 7

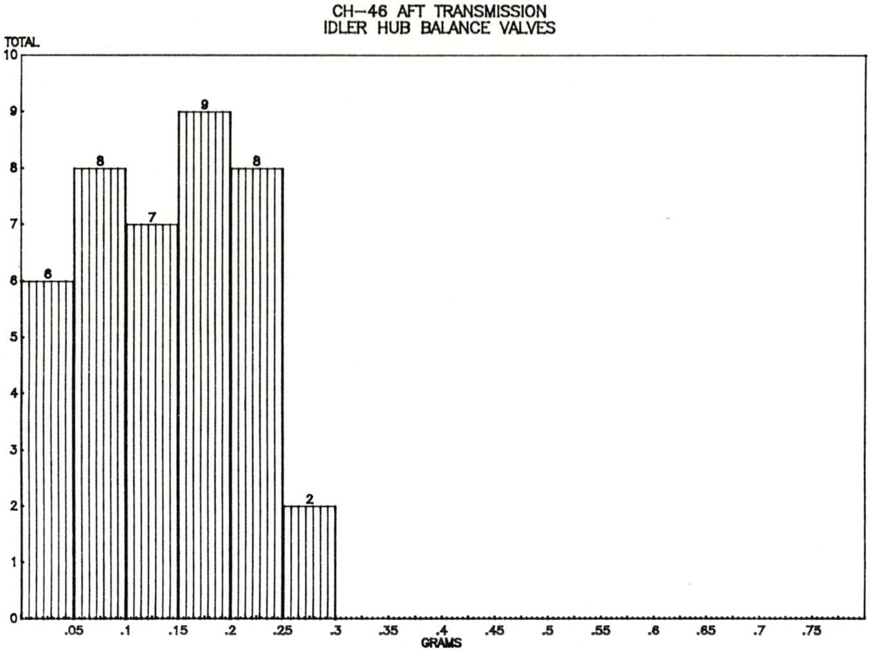

Figure 8

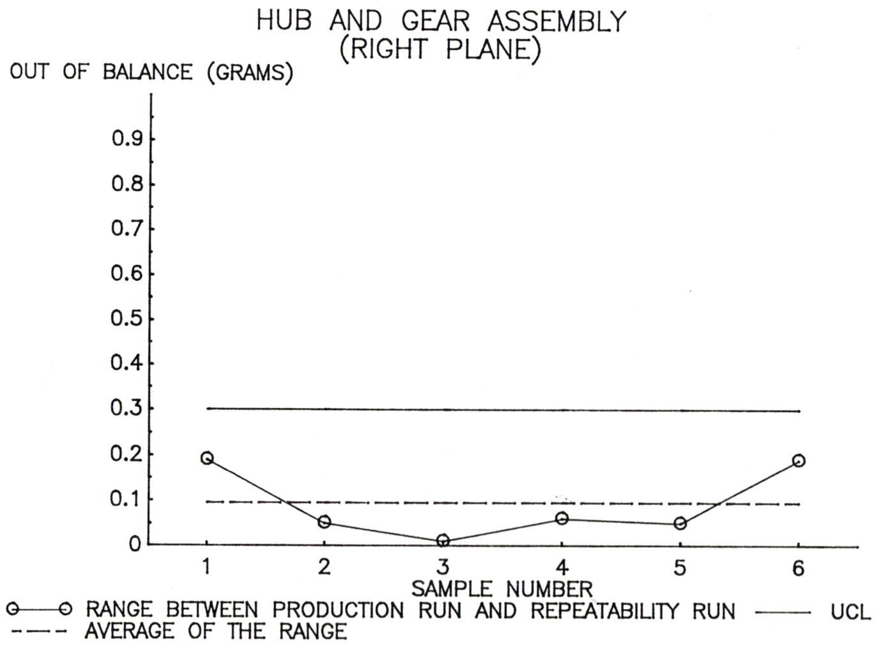

Figure 9

7.3 Productivity Improvement - H-46 Cabin Floorboards

The H-46 Final Assembly Shop was experiencing a high rejection rate on H-46 cabin floorboards. The shop had to return the floorboards to the shop that reworked the floorboards several times before they fit properly, wasting many man-hours and expending more material for this rework. The assembly line asked for assistance in solving this quality problem. The problem was studied and the solution developed was to modify the fixture used to cut and drill the boards to include a pre-set stop for the proper lengths. The result is floorboards of the right length, drilled in the right place, that fit the first time, every time. The floorboard problem was initiated, worked, and implemented all at the shop floor level. The end result is obviously less rework, more time in the shops to produce a quality product, and significantly lower costs.

7.4 T-62-11 and -27 Gas Turbine Compressor Process Improvements

Process problems in the T-62 gas turbine compressor (on board auxiliary power units for aircraft) were indicated by the high number of test cell rejects because of low oil pressure. A study of the rejected units revealed several causes contributing to low oil pressure. These contributing causes were: 75 percent defective rate on relief valve springs assembled into the units, valve covers were not flat, valve seat too deep so that springs could not seat the pressure relief ball, and an unauthorized, substituted plastic pipe elbow fitted in the field in the place of the proper supply system elbow.

In an entirely shop level interdepartmental team made up of artisans and supervisors, solutions were identified and implemented by the team at their own level. The solution included development of capability and testing of all relief valve springs prior to assembly, notifying supply and the vendors of the 75 percent reject rate and following up to assure improved vendor quality performance, machining the valve cover plate flat before installation, rework valve bodies prior to use to correct the valve seating problem, and replacement of the unauthorized plastic elbow. These solutions resulted in a significant drop in test cell reruns required to sell T-62-11 and -27 gas turbine compressors and should result in a more reliable product to the fleet that will require fewer returns for depot overhaul because of low oil pressure. Study is continuing to solve several other test cell reject problems to improve quality.

8 CONCLUSIONS

With the collapse of oil prices worldwide, the U.S. situation is likely to get much worse before it gets better. Global competition will become even more intense as Japan uses its new found bonanza to penetrate even more markets. The Japanese have taught those in other countries, especially the Republic of Korea and Taiwan, the TQC approach to improving quality, productivity, and their competitive position. These countries, in addition to now lower energy costs, have the additional advantage of very low labor rates. To improve our competitive position, U.S. companies must improve quality and productivity while reducing manufacturing/services costs. This is not a question of working harder, but of working smarter and "doing it right the first time, everytime" to reduce scrap and wasted efforts. We can never compete with the Japanese or others that have adopted the TQC approach.

Several of the gurus of TQC state that there cannot be any deviation from the plan. Deming's 14 points for management and 7 deadly sins are examples of the totality of the TQC program. Many, upon hearing these common-sense thoughts and rules, either reject the TQC approach because it doesn't fit the rules, regulations, and laws of their environment, or rally to instantly change their environment. Initially, however, there are many areas within the control of the organization that can be subjected to the TQC process. These internal problems represent over 85 percent of the problems faced by the organization. The TQC statistical techniques can track the other 15 percent and serve as a basis for action based upon data for those creating these environmental problems and serve as a basis for change in these external constraints.

As Deputy Secretary of Defense William H. Taft recently wrote, "The single most important initiative we will pursue in the future will be to extend our concentration on quality. I am advocating a change in our attitude about quality." Secretary Taft, then quoting Dr. Deming, went on to state that, "We can no longer live with commonly accepted levels of mistakes, defects, material not suited for the job, and people on the job who do not know what the job is." The NALC and the NARFs have already begun changing our thoughts on quality and, as a result, to change the way we manage.

9 REFERENCES AND FURTHER READING

Gitlow, H. S., and Hertz, P. T. (1983, September-October). Product defects and productivity. Harvard Business Review, 61, 131-141.

Garvin, D. A. (1983, September-October). Quality on the line. Harvard Business Review, 61, 65-75.

Deming, W. E. (1981-1982, Winter). Improvement of quality and productivity through action by management. National Productivity Review, 1, 12-22.

Tribus, M. (1983, June). Reducing Deming's 14 points to practice. Cambridge, MA: Center for Advanced Engineering Study, MIT.

Tribus, M. (1983, October 6-7). Creating the quality company through company-wide quality control. Cambridge, MA: Center for Advanced Engineering Study, MIT.

Tribus, M. (1982, April). Deming's Way. Cambridge, MA: Center for Advanced Engineering Study, MIT.

Tribus, M. (1982). Productivity...who is responsible for improving it? Cambridge, MA: Center for Advanced Engineering Study, MIT.

Tribus, M. (1983). Deming's redefinition of management. Cambridge, MA: Center for Advanced Engineering Study, MIT.

Tribus, M. (1982). Managing to survive in a competitive world. Cambridge, MA: Center for Advanced Engineering Study, MIT.

Tribus, M. and Tsuda, Y. (1983). Creating the Quality Company. Cambridge, MA: Center for Advanced Engineering Study, MIT.

Metz, E. J. (1984). Managing change: Implementing productivity and quality improvements. National Productivity Review, 3(3), 303-314.

Garvin, D. A. (1984). What does "product quality" really mean? Sloan Management Review, 26(1), 25-43.

AT&T Technologies (1956). Statistical Quality Control Handbook (Section 1). Charlotte, NC: Author.

Siegel, J. C. (1982). Managing with statistical methods. (SAE Technical Paper Series No. 820520). Warrendale, PA: SAE.

Freund, R. A. (1985, January). Education and training in the workplace for statistical process control. Quality Progress, 19-21

Monda, M., Quality Management: An Annotated Bibliography, Quality and Productivity Group, NPRDC.

Introduction to Quality Management: Selected Readings, Quality and Productivity Group, NPRDC.

Houston, A., Shettel-Neuber, J., and Sheposh, J., Management Methods for Quality Improvement Based on Statistical Process Control: A Literature and Field Survey, Quality and Productivity Group, NPRDC, 1986.

Guidelines for Implementing a Quality Improvement System by Quality and Productivity Group, NPRDC.

Action Plans for Implementing Quality and Productivity, Center for Advanced Engineering Study, MIT, 1984, videotape series.

"Military Facility Adopts Total Quality Management" American Productivity Center Case Study 53 on the Naval Air Rework Facility, North Island, San Diego (ISSN 0741-6423)

Productivity Management Frontiers — I, edited by D.J. Sumanth
Elsevier Science Publishers B.V., Amsterdam, 1987 — Printed in The Netherlands

TECHNOLOGY, PRODUCTIVITY AND LABOR RELATIONS

M.G. KOLCHIN and T.J. HYCLAK
College of Business & Economics, Lehigh University
Bethlehem, PA 18015 (U.S.A.)

ABSTRACT
Kolchin, M.G., and Hyclak, T.J., 1987. Technology, Productivity, and Labor Relations.

This paper investigates the importance of labor relations in implementing various forms of technological change by studying four firms who have recently implemented changes in equipment, process, and management style. In all four of these firms, the technological changes introduced resulted in productivity gains and, in each case, labor-management relations were critical.

1 INTRODUCTION

In the recent past, the United States has been faced with a productivity problem that has resulted in a loss of competitiveness in world markets. In response to this problem, U. S. industry has turned to technical solutions, often without regard to the impact on labor relations. As noted by Shaiken (1985) and others, such actions may have disastrous results. Without the support of the workforce in implementing technological change, such changes may be costly and ineffective.

In addition, there is often confusion as to what constitutes technological change. It should be remembered that technology can be defined as the knowledge, processes, tools, and materials necessary to transform raw materials into finished goods and services. With this in mind, technological change can occur in any of these areas -- it does not have to occur only through the installation of new equipment. Additionally, productivity improvement can be obtained in any of these areas and often with less initial cost through changes in process rather than equipment.

This paper investigates the importance of labor relations in implementing various types of technological changes by studying four firms who have recently implemented changes in equipment, process, and management style.

2 TECHNOLOGY AND TECHNOLOGICAL CHANGE

All too often TECHNOLOGY is seen as a panacea to all the productivity ills plaguing U. S. industry. Yet, what is meant by new technology and technological change is not clear. Most frequently people relate the term new technology to new equipment. For instance, when discussing the concept of technology, the image of a highly automated plant comes to mind. However, most organizational theorists define the concept of technology much more broadly. For example, Perrow (1970) defines the concept as pertaining not only to the actual production process but, also, to the processes associated with the procurement of inputs, such as labor, materials, and capital, as well as processes associated with the disposition of finished goods and services. In addition, Perrow would include coordination activities involved in the three phases of input, transformation, and output. As can be seen by this definition of technology, the concept is a much broader one than simply new equipment.

Within this broader context, technology not only includes equipment but, also, such things as shop-floor workflow and management style. With this in mind, investigation of new technology as a means of improving productivity should include the investigation of new shop-floor practices as well as improvement in labor relations. In the U. S., however, companies look to new equipment for productivity improvement. Why is this the case? Cole (1979) suggests the reason for this is the proclivity of industrial engineers, economists, and management to seek technical solutions for productivity problems. They greatly underestimate the potential of harnessing worker cooperation to raise productivity. In essence, they are looking for quick fixes to the organization's lagging productivity.

There are a number of problems associated with this strategy. In some cases, it may simply not be appropriate. For instance, in some mature industries the cost of new equipment precludes its use. More importantly, research on turnaround strategies in mature industries suggests that cost reduction and efficiency strategies are more effective than innovative strategies (Hambrick & Schecter, 1983). One of the more effective turnaround strategies for firms in mature industries was improving employee productivity according to Hambrick and Schecter.

The strategy of looking for technical solutions for productivity problems may also ignore the human factors involved in implementing new technology. This can be disastrous as organizations are sociotechnical systems and a change in the technical systems in the organization must be accompanied by appropriate social changes.

3 WORKER INVOLVEMENT IN IMPLEMENTING NEW TECHNOLOGY

There exists an extensive literature that supports the notion of the critical importance of worker involvement in implementing new technology. For instance, a recent review of this literature by Hyclak and Kolchin (1986) concludes, "...worker involvement in implementing new technology results not only in less resistance to technological change but, more importantly, greater support and commitment from workers for technological change" (p. 1). It would seem that this commitment from workers is absolutely necessary if new technology is to be successfully implemented. Without support of a company's employees, millions of dollars may be spent with resultant long delays and idle equipment.

In order to obtain worker commitment, firms must establish a supportive labor relations climate. One of the ways of accomplishing this is to involve workers in the early stages of any decision to make technological changes. The level of involvement can range from early notification of impending change and efforts to convince workers and the union of the necessity for change to full participation in the decision-making process. The lesser level of participation, early notification, can result in lessened resistance to the announced change while, on the other hand, full participation can result in inducing contributions from workers that will facilitate the implementation process.

Hyclak and Kolchin (1986) suggest that greater worker involvement can have significant benefit in reducing the costs associated with the implementation of new technology. However, there is still a hesitancy by most U. S. firms to involve workers in these decisions. Why is this the case? In all likelihood, this hesitancy comes from the general unwillingness of U. S. industry to look to workers as sources of productivity improvement. As noted earlier, technical solutions are sought for productivity problems. It would seem, however, that, without a supportive labor relations environment, such technical solutions may be counterproductive.

4 IMPORTANCE OF A SUPPORTIVE LABOR RELATIONS CLIMATE

In order to further develop the notion of the importance of a supportive labor relations climate to the successful implementation of technological change, the results of four case studies are presented which examine the relationship between technology, productivity, and relations. These case studies examine technological change at each of the levels described earlier: new equipment, change in workflow, and change in management style. The first case deals with incremental changes in equipment in an existing plant and production line. The second case deals with massive changes in equipment and process in automating an existing plant. The third case deals with a change in workflow process in one building of a multi-building facility which allowed for the introduction of cell technology. The final case deals with a dramatic change in management style that was implemented in conjunction with a move to a new plant.

In all four of these cases, the technological changes introduced resulted in productivity gains and, in each case, labor-management relations were critical.

4.1 Case 1: Incremental Equipment Changes

This case concerns the efforts of a headlamp manufacturer to maintain its competitive position in a mature industry that was experiencing significant product innovation by making incremental changes to existing equipment. In this case, the option of acquiring the latest available technologically advanced equipment was not possible. This was true for two reasons. First, the cost of such acquisition was beyond the resources of the division. Second, massive changes would have likely resulted in severe job displacement and would have been strongly resisted by the union.

Faced with the dilemma of possibly going out of business, the plant manager had to develop some method of keeping up with the competition. This meant decreasing costs and increasing employee productivity. If he were unable to accomplish these objectives, the corporation was seriously considering closing the operation and moving production to its southern operations.

In order to achieve these objectives, the plant manager undertook a two-pronged strategy. First, his engineers and designers worked to adapt existing equipment to make it more productive. In some cases, this would mean automating certain portions of the production line. Such automation would invariably mean the loss of a small number of jobs. Knowing this could mean trouble with the union, the plant manager worked on the second part of his strategy, maintaining a good labor-management climate.

In implementing the second part of his strategy, the plant manager was fortunate that the plant had a history of excellent labor relations. It was now his task to insure that this climate was maintained. The plant manager did this by keeping the union informed of changes as they were being developed. This was accomplished by monthly meetings with the local president to inform her of the company's current market position and profitability. In these meetings, the plant manager stressed the importance of maintaining market share to avoid closing the plant. In order to maintain market share, it was stressed to the local president that some automation would be required which would result in the loss of a few jobs. The plant manager and local president agreed that, in order to diminish the impact of these changes, temporary employees and employees with low seniority would be used in areas of the plant that were likely to be automated. As these areas were automated, these employees would face temporary layoffs or act as floaters and be moved to other areas in the plant that were likely candidates for automation as production warranted. This was designed to make the impact of the changes less noticeable and disruptive and to keep the core of the workforce, people who had been working in the plant for over twenty years, largely unaffected.

The local president would meet with the shop stewards after meeting with the plant manager and keep them abreast of what changes were occurring. By keeping the union informed and providing a measure of job security for the core of the workforce, the plant manager was able to gain the support of his workers in making the changes necessary to remain competitive. The results of this support were dramatic. The area where the effects of the changes were most noticeable was in reducing scrap. The scrap rate dropped from over 14 percent to just over five percent in just one year, a period that included four months in a row where plant records were achieved. This performance was so dramatic that not only was plant kept in business but the corporation even moved some of its southern plant's operations back to this older plant.

4.2 Case 2: The Automation of an Existing Plant

The second case deals with the Canadian subsidiary of one of the big three auto makers. In this case, it was decided that this subsidiary would begin to build a new product that was to be introduced in the near future. In order to competitively build the best quality new product possible, the corporation felt it was necessary to completely renovate the existing facility and automate the production process.

Here again, corporate management was concerned that such a drastic change would evoke strong union resistance and developed a strategy to insure employee support of such changes. Like the previsous case, early notification was a key element of this strategy. Two years before the anticipated changes, the union leaders were asked for their input concerning the proposed change. Initially, the union exhibited some hesitancy concerning the change. Their concerns centered around the loss of jobs and limiting production to a single product. The corporation argued that, while the proposed automation of the plant would result in some job loss, it was their belief that the productivity improvements made possible by the change would mean increased market share and improved profitability for the plant. Based on these anticipated outcomes, the union was assured that the job loss would be minimal and, over the long run, imperceptible. Also, the corporation argued that they would not be investing in excess of $400 million if they were not certain of success in this venture.

The corporation's enthusiasm and assurances and its eagerness to include the union by notifying them at a very early date seemed to have the desired effect. The union worked with the company to insure the successful conversion of the plant.

Another element of this case which demonstrated the importance of good labor relations was the extensive training undertaken by the corporation to insure that the converted plant would be up to full capacity as soon as possible. Again, two years out from the actual conversion date, the corporation set up a training program that would familiarize all plant employees with the new equipment that would be in the plant after its conversion. As an example of the type of training undertaken, a robot lab which included each type of robot that would be part of the automated production process was set up in a cafeteria and all employees were exposed to the equipment. This allowed the workforce to see the robots firsthand and to become familiar with them before they were actually installed on the production line. In addition, those employees who were to have the responsibility for programming, operating, and maintaining the robots were given extensive training in robotics. In all, this corporation expended close to a million manhours in training before the plant was actually converted.

These actions undertaken by the corporation resulted not only in reducing resistance to this major technological change but actually induced active employee support for the change. The evidence of this support can be demonstrated by the fact that the actual conversion of this plant, including the complete gutting of the existing building and the installation of 125 robots and over ten miles of new conveyor lines, took only 17 weeks and the newly automated plant was up to a two-shift operation only two months after the conversion was completed. This was a truly remarkable feat that could have only been accomplished with the active support of the union and the workforce. The end result of this conversion was the most productive automobile plant at the time. The success of the plant and the product it was building actually resulted in minimal job loss from the original operations and the addition of almost 500 jobs by including the painting operations in the newly converted plant. In this case, good labor relations resulted in benefit not only to the corporation but to the workers and union as well.

4.3 Case 3: The Implementation of Cell Technology

The third case concerns a company that was faced with the loss of a production line unless it could improve productivity. In this case, the company felt that a more flexible manufacturing process was the key to achieving greater productivity. However, before the company could implement a change in shop-floor workflow, a change in union work rules was required.

The particular type of technology involved here is called the manufacturing cell which groups processes by product line. In the cell, workers rotate through the different operations ncessary to make the product which, in this case, is an impeller for pumps. In contrast to the traditional line process for making these impellers, cell technology allowed the company to produce finished product for approximately 40 to 50 percent less in wage costs. This reduction in labor costs resulted from the elimination of setups and the elimination of an archaic incentive system which encouraged excessive overtime. All employees in the cell are paid on a straight eight-hour basis. In this particular cell, there are four operations: first-side machining, second-side machining, deburring and hand operations, and balancing. These four operations are a substantial reduction from the previous ten to twelve operations required from a more traditional line process. In addition to the overall 40 percent reduction in total process costs, there is also a significant reduction in the amount of rework necessary and a greater flexibility in the process that allows the company to operate in a manufacture-to-order environment where, previously, they operated in a assemble-to-order environment. This latter benefit resulted in closer scheduling and reduced inventory.

However, in order to obtain the benefit of cell technology, the company needed to get the union to agree to changes in existing work rules. In essence, there was a need to reduce the number of job classifications from 50 to four. The company secured the local union's support in obtaining these modifications by informing them of the company's plight. The general manager held state-of-company meetings where he disclosed the company's position in the market and its profit picture. He was successful in convincing local union members of the dire plight of the company and the necessity for improving productivity of the plant. The union did agree to modify the work rules in the building where this manufacturing was to be housed. In return, the company agreed to increase the job grade for the positions in the cell. It was their belief that the increased productivity from the change would more than make up for the increased wage rate and, in fact, this was exactly what occurred.

In this case, not only was a different type of technological change implemented, that is, a non equipment one, but, again, the importance of securing worker support was demonstrated. The company involved in this instance did not have the luxury of completely automating the process as it was in a mature industry which required drastic cost reductions in order for it to survive. The supportive local labor environment was critical to making these reductions. An additional benefit of the change to manufacturing cells was the increased employee satisfaction of those operators in the cells. The company and the union have been so pleased by the success of the cell that they have agreed to expand cell technology to other buildings.

4.4 Case 4: Implementing Team Management

The final case reported here deals with a change in management style from a traditional authoritarian style to team management. Team management is a form of worker participation whereby team members are responsible for task-related decisions concerning their areas of responsibility. Under this concept, teams are formed along functional lines around the production process and support services found in the plant.

The company in this case was also in a mature industry and was experiencing significant productivity problems in its existing plant. A decision was made to move one product line from this plant to a new facility. In order to increase plant productivity, it was also decided to experiment with team management. The plant manager assigned to the new facility was a proponent of a more participatory style of management and the move to a new facility was seen as a golden opportunity to test his ideas. Again, as in two of the other cases reported above, the acquisition of new technology was not feasible. The product line in question, oilwell cable, was a mature one and, in order for the company to remain competitive in its world markets, it needed to refocus its efforts on a single product line and to engage in significant cost reduction. The company in question was the industry leader and wanted to maintain this position by being the low cost producer. Since the technological process involved in producing cable was relatively stable, the company was forced to look to increased labor productivity and reduced scrap to maintain its competitive position. Team management was thought to be the key to increasing labor productivity at the new facility.

Team management is an example of a management technology. It is not a quick-fix solution to productivity problems as is often thought to be the case in the acquisition of new technology. Team management requires a complete change in philosophy with regard to labor-management relations. As such, there are considerable start-up costs involved in switching from more traditional styles of management to worker self control as both parties, labor and management, become more comfortable with their new roles. There is also a need for a more cooperative and trusting work environment to evolve in order for the concept to be effective.

This company's experience with implementing team management was no different than could be expected with such a dramatic switch in management style. In the beginning, there was a good deal of mistrust among employees regarding management's motives. Also, when starting up the new facility, only one technical person was brought from the old plant. The rest of the people hired had little experience with the production process involved in making wire cable. As a result, there was a lot of frustration with a resultant high level of turnover. The turnover rate of 12 percent in the first two years of the new facility compared to a national average of 3.8 percent during the comparable period.

These problems highlight the need to treat any organization as a socio-technical system. In this particular case, the plant manager was so busy concentrating on the social requirements of the system that the technical aspects were ignored. In order to address this problem, a compensation system was developed that encouraged employees to master various pieces of equipment in the plant. This action seemed to have the desired effect for the division became profitable by the end of its second year of existence. Turnover dropped from an excess of 12 percent to a range of two to four percent which was more in line with the national average for manufacturing firms. More impressive was the absentee rate which hovered in the range of 2.5 to three percent during the next four years. The national average during the comparable period was closer to 6.5 percent. Productivity was improving steadily as well. This division now enjoyed the highest productivity of any plant in the oilwell cable industry.

5 CONCLUDING COMMENTS

Each of these cases demonstrates the importance of a supportive labor-management relations climate to the effective implementation of changes in technology. In the first three cases, the changes undertaken would not have been possible without the support of the union. It is important to note here that the union in this case is the local union and not the national. In fact, in two of the cases, the auto maker and the pump manufacturer, the national union was not at all enthusiastic about the proposed changes. In both cases, it was the strong relationship with the local that enabled the firm to make the changes they wanted. This suggests that, when talking about labor relations, it is the local plant environment that is important not relations with the national union.

Also, labor relations is a much broader concept than union relations. It includes the direct relationship between managers and shop-floor workers. This is particularly true where there is a change in management style, such as in the case of the switch to cell technology or in the case of team management, where the shop-floor worker is being asked to take on responsibility of self management. A cooperative and trusting atmosphere must exist or be developed before these types of changes can occur.

These cases also highlight the necessity to deal with both the social and technical aspects of the work environment in making technological changes. Both the auto maker and wire cable manufacturer cases demonstrate the importance of adequate and appropriate training in insuring effective changes in technology. It is interesting to note that the auto maker has engaged in an additional $460,000 in training since the conversion in order that they are able to obtain the most from the equipment in the plant.

All of these cases show the value of obtaining employee input and participation in the process of technological change. Such involvement results not only lessened resistance to the change but also results in active support of the change. This support insures the greatest likelihood of successful technological change.

The last two cases also demonstrate the value of so-called non technical solutions to productivity problems. This is particularly important since many firms in the U. S. do not have the resources necessary to acquire new equipment and must seek out other solutions to their lagging productivity. Such solutions may result in significant productivity gains without the expense of new equipment. One caution must be addressed in these types of technological changes in that the results from such changes are not immediately forthcoming. Such changes require a good deal of patience on the part of management if they are to have a chance of being successful.

A very tentative conclusion that may be drawn from these case studies is that productivity may be improved by improving or maintaining good labor-management relations regardless of the type of technological change employed. While these companies represent a relatively small sample, several studies of the automobile industry seem to support this conclusion. For instance, Katz, Kochan and Gobeille (1983) present statistical evidence that firm investments in better labor relations via quality-of-work-life programs paid off in improved worker attitudes. Also, the results of a study by Norsworthy and Zabala (1985) demonstrate that such worker attitudes are significant determinants of productivity in the automobile industry. Given the dramatic technological changes that are occurring in the automobile industry, it would seem that the attention being paid to improving labor relations through such QWL programs is highly justified. The results presented from the four case studies in this paper would suggest that such efforts are generalizable to a larger population of firms in the U. S.

6 REFERENCES

Cole, R.E., 1979. Work, Mobility, and Participation: A Comparative Study of American and Japanese Industry, University of California Press, Berkeley, California.

Hambrick, D.C. and Schecter, S.M., 1983. Turnaround strategies for mature industrial-product business units. Academy of Management Journal, 26(2):231-248.

Hyclak, T.J. and Kolchin, M.G., 1986. Worker involvement in implementing new technology. Technovation, 4(2):143-151.

Katz, H.C., Kochan, T.A. and Gobeille, K.R., 1983. Industrial relations performance, economic performance, and qwl programs: An interplant analysis. Industrial and Labor Relations Review, 37(1): 3-17.

Norsworthy, J.R. and Zabala, C.A., 1985. Worker attitudes, worker behavior, and productivity in the U.S. automobile industry, 1959-1976. Industrial and Labor Relations Review, 38(4):544-557.

Perrow, C.B., 1970. Organizational Analysis: A Sociological View, Brooks/ Cole Publishing Company, Belmont, California.
Shaiken, H., 1985. The automated factory: The view from the shop floor. Technology Review, 88(1):16-27.

7 ACKNOWLEDGEMENTS

The authors gratefully acknowledge the financial support of both the NET Ben Franklin Advanced Technology Center and the Fairchild-Martindale Center for the Study of Private Enterprise of Lehigh University which enabled the authors to conduct the case studies reported herein.

Productivity Management Frontiers — I, edited by D.J. Sumanth
Elsevier Science Publishers B.V., Amsterdam, 1987 — Printed in The Netherlands

AN EMPIRICAL STUDY OF HOW SOME U.K. ORGANISATIONS IMPLEMENT PRODUCTIVITY IMPROVEMENTS

W.N. SHAW
Department of Business Organisation, Heriot-Watt University, 31 Grassmarket, Edinburgh EH1 2HT, Scotland,U.K.

ABSTRACT

This paper reports some of the key findings about implementing productivity changes based on an in-depth study of the role of the internal productivity/efficiency departments in 34 U.K. commercial, industrial and public sector organisations. The paper examines a range of real life issues faced in operating such a department or unit successfully. Issues dealt with include; the type of backing given by top management; performance measures for departments; links with clients; sources of projects or tasks; future areas of productivity related work; definition and diagnosis of problems; implementation of solutions and finally, power, influence and organisational politics.

INTRODUCTION

In the U.K., internal productivity/efficiency departments or units have existed in most industrial sectors since the late 1950's or early 60's. The effectiveness of these units varies from industry to industry and within industries. Increased attention has been focused on these units since 1980 when the present U.K. Conservative government began to emphasise the need for achieving greater value for money from central and local government. Internal productivity practitioners, within the U.K., are usually located in management services (MS) departments or their equivalents e.g. industrial engineering, productivity services, organisation and methods (O & M) or internal consultancy.

The data for this paper has come from over 100 structured interviews with chief executives, line managers and management services managers. In addition information was obtainedfrom questionnaires and documents of various kinds. Although the research took place in 34 U.K. commercial, industrial and public sector organisations the bulk of the data came from 28 of them. The organisations cover a wide range of sizes (from 250 to 30,000 employees) and provide a variety of products and services. Complete details of the background to the research project along with a description of the characteristics of the management services units and their managers are contained in a companion paper presented at the conference (Shaw, 1987).

WORKING RELATIONSHIPS

A number of writers have focused on the issue of working relationships in organisations. Handy (1985), for example, argues that our interactions or relationships with people are affected to a large extent by the way we perceive them, the things they say and what they do. Personality, inter-personal skills and the structure of the situation, as Handy points out, are other factors that affect our working relationships.

In discussing the influence process between specialists and executives Pettigrew (1974) takes the argument a stage further by declaring that difficulties between specialists and executives cannot be attributed just to incompatible personalities, faulty communications and differences in values and attitudes. He believes these to be the surface manifestations of the problem, whilst deeper structural issues are at work. Pettigrew goes on to list these deeper issues as being the design of organisational structures, the conflicts of interest which emanate from inappropriate structural designs, the role problems of specialists and the internal political issues raised amongst those supposed to co-operate in producing organisational change. Mumford and Pettigrew (1975) claim that when more than one group wishes to have responsibility for a set of tasks there are likely to be power battles over where the boundaries of each occupation are to be drawn. Kaplan (1984) after synthesising recent literature on the subject argues that a network of relationships is the complex and richly textured setting in which most managers work.

A discussion of relationships raises many relevant and important questions for management services. For example, what type of backing does senior management give to MS units? What are the criteria by which clients accept or reject the recommendations made by units? What are the activities and issues over which MS units have the greatest influence?

Before commenting on the backing given by top management to management services it is necessary to look at the context in which this backing is given. It was clearly established from the study that the work objectives or goals set by top management for management services were both vague and general. Some 50% of the 28 MS managers interviewed were never given specific work objectives by their superiors, 25% of them were given objectives occasionally and the remaining 25% had objectives set regularly. Turning to financial goals, only 21% of the managers were set any kind of goal in this area and these were generally cost reduction targets. Most management services units were not entirely left to their own devices in that 75% of them reported to senior or top management on either a monthly or quarterly basis.

An important factor in the success or otherwise of management services work was found to be the extent to which senior management were seen to be backing and supporting the MS effort within an organisation. All but one of the MS managers were "completely or reasonably" happy with the backing given to them and their units by top management. Allied to the backing given by senior management is the way that this same group measure the effectiveness of MS effort. It can be seen from Table 1 that although a wide range of measures were used they were all of a qualitative nature. In nine organisations, virtually one third of the sample, the MS manager either did not know what measures of effectiveness were used or no measures of effectiveness existed. It should be noted, however, that a number of senior management groups used more than one factor for measuring effectiveness.

TABLE 1

The measures by which senior management assess the effectiveness of MS units (N = 28)

Measures of effectiveness	No. of mentions
Produce cost effective solutions to problems	13
Ensure the smooth running of the organisation	5
Don't know	5
Speed of response to requests for work	4
The quality of solutions provided	4
None/no formal measure	4
Dealing effectively with innovation and change	3
The amount of work waiting to be carried out	2

Source: Interviews

The reason why some organisations have no formal measures is partially explained in the following comment made by a productivity services manager from a chemical manufacturing company,

> "This is a small factory and the work climate is relatively informal because we liaise closely with the works management and others, hence there is no real need for measures of effectiveness."

In general, the data about measures of effectiveness supports the argument advanced by Mumford and Pettigrew (1975) that top management do not, provide their specialists with a clear picture of the tasks they are expected to perform and indeed the nature of the specialist role has often not been thought through by them in any meaningful way.

The creation and continued survival of a management services unit is based upon undertaking work for others. The MS managers were asked to indicate the main sources from which their units obtained, or were given projects, assignments or tasks. The responses are shown in Table 2 where it can be seen that senior management appear to have the greatest influence on the work that MS units undertake. What is surprising, however, is the large amount of self selection of work by the units. This is generally carried out by the MS manager alone or in conjunction with his staff. The Table does not indicate the fact that most MS units obtained work from more than one source. Public

TABLE 2

Sources of projects, assignments or tasks for MS Units (N = 28)

Source	No. of times mentioned
From senior management	22
Self selected by unit	18
From line managers	13
From committees	10
Other	6

Source: Interviews

sector organisations made the most use of committees in allocating work to MS, whereas the largest number of responses in the line manager category came from industrial organisations. A further aspect of undertaking work for others is what criteria do they, the clients, use for accepting or rejecting recommendations made to them by an advisory unit such as management services. It is evident from the results shown in Table 3,

TABLE 3

Criteria by which clients* accept or reject recommendations made by MS units (N = 28)

Criteria for acceptance or rejection	No. of mentions
Financial benefits e.g. "savings in general" "work within budgets" "reduce costs" "cost effective solutions"	18
Improvement in services e.g. "value for money" "improve quality of service" "utilisation of resources"	13
Practical commonsense solutions e.g. "do recommendations make sense" "can recommendations be practically implemented" "credible solutions"	9

Source: Interviews

* Clients include top, middle and lower management

that clients were looking first for financial benefits to their operations such as reduction in costs or being able to work within budgets. Next they hoped to achieve an improvement in the services that they provided by, for example, giving better value for money and/or an improved quality of service. Lastly, clients expected recommendations to be practical and based on sound commonsense. Often more than one criteria was used by clients in making decisions about recommendations. One other issue that was found to have a direct bearing on the nature of the relationships with clients was the extent to which the management services function liaised with other specialist advisory functions. It can be seen from Table 4 that the main links were with personnel and computer services/data processing. However, it should be noted that in some instances MS units already contain some of these specialist functions.

TABLE 4

The specialist functions with which MS units have the greatest dealings (N = 28)

Specialist function	No. of mentions
Personnel	16
Computer Services/Data processing	15
Finance	8
Technical	8
Marketing	3

Source: Interviews

ORGANISATIONAL PROCESSES

The second part of the paper concentrates on how certain organisational processes impinge on the workings of internal management services units. The processes examined include the definition and diagnosis of problems, the implementation of solutions and the impact of power, influence and organisational politics on the successful achievement of productivity improvements. A number of writers such as Doris (1981) have argued that after some time in a job, specialist advisory staff such as management services come to realise that their effectiveness is limited not by knowledge of their own technical field, but by organisational and political factors, in the settings in which they operate. This is often because they have little direct control over events and must make progress by influencing others.

Much has been written, especially in textbooks, about the use of various techniques associated with productivity improvement. However little has been said, until comparatively recently, about the definition or diagnosis of problems in this area. Kilman and Mitroff (1979) when talking about consulting interventions claim that surprisingly little is known or understood about the process by which problems are defined. More often than not, it seems that individuals assume that their view of the world, often from their specialism, defines the essence of the problem. Tichy (1975) expands this point somewhat by arguing that like blind men, change agents seem to have limited perspectives, not only due to their diagnostic frameworks, but also very probably due to real world constraints of time, money and access to certain information. Hence their diagnoses are partial because they only look at certain aspects of complex social systems. In an article outlining the range of diagnostic methods available to an internal consultant Phillips (1980) states that correct diagnosis is vital to the success of any assignment or intervention. One further aspect of problem definition and diagnosis is depth of intervention. Harrison (1970) suggests that there are two criteria for this, first, to intervene at a level no deeper than that required to produce enduring solutions to the problems in hand; and second, to intervene at a level no deeper than that at which the energy resources of the client can be committed to problem solving and to change. From an analysis of the data it was established that there was no standard procedure used for defining and diagnosing problems.

It used to be generally accepted that at the other end of the problem solving or project continuum from definition and diagnosis was implementation. Increasingly the view now being put forward is that implementation is a process that runs through all phases of a project from beginning to end (Ginzberg, 1978). Nevertheless implementation is seen as a process which involves human participants, social interaction, organisation structure and management change - in short, a complex behavioural process (Shultz and Slevin, 1975). In its review of the results of U.K. government efficiency studies since 1979 and how they were implemented the Efficiency Unit (1985) found that it took at least two years and often as long as five to implement savings. The Unit was asked to find out why this was the case and to suggest how implementation could be improved. As a result of their analysis, the Unit concluded that it was not the scrutiny technique which needed attention but the wider process of achieving change into which it fitted. It suggested improvements in five areas. These were, greater top management support and commitment; implementation being the responsibility of only one person; a greater sense of urgency during implementation; a maximum of 2 years for full implementation and lastly, political support from Ministers. As the Unit notes in its report,

> "Effective implementation means both getting savings quickly and also ensuring that change is real and lasting."

Most MS units in the study saw implementation as a dynamic ongoing process commencing at the start of a project. Rarely, if ever, was it seen as a "once off" at the end of a project.

Most management services managers would no doubt accept the argument that in trying to operate a unit it is impossible if not unwise to completely ignore issues about power, politics and conflict within their organisation. Almost all management services managers and practitioners interviewed in this research indicated that they had little formal power as such but that in certain circumstances their influence could be significant. They reported having the greatest influence with issues related firstly to staffing, such as setting work standards, and deciding staffing levels and secondly in developing operating systems and procedures (Table 5). Allied to these was that in the service organisations such as banks, insurance companies and local

TABLE 5

Activities and/or issues where MS units have the greatest influence (N = 28)

Activity/issue	No. of mentions
Staffing levels	18
Operating systems and procedures	11
Selection of certain types of equipment	11
Making financial savings	10
The work of key operating departments	9

Source: Interviews

authorities many O&M/MS units have a considerable say in the selection of office equipment. The same was true in certain manufacturing companies where the IE department in some cases had the power of veto over equipment purchases. It is perhaps a little surprising that given today's climate in the UK for obtaining efficiency and value for money, the making of financial savings do not rate more highly.

A discussion of influence leads not unnaturally to an examination of the causes of conflict between units and their clients. Although conflicts did arise in relation to staff and manning levels, this was not the most contentious area. The major conflicts seem to revolve around two key issues namely the resistance to change shown by either individuals or groups and the often difficult working relationships between a particular department and the management services unit.

A detailed analysis of influence, power and organisational politics in management services has been completed by the author (Shaw, 1986). His paper deals with the importance of understanding these inter-related concepts, their relevance to management services, the characteristics of those involved in organisational politics, the situations most permeated by power, influence and politics and lastly boundary management, a strategy whereby management services people can deal effectively with these issues. Many writers, such as Handy (1985), Mintzberg (1983), Newman (1979) and Kotter (1977) have emphasised the importance of power, influence and politics. As Handy notes, organisations can be looked at as a fine weave of influence patterns whereby individuals or groups seek to influence others to think or act in particular ways. He believes that for anyone to neglect or by-pass the political reality of organisations would be misleading or blind or both, since all organisations of any size are political systems. In all organisations, he argues, there are individuals and groups competing for influence or resources, there are differences of opinion and of values, conflicts of priorities and of goals. There are pressure groups and lobbies, cliques and cabals, rivalries and contests, clashes of personality and bonds of alliance. The concensus view of respondents, chief executives, line managers and MS managers was that MS units did have power and were able to exert influence, sometimes to a considerable degree. As the MS manager in a local authority said,

> "Power produces the climate of the organisation and the rules of the game."

With regard to organisational politics the predominant view of most, but not all, was that management services people should not get involved in politics but equally should be aware of its existence and impact on the work that they do. Views from both ends of the spectrum can be seen in the following statements made by respondents.

> "This (productivity services) unit brings about change by being devious if necessary, the soft sell, politicing like hell and having an excellent formal and informal communication network." (Productivity Services Manager of a small privately owned public utility.)

At the other end of the spectrum is the following statement by the organisation and methods manager from an insurance company,

> "O and M should not take sides, we should not get involved in politics, we should offer a service and not impose ourselves. We must be completely impartial, discreet and careful how we say things."

The concept of social network analysis (Tichy et al, 1979) was used in the research to show how MS people not only crossed many boundaries within their organisation but also had links with many individuals and groups external to the organisation. Internally MS managers and practitioners acted as boundary spanners between different functional groups and between different hierarchial levels. The research demonstrated that establishing a boundary management strategy was a prerequisite for dealing effectively, with amongst other things, politics, power and influence in organisations.

When asked about the most significant recent changes in the way units operated the responses, as one would have expected, varied considerably. Nevertheless certain changes had occurred more frequently than others, these were firstly the increasing breadth of work being undertaken by many units, secondly the rapidly increasing amount of work being undertaken in the field of new technology and finally the improved standing of the unit within the organisation. Respondents were then asked to predict future trends in productivity work over the next few years. Many of the replies were inter-related and dealt either with some aspects of new technology or computer systems.

CONCLUSION

This paper has identified how certain critical issues and organisation processes had an impact on the achievement of successful productivity improvements in a number of U.K. organisations. In particular the following key points are worth restating;

- Top management control management services in an informal and fluid manner. This results in the objectives for most MS managers being vague and unspecific.
- There is generally a very high level of support given to MS by top management though their measures of effectiveness, where they exist at all, were found to be almost exclusively qualitative and subjective.
- The critical processes for the effective functioning of an MS unit are an awareness of organisational politics and boundary management and the ability to use power and influence as and when appropriate.

REFERENCES

Doris, G., 1981. Increasing the Effectiveness of the Management Services Professional. Management Services, Vol. 25, No. 6: 14-16.

Efficiency Unit, 1985. Making Things Happen. A report on the implementation of government efficiency scrutines. London: HMSO.

Ginzberg, M.J., 1978. Steps towards more effective implementation. Interfaces, Vol. 8, No. 3: 57-64.

Handy, C.B., 1985. Understanding Organisations. Penguin, Harmondsworth.

Harrison, R., 1970. Choosing the depth of organisational intervention. Journal of Applied Behavioural Science, Vol. 6: 181-202.

Kaplan, R.E., 1984. Trade Routes: The Manager's Network of Relationships. Organizational Dynamics, 37-52.

Kilman, R.H., & Mitroff, I.L., 1979. Problem Defining and the Consulting Intervention Process. California Management Review, Vol. XXI, No. 3: 26-36.

Kotter, J.P., 1977. Power dependence and effective Management. Harvard Business Review, Vol. 56, No. 4: 125-136.

Mintzberg, H., 1983. Power in and around organisations, Prentice-Hall, Englewood Cliffs.

Mumford, E., & Pettigrew, A., 1975. Implementing Strategic Decisions, Longman.

Newman, W.H., 1979. Company Politics: Unexplored Dimensions of Management. Journal of General Management, Vol. 5, No. 1: 3-12.

Pettigrew, A.M., 1974. The Influence Process between Specialists and Executives. Personnel Review, Vol. 3, No. 1: 24-30.

Phillips, A., 1980. Diagnosis in Investigations. Management Services in Government, Vol. 35, No. 2: 76-80.

Schultz, R.L., & Slevin, D.P., 1975. Implementing Operations Research/Management Science. American Elsevier, New York.

Shaw, W.N., 1986. Power, Politics and Influence in Management Services. Proceedings of the Institute of Management Services. Summer School, St. John's College, Cambridge.

Shaw, W.N., 1987. Productivity Management, An in-depth study of selected U.K. organisations. International Conference on Productivity Research, University of Miami.

Tichy, N.M., 1975. How different types of change agents diagnose organisations. Human Relations, Vol. 28, No. 9: 771-799.

Tichy, N.M., et al 1979. Social Network Analysis for organisations. The Academy of Management Review, Vol. 4, No. 4: 507-521.

LATE SUBMISSION

Productivity Management Frontiers — I, edited by D.J. Sumanth
Elsevier Science Publishers B.V., Amsterdam, 1987 — Printed in The Netherlands

AN OPERATION CONTROL STRATEGY OF FMS PROVIDING HIGH PRODUCTIVITY AND QUALITY

Paul G. Ranky, Visiting Associate Professor, Department of Mechanical Engineering and Applied Mechanics, The University of Michigan, Ann Arbor, MI 48109, USA

ABSTRACT

Every leading institution and manufacturing company in the world is dealing with some aspects of flexible automation, because many researchers and managers in industry have found that the concept of designing and making goods and products on order, rather than for stock at high quality, is the key issue in order to cut down manufacturing costs and lead time, and eventually to stay in business in the future. The important fact to realize is that demand, as well as available technology are changing at an exponentially growing rate, but money is tight and worldwide competition is tough; thus companies need to react to changes much faster and in a more flexible way than in the past, not only when designing new products, but also when manufacturing them. The purpose of the paper is to summarize those FMS design and dynamic operation control rules which must be followed and/or programmed when designing and operating flexible automation systems in order to achieve high productivity and quality.

1. INTRODUCTION

Computer Integrated Manufacturing (CIM) is concerned with providing computer assistance, control, and high level integrated automation at all levels of the manufacturing industries, including the business data processing system, CAD, CAM, and FMS by linking islands of automation into a distributed processing system. The technology applied in CIM makes intensive use of distributed computer networks and data processing techniques, Artificial Intelligence and Database Management Systems. FMS in this aspect plays the role of a highly efficient and "ready to react to random requests" manufacturing facility (e.g. machining, test, assembly, welding, etc.) in this "total concept". (Ref. 1, 2, 3)

CIM technology offers a range of benefits to managers too, when making decisions. In such cases computing and human decision-making skills cannot be ignored either, but:

- if the decision-makers in the organization have up-to-date information on the requested process and data,
- if there is the possibility of controlling and analyzing large amount of business and technical data by computers,
- if design and production engineering are integrated; in other words, if the designers can obtain guidance and

feedback from databases, and knowledge bases of expert systems,

- if manufacturing in its broad sense (i.e. not only machining, but also inspection, test, assembly, etc.) is done mainly by FMS producing the necessary workparts on order rather than for stock

then the company will have a better chance for survival in the age of CIM.

2. COMPUTER INTEGRATED MANUFACTURING (CIM) AND FMS DESIGN CONCEPTS

The aim of CIM is to let the advanced information processing technology penetrate into all areas of the manufacturing industry in order to:

- make the total process more productive and efficient,
- increase product reliability and
- decrease the cost of production and maintenance relating both to the manufacturing system as well as to the product.
- Reduce the number of hazardous jobs and increase the involvement of well educated and able humans in the manufacturing activity and design.

In comparison with FMS, CIM is mainly concerned with the information processing tasks at all levels of the factory and its management, whereas FMS provides the essential computer controlled manufacturing tools and systems for CIM to execute the computer generated plans and schedules that take account of a total system rather than just one cell or shop. (Ref. 1, 12, 14)

One could also say the CIM integrates "FMS islands" with the overall computer network, the business system, with different design and manufacturing databases of the company and allows the optimization of the data flow and eventually all activities at a much broader level than FMS using distributed processing techniques. (Figure 1)

3. FLEXIBLE MANUFACTURING SYSTEMS (FMS)

The latest research and FMS implementation results suggest that instead of the "traditional" hierarchical control architecture, FMS cells and workstations, from a control point of view, should be designed as nodes in a distributed computer network. Today, when the most up-to-date CNC (Computer Numerical Control), robot, Co-ordinate Measuring Machine (CMM), etc. controllers offer the processing and communication power of a 16-32 bit micro-computer linking machine controllers into a local area

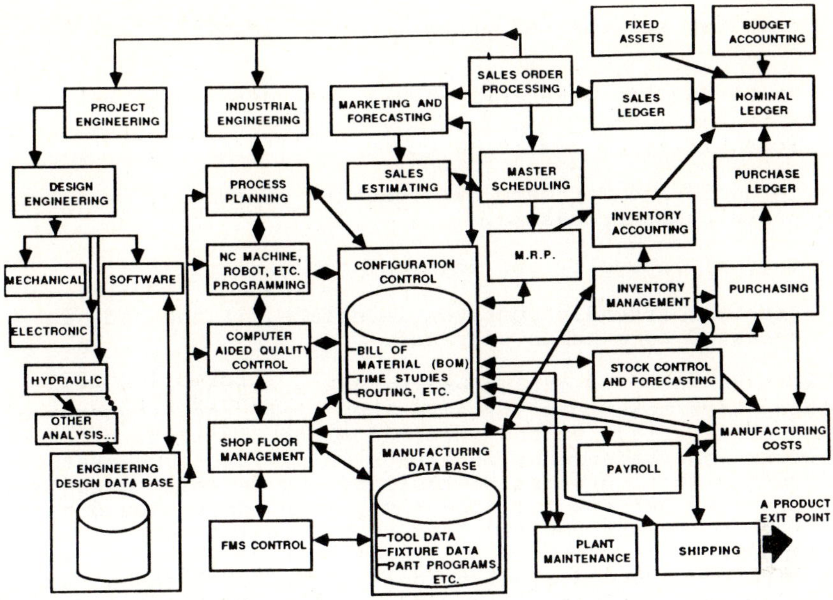

Figure 1. Computer Integrated Manufacturing System (CIM) concept integrating four major modules. The Business Data Processing System, CAD, CAM, and FMS. As the figure indicates distributed database management and computer network as well as expert learning and decision-making systems are important modules of this "total concept." (Adapted from Reference 1)

network (LAN) similarity to, for example graphics workstations in a CAD-LAN (Computer Aided Design Local Area Network) it is a reality now and as soon as MAP (the Manufacturing Automation Protocol) becomes a world-wide standard, will be a must for all hardware and soft- ware manufacturers. This is an important trend that will change the control architecture of all new automation systems and will dominate this business during this and the forthcoming decades. (The important point is that one cannot perform valuable manufacturing system design, operation control and modelling research work without understanding and contributing to such dramatic system control developments as the one mentioned above).

3.1 Principles of Distributed Processing

In a distributed processing system different computers and controllers, machines robots, CMMs, etc. must communicate with each other and process each other's data, and/or execute each other's programs.

Because communication takes time and costs money, low cost and efficient processing and storage methods should be used. This requires a number of conditions to be able to fulfill at each node of the computer network, of which the most important are as follows:

- Compatibility with the "outside world".
- Possibility of real-time communication and data update.
- Hardware and software architecture capable of performing distributed processes.
- Logical integrity and physical modularity.
- Safe and friendly operator interface, if the node is operated by a human operator.

"Well designed" distributed systems are generally more reliable than other solutions, because they limit the frequency of failures by distributing the workload among different processors and because some redundancy is always built into such systems. If each node has local processing facility and is able to run without the distributed network for some time, then there should be enough time either to repair the fault or to switch the control to the other computer. In distributed systems hardware and software should work together in a balanced way. Software tasks should be structured into standalone procedures and functions. In most cases the tasks to be performed are clear, and lend themselves to "natural demodularization." Software modules should be self contained, thus global variables should be avoided as much as possible. As a general rule, the FMS software should be designed, coded, tested and implemented as any other "well engineered" product.

3.2 FMS Design Rules

FMS may be defined as a system dealing with high level distributed data processing and automated material flow using cells (i.e. computer controlled machines, assembly robots, inspection machines and so on), together with computer integrated materials and storage systems. In this sense FMS is a production know-how offering "on-order" rather than "on-stock" manufacturing facility for a large variety of components. [Ref 1, 2, 4, 6, 8, 9]. In data processing terms an FMS "cell" is a node in the distributed control system, or in a LAN (Local Area Network).

To be able to achieve "truly" random production and the above listed economic benefits, the FMS system architecture must conform to the following basic design and operating rules:

- Utilize highly automated and programmable cells, incorpoating powerful controllers and self diagnostic systems, able of changing their tools and parts, preferably unmanned, and keeping in touch with a central computer or perferably with a LAN node controller from where the production plan, the part programs and further necessary data arrive and are fed back.
- Link these cells into a system by providing preferably direct access, or random material handling systems (e.g. AGV), rather than serial access (e.g. conveyor line) between them.

- Create a part, tool and pallet (fixture and clamping device) storage facility (eg. a warehouse), and

- Provide high level computer control "inside" and "outside" the system, based on a distributed processing system, on databases and on the necessary links to other subsystems, such as CAD, CAM, and the business data processing system.

- Ensure that if any cells break down the production planning and control system can reroute and reschedule production. In other words, design in the system not only programmable production facilities, but also ensure that the routing of parts can be dynamically altered, (the source of this loading as well as the real-time schedule should be a production rule base, discussed later).

Generally, requirements set against the FMS materials handling system include part transportation and storage of workpieces, empty pallets, auxiliary materials, wastes, fixtures and tools.

Material handling systems in FMS provide palletized component transfer between cells and include:

- Automated Guided Vehicles (AGV),

- conveyors,

- industrial robots,

- special purpose pallet changing and pallet transportation systems (e.g. automated pallet carrier on rails, towed carts, shuttle carts, etc.)

- forklift trucks and other solutions.

Since computer controlled material handling systems offer a wide range of solutions in FMS and because the inadequately designed materials handling system can limit an otherwise highly flexible manufacturing environment, it is important to provide random access material transportation systems when designing the FMS.

4. CIM LINKS FOR OPTIMAL FMS OPERATION CONTROL

Let us extend the already listed principles with some new rules regarding FMS design and operation control and summarize them in order to understand more details of the ways FMS systems should be designed, operated and controlled in order to achieve high level of reliability and productivity.

4.1 The Most Important FMS Operation Control Activities

The most important operation control activities are shown in Figure 2 identifying three levels at which simulation and optimization is required prior to, or during FMS part manufacturing. The three levels being:

- The factory level, or business level handled by the business system of CIM,

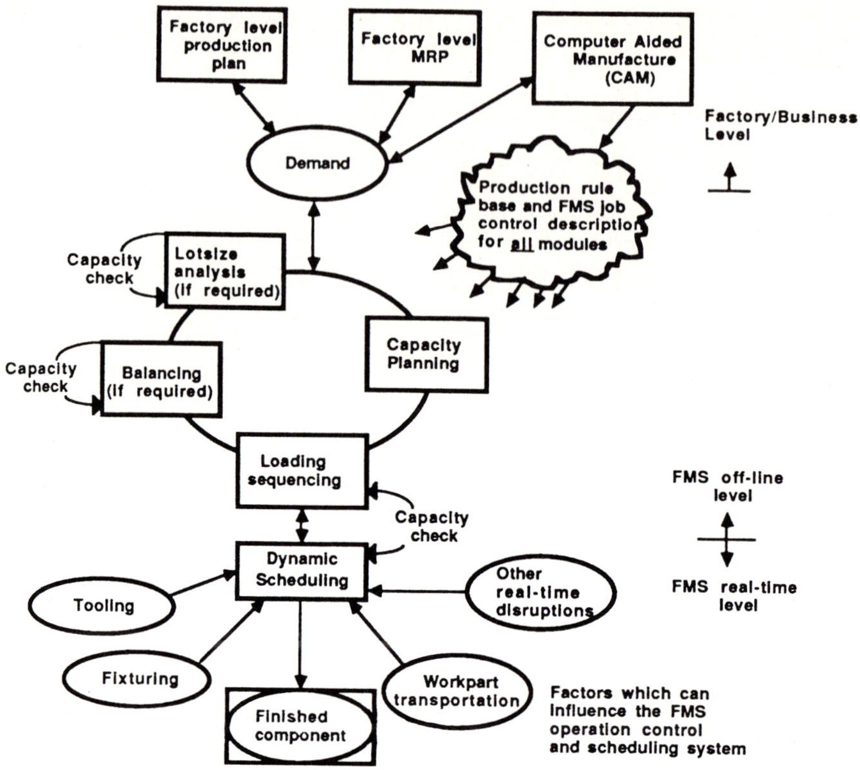

Figure 2. FMS operation control tasks and their links to the CIM modules.

- The FMS off-line level, representing simulation and optimization activities prior to loading a batch or a single component on the FMS, handled sometimes by the CAM system, sometimes by the FMS part programming computer, and

- The real-time controlled level, handled by the FMS operation control system, representing a situation where the parts are already physically as well as logically in the DNC controlled environment.

One can also see from this Figure, that to satisfy the demand generated by the business systems modules of CIM (i.e. the MRP and the MPS programs) the major task is the selection of the appropriate part mix for a defined period of time, which can be as short as a few hours, a shift (normally eight hours) a day or a few days.

Since almost "everything relates to everything" in CIM, the emphasis in our discussion is put naturally on integration and on the fact that static production management systems (thanks to the factory wide data logging facilities and the computer networks) are becoming more and more real time, thus the available data to be processed by the different modules of the business system is more accurate.

4.2 FMS Loading Sequencing

Loading sequencing can be a simulation and/or optimization method to establish the best order in which parts should be processed by the system in a short period of time, e.g. few hours, or up to the length of a shift.

If parts arrive at the FMS in random order, in very small batches, or as single components and if they must be executed as soon as they arrive there is not much point in selecting the appropriate part mix, nor in establishing an optimum loading sequence, because there is simply no time for that. However, in the case of most FMSs parts arrive in a fairly well planned order, there is time to load sequence and then it is essential not only to utilize the system at a level close to 100%, but also to process the parts without delay.

The input to the loading sequencing program can be the MRP, or directly the MPS output, or alternatively if higher level optimization is possible, the result of the part mix evaluation procedure using lotsize analysis and operation balancing with capacity checks.

4.3 Batchsize Analysis

If there are sufficient number of different batches of components and if the time span allows (i.e. due dates are not too close) it is worth while doing a batchsize analysis to group the components into economic batches and to establish economic cycle times in particular if there are resource constraints such as machine or fixture etc. availability.

4.4 Balancing

Balancing can be useful in particular if there are many operations of different length are to be done on the part, and if the resources allow diffferent operations to be performed in a random order at no additional cost (e.g. if the machines have an adequate selection of tools in their magazines, or if robots can automatically change their hands, if machines and assembly robots can change parts automatically, etc.). Balancing has a potential not only in assembly but also in the machining oriented and other FMS too, when one or more cells become frequently overloaded.

4.5 Capacity Planning

To avoid unrealistic planning, capacity planning is important too, but capacity checking is even more important shortly after the above discussed processes in order to ensure that the established part mix can be processed on a given system. (This process is indicated by feedback loops at the three major modules of the off-line process in Figure 2).

FMS must be able to react to real-time changes; to mention a few, what should happen to its operation control

- if the processing priorities of a part or a batch must be changed?
- if a part, or a batch must be deleted from the schedule and removed from the system?
- if a cell breaks down?

This is an area where in order to ensure maximum productivity fast "secondary optimization" is required using a dynamic FMS scheduling program.

5. THE FMS DYNAMIC SCHEDULING CONCEPT

To implement the maximum level of flexibility in the dynamic scheduling system, the following strategy is used:

The FMS part programmer knows what processes he wants to carry out on the workpiece and in what possible variations these processes could be performed. He designs the processes and their possible orders and documents the results following the "variable route" part programming method which is a "production rule base" using an FMS job description language, describing the order and structure of the various operations required, the tooling and fixturing needs and the necessary pointers to different FMS data bases. (Figure 3)

On the basis of this production plan, the already discussed capacity planning, balancing and other optimization programs can be run in the desired order. (Figure 2). The result of this analysis is an FMS part loading sequence that is optimum when established, but can be changed if the real-time circumstances change. Whatever changes there are, the production knowledge base must be consulted before finalizing process execution plans, since it contains all necessary knowledge that results of the fabrication of each individual part. Rules at the operation level programs in this knowledge base can follow a format that is similar to this:

```
IF  [Condition true]

THEN PERFORM [Operation code]  WITH  [Tool code]

                               ON MACHINE [Preferred list of
                                           machine IDs]
```

Note that in this rule structure "Condition true" relates to real-time disturbances, the "Operation code" normally means a CNC

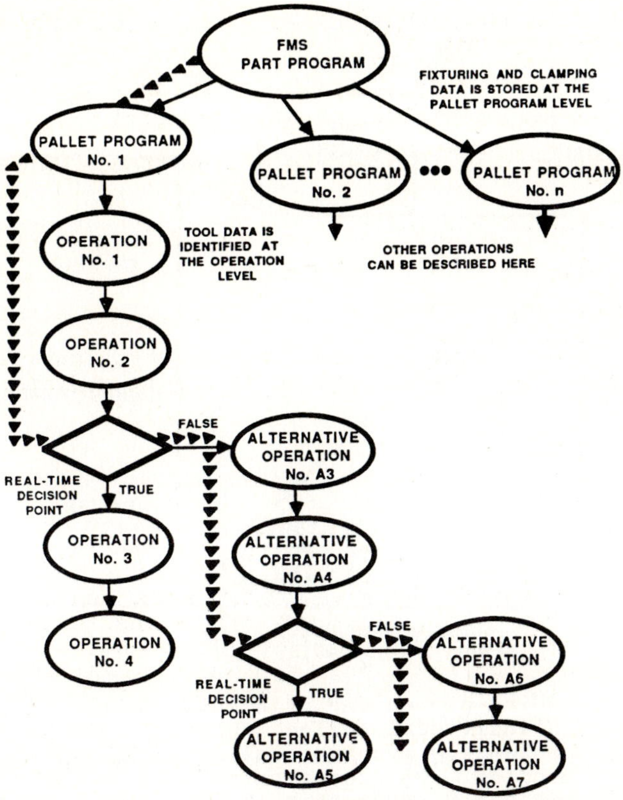

Figure 3. The "variable route" FMS part programming concept and program structure describing the job for the FMS allowing alternative operations to be taken according to the real-time status of the system.

macro call, or a robot program subroutine call, the "Tool code" is an ID of a tool described in the tool data base, and finally the "Preferred list of machine IDs" is a not always necessary, but often useful help from the process planner (human or machine assisted) to the system because it helps to choose from a possible list of machines for which the part should be routed real-time. If provided, this list gives the FMS dynamic sched-scheduling program the chance to choose the "best-fit machine " not only considering the process planner's performance, but also such real-time related aspects as: machine availability (i.e. up/down, if down back when?, etc.) and the input part queue size currently known (and or scheduled) for that particular machine.

Once a part is loaded on an FMS it can be (and should be) dynamically scheduled. (Figure 4). (Note the differences between FMS loading sequencing and the job shop, in which scheduling is done off-line, rather than real-time, thus the difference between the "real-world situation" and the "desired, dream world situation" gradually increases, meanwhile the operation control of the job shop falls apart and "panic actions" must be taken by able "human rescuers").

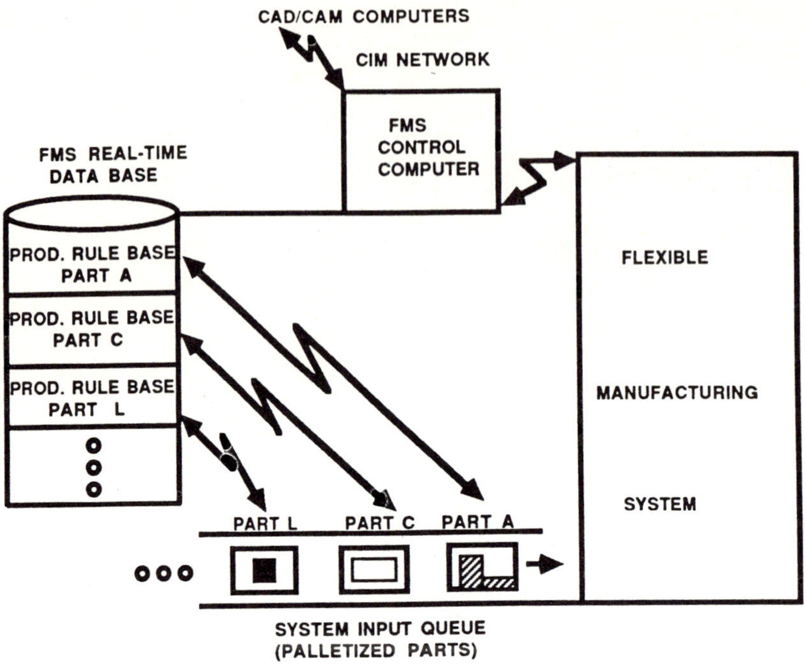

Figure 4. FMS loading sequencing principle, using the production rule base for each individual pelletized part. (In this concept the FMS loading sequencing program considers the FMS as one, multifunctional machine. Of course parts will be dynamically re-scheduled as necessary after this at cell level in real-time)

Dynamic scheduling should be done both at cell and FMS system levels. Figure 5 indicates FMS cells that can change parts and tools in a random order, are linked together by an AGV network offering random visits on behalf of the material handling system. Cell controllers in this model are linked to a LAN, as discussed before. Each part in this system is put on a standard pallet and has access to the FMS production rule base. The selection of operation alternatives should be performed real-time by the FMS dynamic scheduler and the FMS control system.

Scheduling rules, such as SPT (Shortest Processing Time), WSPT (Weighted Shortest Processing Time), EDD (Early Due Date) etc., can be used at all level assuming zero slack time. Once a part is ready to be loaded on the next programmed cell, scheduling must be performed at the selected cell (or machine) for the job. This re-scheduling process must be repeated every time a new part is put into the pallet pool of the new programmed cell) since cell schedulers can be as fast as 2-8 seconds, parts will not suffer any delay because of being rescheduled so frequently. (Figure 5)

Using the above described procedure and concept, the production of the FMS can be maximized, maintaining high level quality, since new operators are not performed on the part while the current ones have not been checked.

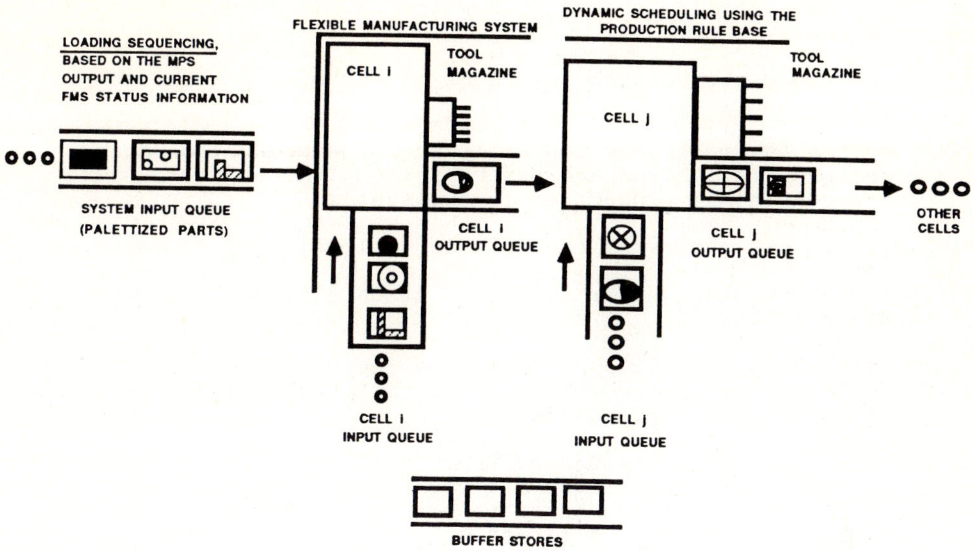

Figure 5. Dynamic scheduling principles in cellular flexible manufacturing. In this concept each part on each cell can be dynamically re-scheduled, just before processing. The principle is that of the part is not available only simulation is performed. Once the part is at the cell, a very fast cell schedules decides the processing order on the basis of the production rule base (attached to each part) and by evaluating high level real-time interrupts if their are any.

The dynamic scheduling system should incorporate procedures such as DELETE workpiece from any production route and/or the buffer store, WAIT with the production of a particular workpiece, because the system is not capable of accepting further components for the time being, ASK the status of a component, REARRANGE any production route in real-time because of a break down or maintenance, and so on.

The WAIT procedure for example could be implemented using this algorithm:

IF the production route can be loaded with further components (in other words if the queue is not yet full) and/or if there is a free place in the buffer stores,

THEN the part should be placed logically and physically as soon as possible on the production route

ELSE the part should wait in the automated warehouse, or, if already delivered to the system, in a free buffer store

As a further example, let us see how the DELETE algorithm could be designed:

IF the processing of the workpiece in any of the possible production routes is deleted

THEN remove the workpiece logically from the production queue and physically from the buffer store without interrupting the continuous production on the system.

Finally, the ASK algorithm could be as follows:

IF the dispatcher and/or subsystem sends a request for workpiece status

THEN list those FMS cells and selected routes which the component has passed by and also those which are planned to be met in the current schedule.

It is clear from these algorithms that for each possible manufacturing route, two lists are required: a waiting list of workpieces to be manufactured on each selected route, (gained as a result of frequent simulation of the system) and the current status description of the availability of the system (gained using real-time sensors). Normally components waiting to be manufactured via the same route in the FMS should be taken one-by-one from the front of each queue (i.e. buffer store), until there are no components in the queue, but the rearrangement of components in queue should also be allowed using the SPT, WSPT, EDD or other scheduling rules. This feature is important if a job must be processed at the maximum possible speed, or in "panic situations" where the system needs to recover fast.

Having allowed for the real-time alternation of queues, the program must keep track of how much a production route can be loaded with workparts and must also know the number of workparts already accepted for production for each route.

The above outlined procedures and dynamic FMS operation control principles can be tested by simulation. Figures 6 and 7 illustrate some animated graphics output, using ROBCAD [Ref.13, 14, 15]. The objects (i.e. robots, machine tools, etc) in these simulation programs are not only animated, but controlled using a high level, task oriented language. This technique allows realistic planning and design of the simulation model, as well as the generation of the operation and software such that if the results of the simulation are acceptable, the operation control software can be downloaded to the real robots, machine tools, etc.

6. SUMMARY

To summarize, the operation control method to be used in FMS should be multi-level and dynamic. This means that loading sequencing and operation control in FMS should rely on information sources provided by different subsystems of CIM, thus there should be an overall planning level and a dynamic, or real-time level otherwise the potentially ofered benefits of the FMS technology cannot be acieved.

Currently further research is carried out in formally defining the FMS job control language and a generic production rule base description method, generally for human users and in a computer readable, high level language format for systems analysts and for the FMS real-time scheduling program. [Reference 13].

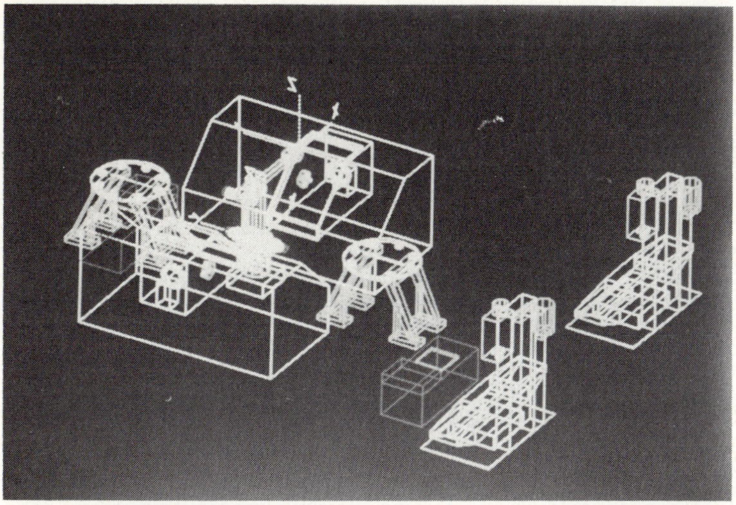

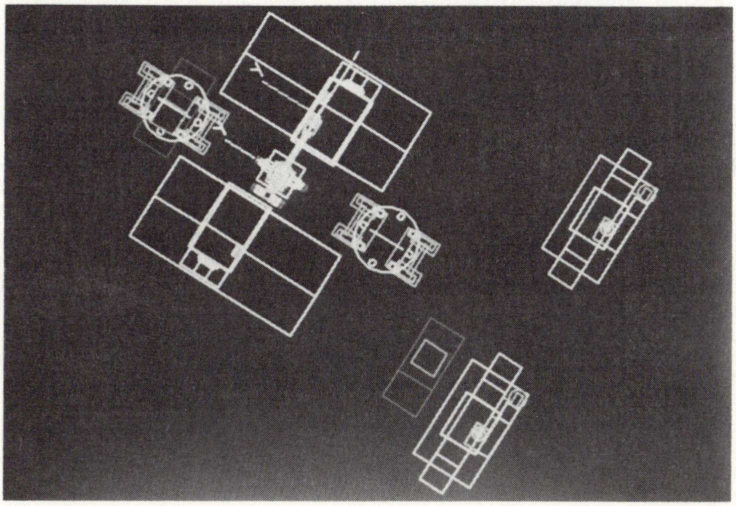

Figures 6 and 7. The photographs show the graphics simulation of a small size FMS. Parts are delivered and docked by AGVs to the illustrated AGV docking stations. Cylindrical part magazines index and the robot as well as the AGVs can access them real-time. The other two machines on the right hand side are CNC milling machines, loaded/unloaded by the same AGVs used for the CNC lathes. [Ref. 12]

7. ACKNOWLEDGEMENTS

The author wishes to express his thanks to the University of Michigan, College of Engineering for supporting this research.

8. REFERENCES AND FURTHER READING:

[1] Paul G. Ranky: Computer Integrated Manufacturing, Prentice-Hall International, 1986, p. 528.

[2] Paul G. Ranky: The Design and Operation of FMS, IFS (Publications) Ltd. and North Holland, 1983, p. 348.

[3] Paul G. Ranky and Philip H. Francis: Design and Solid Model Simulation of Generic Assembly Cells and Systems, Japan-USA Flexible Manufacturing Systems Symposium, Osaka, Japan, 1986.

[4] Paul G. Ranky: Tool Management Tasks in Flexible Manufacturing Systems (FMS), NAMRC XIV (North American Manfacturing Research Conference), Minnesota, 1986.

[5] M. Horvath and A. Narkin: Operating Sequence Planning Using Optimization Concepts and Logic Programming, IFIP/IFAC World Congress, Budapest, 1984, pp. 103-156.

[6] U. Rembold and R. Dildmann (ed.): Methods and Tools for Computer Integrated Manufacturing, Springer Verlag, 1984.

[7] David A. Bourne and Mark S. Fox: Anonymous Manufacturing: Automating the Job-Shop, Computer, September, 1984, pp. 76-79.

[8] R. K. Sinha and R. H. Hollier: A review of production control problems in cellular manufacture, Int. J. Prod. Res. 1984, Vol. 22, No. 5, pp. 773-789.

[9] J. Erschler, D. Leveque, F. Roubellat: Periodic Loading of Flexible Manufacturing Systems, Advances in Production Management Systems, Elsevier Science Publ., IFIP 1984.

[10] Paul G. Ranky: The FMS Software Library, User and System Manuals, 1983-1984.

[11] Paul G. Ranky: "Robot Tool Management for Flexible Automation Systems, End-of-Arm Tooling and Tool Design Seminar, SME, Detroit, 1986, USA.

[12] Paul G. Ranky: "End of Arm Tool Management System for Robotized Assembly Lines," Robots West Conference, Long Beach, CA, 1986, USA.

[13] Paul G. Ranky: "A Program Prospectus for the Simulation, Design and Implementation of Flexible Assembly and Inspection Cells," International FMS Conference at the University of Michigan, Ann Arbor, 1986, USA.

[14] Paul G. Ranky: "Robotic Assembly Cell Design by Solid Model Simulation," Industrial Affiliates Review, CRIM, Center for Research on Integrated Manufacturing, University of Michigan, 1986, USA.

[15] ROBCAD User and System Manuals, Techn omatix, ROBCAD, Detroit, 1985-86, USA.

Author Index